KV-638-857

THE FUTURE OF THE LAND

THE FUTURE OF THE LAND

MOBILISING AND INTEGRATING KNOWLEDGE FOR LAND USE OPTIONS

Edited by

LOUISE O. FRESCO

LEO STROOSNIJDER

JOHAN BOUMA

HERMAN van KEULEN

Wageningen Agricultural University,
The Netherlands

JOHN WILEY & SONS
Chichester · New York · Brisbane · Toronto · Singapore

Telephone National Chichester (0243) 779777
International (+44) (243) 779777

Other Wiley Editorial Offices

John Wiley & Sons, Inc., 605 Third Avenue,
New York, NY 10158-0012, USA

Jacaranda Wiley Ltd, 33 Park Road, Milton,
Queensland 4064, Australia

John Wiley & Sons (Canada) Ltd, 22 Worcester Road,
Rexdale, Ontario M9W 1L1, Canada

John Wiley & Sons (SEA) Pte Ltd, 37 Jalan Pemimpin #05-04,
Block B, Union Industrial Building, Singapore 2057

Library of Congress Cataloging-in-Publication Data

The Future of the land : mobilising and integrating knowledge for land use options / edited by L. O. Fresco . . . [et al.].
p. cm.
"An international interdisciplinary conference entitled: 'The future of the land, mobilising and integrating knowledge for land use options' was held in Wageningen [Netherlands] from August 22–25, 1993"—Pref.
Includes bibliographical references and index.
ISBN 0 471 95017 3
1. Land use, Rural—Planning—Congresses. 2. Land use—Information services—Congresses. 3. Land use—Environmental aspects—Congresses. 4. Sustainable development—Congresses.
I. Fresco, Louise.
HD105.F88 1994
333.76′13—dc20 94-2446
CIP

British Library Cataloguing in Publication Data

A catalogue record for this book is available from the British Library

ISBN 0 471 95017 3

Typeset in 10/12pt Times by Dobbie Typesetting Limited, Tavistock, Devon
Printed and bound in Great Britain by Bookcraft (Bath), Avon

Contents

List of Contributors

L. Agustina, INRES Research Institute, University of Brawijaya, Jl. Majen Haryono 169, Malang-65420, Indonesia
D. Ah Koon, Mauritius Sugar Industry Research Institute, Réduit, Mauritius
R. Alfaro, Atlantic Zone Programme, Estación Experimental Los Diamantes, Apartado 224, 7210 Guápiles, Costa Rica
Z. Alikhani, Department of Agricultural Engineering, McGill University, Macdonald Campus, Montreal, Canada
W. Andriesse, The Winand Staring Centre for Integrated Land, Soil and Water Research, Department of International Cooperation, PO Box 125, 6700 AC Wageningen, The Netherlands
E. Asteraki, IGER, North Wyke, EX20 2SB, UK
J. Bachacou, INRA C.R.F. Champenoux, 54280 Seichamps, France
A. K. Bandyopadhyay, Central Agricultural Research Institute, Post Bag No. 181, Port Blair 744101, Andamans, India
C. J. Barr, Institute of Terrestrial Ecology, Merlewood, Grange-over-Sands, Cumbria LA11 6JU, UK
K. J. Beek, ITC, PO Box 6, 7500 AA Enschede, The Netherlands
M. Benoit, INRA Station, S.A.D., 88500 Mirecourt, France
J. J. E. Bessembinder, Atlantic Zone Programme, CATIE-WAU-MAG, Apartado 224, 7210 Guápiles, Costa Rica
E. Y. H. Bobobee, Agricultural Engineering Department, University of Science and Technology, Kumasi, Ghana
L. Boerboom, International Service for National Agricultural Research (ISNAR), PO Box 93375, 2509 AJ Den Haag, The Netherlands
C. H. Bonte-Friedheim, ISNAR, PO Box 93375, 2509 AJ The Hague, The Netherlands
J. Bouma, Department of Soil Science and Geology, Wageningen Agricultural University, Duivendaal 10, 6701 AR Wageningen, The Netherlands
B. A. M. Bouman, Research Institute for Agrobiology and Soil Fertility (AB-DLO), PO Box 14, 6700 AA Wageningen, The Netherlands
P. Bowling, Institute of Grassland and Environmental Research, Aberystwyth, SY23 3EB, UK
R. Brinkman, Land and Water Development Division, FAO, via delle Terme di Caracalla, Rome 00100, Italy
M. C. Bronsveld, ITC, PO Box 6, 7500 AA Enschede, The Netherlands

R. G. H. Bunce, Institute of Terrestrial Ecology, Merlewood, Grange-over-Sands, Cumbria LA11 6JU, UK
S. Chadrapatya, IBSRAM, PO Box 9-109, Chatuchak, 10900 Bangkok, Thailand
L. Chitoli, Mabouya Valley Development Project, Ministry of Agriculture, Land, Fisheries and Forestry, St Lucia
M. Chung, Mauritius Sugar Industry Research Institute, Réduit, Mauritius
S. Cissé, Etude sur les Systèmes de Production Rurales en 5ème Région (ESPR), Sévaré, Mali
D. B. Dalal Clayton, International Institute for Environment and Development, 3 Endsleigh Street, London WC1H 0DD, UK
N. Danby, FARD, Talybont-on-Usk, Brecon, LD3 7YN, UK
M. de Bakker, Department of Environmental Sciences, Professor H. C. van Hall Institute, PO Box 17, 9700 AA Groningen, The Netherlands
J. A. Deckers, Institute of Land and Water Management (K. U. Leuven), Vital Decosterstraat 102, 3000 Leuven, Belgium
G. H. J. de Koning, Research Institute for Agrobiology and Soil Fertility (AB-DLO), PO Box 14, 6700 AA Wageningen, The Netherlands
D. L. Dent, School of Environmental Sciences, University of East Anglia, Norwich, Norfolk NR16 1HT, UK
J. Deville, Mauritius Sugar Industry Research Institute, Réduit, Mauritius
C. T. de Wit, Abersonlaan 31, 6703 GE Wageningen, The Netherlands
J. Dijsselbloem, Social Democratic Party (PvdA), PO Box 20018, 2500 EA Den Haag, The Netherlands
C. Dreiser, Mozart Str. 7, 6920 Sinsheim, Germany
M. Dzatko, Soil Fertility Research Institute, Gagarinova 10, Bratislava, Slovakia
W. R. Eason, Institute of Grassland and Environmental Research, Aberystwyth, Dyfed SY23 3EB, UK
S. Efdé, Department of Soil Science and Geology, Wageningen Agricultural University, Duivendaal 10, 6701 AR Wageningen, The Netherlands
J. W. Erdman, Euroconsult, PO Box 441, 6800 AK Arnhem, The Netherlands
J. Feyen, Institute of Land and Water Management (K. U. Leuven), Vital Decosterstraat 102, 3000 Leuven, Belgium
L. O. Fresco, Department of Agronomy, Wageningen Agricultural University, PO Box 341, 6700 AH Wageningen, The Netherlands
E. K. Gill, Institute of Grassland and Environmental Research, Aberystwyth, SY23 3EB, UK
P. Goldsworthy, International Service for National Agricultural Research (ISNAR), PO Box 93375, 2509 AJ Den Haag, The Netherlands
P. Gosseye, Research Institute for Agrobiology and Soil Fertility (AB-DLO), PO Box 14, 6700 AA Wageningen, The Netherlands
A. Grillet, Institute of Land and Water Management (K. U. Leuven), Vital Decosterstraat 102, 3000 Leuven, Belgium
R. J. Haggar, Institute of Grassland and Environmental Research, Aberystwyth, SY23 3EB, UK
J. W. M. Hardon, Euroconsult, PO Box 441, 6800 AK Arnhem, The Netherlands

W. B. Harms, DLO Winand Staring Centre for Integrated Land, Soil and Water Research, PO Box 125, 6700 AC Wageningen, The Netherlands
W. J. M. Heijman, Department of General Economics, Wageningen Agricultural University, PO Box 8130, 6700 EW Wageningen, The Netherlands
H. Hetsen, Department of Physical Planning and Rural Development, Wageningen Agricultural University, Generaal Foulkesweg 13, 6703 BJ Wageningen, The Netherlands
M. Hidding, Department of Physical Planning and Rural Development, Wageningen Agricultural University, Generaal Foulkesweg 13, 6703 BJ Wageningen, The Netherlands
J. Houssait-Young, IGER, North Wyke, EX20 2SB, UK
D. C. Howard, Institute of Terrestrial Ecology, Merlewood, Grange-over-Sands, Cumbria LA11 6JU, UK
L. Hubrechts, Institute of Land and Water Management (K. U. Leuven), Vital Decosterstraat 102, 3000 Leuven, Belgium
E. J. Huising, PO Box 339, 6708 CR Wageningen, The Netherlands
H. Huizing, ITC, PO Box 6, 7500 AA Enschede, The Netherlands
D. M. Jansen, Atlantic Zone Programme, Estación Experimental Los Diamantes, Apartado 224, 7210 Guápiles, Costa Rica
J. M. L. Jansen, The Winand Staring Centre for Integrated Land, Soil and Water Research, PO Box 125, 6700 AH Wageningen, The Netherlands
I. Jhoty, Mauritius Sugar Industry Research Institute, Réduit, Mauritius
D. Jones, Institute of Grassland and Environmental Research, Aberystwyth, SY23 3EB, UK
R. H. G. Jongman, Department of Physical Planning and Rural Development, Wageningen Agricultural University, Generaal Foulkesweg 13, 6703 BJ Wageningen, The Netherlands
Z. Karacsonyi, Environmental Management Workgroup, Debrecen Agricultural University, Faculty of Agronomy, pf. 36., 4015 Debrecen, Hungary
A. K. Kassam, The Secretariat of the Technical Advisory Committee to the CGIAR, Research and Technology Development Divison, FAO, Via delle Terme di Caracalla, 00100 Rome, Italy
Ph. A. Kips, The Winand Staring Centre for Integrated Land, Soil and Water Research (SC-DLO), PO Box 125, 6700 AC Wageningen, The Netherlands
J. P. Knaapen, The Winand Staring Centre for Integrated Land, Soil and Water Research, PO Box 125, 6700 AC Wageningen, The Netherlands
K. W. Knickel, Institut fuer laendliche Strukturforschung, Johann Wolfgang Goethe-Universitaet, Zeppelinallee 31, 60325, Frankfurt/Main, Germany
S. A. Koram, Agricultural Engineering Department, University of Science and Technology, Kumasi, Ghana
S. B. Kroonenberg, Department of Soil Science and Geology, Wageningen Agricultural University, Duivendaal 10, 6701 AR Wageningen, The Netherlands
M. J. Kropff, International Rice Research Institute (IRRI), Los Banos, PO Box 933, 1099 Manilla, The Philippines
H. J. Krüger, Soil Conservation Research Project Ethiopia, PO Box 2597, Addis Ababa, Ethiopia
N. Kyei-Baffour, Agricultural Engineering Department, University of Science and Technology, Kumasi, Ghana

F. B. Labib, Soils and Water Use Department, National Research Centre Dokki, Cairo, Egypt
M. Latham, IBSRAM, PO Box 9-109, Chatuchak, 10900 Bangkok, Thailand
J. A. Laws, IGER, North Wyke, EX20 2SB, UK
F. Le Ber, INRA C.R.F. Champenoux, 54280 Seichamps, France
C. A. Madramootoo, Department of Agricultural Engineering, McGill University, Macdonald Campus, Montreal, Canada
S. Meerman, Raw Materials, AVEBE, PO Box 15, 9640 AA Veendam, The Netherlands
M. Mitchell, Department of Economics, Marketing and Management, SAC-Auchincruive, Ayr, Ayrshire KA6 5HW, UK
M. Molenaar, Department of Surveying, Photogrammetry and Remote Sensing, Wageningen Agricultural University, Hesselink van Suchtelenweg 6, 6703 CT Wageningen, The Netherlands
M. C. Muhar, INRA Station S.A.D., 88500 Mirecourt, France
J. A. Ngailo, National Soil Service, PO Box 5088, Tanga, Tanzania
A. Nieuwenhuyse, Atlantic Zone Programme, CATIE-WAU-MAG, Apartado Postal 224-7210, Guápiles, Costa Rica
Nurpilihan, Agricultural Faculty of Padjadjaran University, Jalan Raya Sumedang, Jatinangor, West Java, Indonesia
R. J. Olembo, United Nations Environment Programme, PO Box 47074, Nairobi, Kenya
M. Omakupt, Department of Land Development, Land Use Planning Division, Phaholyothin Road, Chatuchak, 10900 Bangkok, Thailand
S. Panichapong, TA&E Consultants Co., Ltd., 3154 Drive in Centre, Ladprao Road Soi 130, 10240 Bangkapi, Bangkok, Thailand
T. W. Parr, Institute of Terrestrial Ecology, Merlewood, Grange-over-Sands, Cumbria LA11 6JU, UK
S. Patinavin, Department of Land Development, Land Use Planning Division, Phaholyothin Road, Chatuchak, 10900 Bangkok, Thailand
C. Patrick, Department of Agricultural Research, Private Bag 0033, Gaborone, Botswana
F. W. T. Penning de Vries, Research Institute for Agrobiology and Soil Fertility (AB-DLO), PO Box 14, 6700 AA Wageningen, The Netherlands
R. Rabbinge, Department of Theoretical Production Ecology, Wageningen Agricultural University, Bornsesteeg 65, 6708 PD Wageningen, The Netherlands
I. S. Rahim, Soils and Water Use Department, National Research Centre Dokki, Cairo, Egypt
F. N. Reyniers, Water Management Research Unit, CIRAD-CA, BP 5053, 34032 Montpellier, France
C. Ricaud, Mauritius Sugar Industry, Réduit, Mauritius
R. B. Ridgway, Natural Resources Institute, Central Avenue, Chatham Maritime, Kent ME4 4TB, UK
J. E. Roberts, Institute of Grassland and Environmental Research, Aberystwyth, SY23 3EB, UK
R. P. Roetter, Department of Geography, University of Trier, PO Box 3825, 5500 Trier, Germany
D. G. Rogers, FARD, Kennford, Exeter, Devon EX6 7XR, UK

N. Röling, Department of Communication and Innovation Studies, Wageningen Agricultural University, Hollandseweg 1, 6706 KN Wageningen, The Netherlands
J. Roos-Klein Lankhorst, DLO Winand Starling Centre for Integrated Land, Soil and Water Research, PO Box 125, 6700 AC Wageningen, The Netherlands
G. Russell, Institute of Ecology and Resource Management, The University of Edinburgh, West Mains Road, Edinburgh EH9 3JG, UK
B. Saengwan, TA&E Consultants Co., Ltd., 3154 Drive in Centre, Ladprao Road Soi 130, 10240 Bangkapi, Bangkok, Thailand
A. Sajjapengse, IBSRAM, PO Box 9-109, Chatuchak, 10900 Bangkok, Thailand
S. Sappong, Department of Agricultural Economics and Business, University of Guelph, Guelph, Ontario, Canada N1G 2W1
C. E. Schaefer, Soil Science Department, University of Reading, London Road, Reading, Berkshire RG1 5AQ, UK
R. A. Schipper, Department of Development Economics, Wageningen Agricultural University, Hollandseweg 1, 6706 KN Wageningen, The Netherlands
J. W. Schultink, SEAMEO-TROPMED/GTZ, University of Indonesia, PO Box 3852, Jakarta 10038, Indonesia
S. Senol, Department of Soil Science, Cukurova University, 01330 Adana, Turkey
R. J. Sevenhuysen, Atlantic Zone Programme, Estación Experimental Los Diamantes, Apartado 224, 7210 Guápiles, Costa Rica
K. M. Severin, Mabouya Valley Development Project, Ministry of Agriculture, Land, Fisheries and Forestry, St Lucia
J. M. Shaka, National Soil Service, PO Box 5088, Tanga, Tanzania
R. D. Sheldrick, IGER, North Wyke, EX20 2SB, UK
J. Simpson, Forestry Authority Research Division, Midlothian EH25 9SY, UK
K. Smets, Institute of Land and Water Management (K. U. Leuven), Vital Decosterstraat 102, 3000 Leuven, Belgium
A. B. Smit, Department of Agronomy, Wageningen Agricultural University, PO Box 341, 6700 AH Wageningen, The Netherlands
J. H. J. Spiertz, Research Institute for Agrobiology and Soil Fertility (AB-DLO), PO Box 14, 6700 AA Wageningen, The Netherlands
P. Spijkers, Department of Agriculture and Regional Development, DHV Consultants BV, PO Box 1399, Amersfoort, The Netherlands
V. S. Stolbovoy, GIS Laboratory, Dokuchaev Soil Science Institute, 109017, Pyzhevsky 7, Moscow, Russia
J. J. Stoorvogel, Atlantic Zone Programme, CATIE-WAU-MAG, Apartado Postal 224-7210, Guápiles, Costa Rica
L. Stroosnijder, Department of Irrigation and Soil and Water Conservation, Wageningen Agricultural University, Nieuwe Kanaal 11, 6709 PA Wageningen, The Netherlands
G. H. A. Te Braake, Reconstruction Committee, Engelsekamp 6, Groningen, The Netherlands
S. Thijsen, Grontmij, Nieuwe Stationstraat, 9750 AC Haren, The Netherlands
G. W. van Barneveld, Department of Agriculture and Regional Development, DHV Consultants BV, PO Box 1399, Amersfoort, The Netherlands
D. van den Broucke, Institute of Land and Water Management (K. U. Leuven), Vital Decosterstraat 102, 3000 Leuven, Belgium

P. C. van den Noort, Department of Agricultural Economics and Policy, Wageningen Agricultural University, Hollandseweg 1, 6706 KN Wageningen, The Netherlands
P. van der Molen, Department of Human Nutrition, Wageningen Agricultural University, PO Box 8129, 6700 EV Wageningen, The Netherlands
B. J. A. van der Pouw, The Winand Staring Centre for Integrated Land, Soil and Water Research (SC-DLO), PO Box 125, 6700 AC Wageningen, The Netherlands
R. van der Veeren, Department of General Economics, Wageningen Agricultural University, PO Box 8130, 6700 EW Wageningen, The Netherlands
M. van der Velden, Institute of Land and Water Management (K. U. Leuven), Vital Decosterstraat 102, 3000 Leuven, Belgium
F. van der Wal, National Soil Service, PO Box 5088, Tanga, Tanzania
C. A. van Diepen, The Winand Staring Centre for Integrated Land, Soil and Water Research (SC-DLO), PO Box 125, 6700 AC Wageningen, The Netherlands
N. van Duivenbooden, Department of Agronomy, Wageningen Agricultural University, PO Box 341, 6700 AH Wageningen, The Netherlands
M. K. van Ittersum, Department of Theoretical Production Ecology, Wageningen Agricultural University, Bornsesteeg 65, 6708 PD Wageningen, The Netherlands
H. van Keulen, Research Institute for Agrobiology and Soil Fertility (AB-DLO), PO Box 14, 6700 AA Wageningen, The Netherlands
H. C. van Latesteijn, Netherlands Scientific Council for Government Policy, PO Box 20004, 2500 EA Den Haag, The Netherlands
A. C. J. van Leeuwen, Atlantic Zone Programme, Estación Experimental Los Diamantes, Apartado 224, 7210 Guápiles, Costa Rica
H. N. van Lier, Department of Physical Planning and Rural Development, Wageningen Agricultural University, Generaal Foulkesweg 13, 6703 BJ Wageningen, The Netherlands
J. H. van Niejenhuis, Department of Farm Management, Wageningen Agricultural University, Hollandseweg 1, 6706 KN Wageningen, The Netherlands
J. van Orshoven, Institute of Land and Water Management (K. U. Leuven), Vital Decosterstraat 102, 3000 Leuven, Belgium
T. van Rheenen, Departments of Development Economics and Theoretical Production Ecology, Wageningen Agricultural University, Hollandseweg 1, 6706 KN Wageningen, The Netherlands
J. van Valckenborgh, Institute of Land and Water Management (K. U. Leuven), Vital Decosterstraat 102, 3000 Leuven, Belgium
W. van Vuuren, Department of Agricultural Economics and Business, University of Guelph, Guelph, Ontario, Canada N1G 2W1
J. van Zijl, Social Democratic Party (PvdA), PO Box 20018, 2500 EA Den Haag, The Netherlands
E. R. Veeneklass, The Winand Staring Centre for Integrated Land, Soil and Water Research (SC-DLO), PO Box 125, 6700 AC Wageningen, The Netherlands
T. Vetter, GTZ, QRDP, PO Box 18, Marsa Matruh, Egypt
A. C. Vlaanderen, TAUW Infra Consult, Curieplein 7, 7242 KH Lochem, The Netherlands
J. P. Watson, Department of Crop Science, University of Venda, PO Box 810, Louis Trichardt 0920, South Africa

V. Watson, Centro Científico Tropical, Apartado 8-3870 CP 1000, San José, Costa Rica
U. Wiesmann, Group for Development and Environment, Institute of Geography, Haellerstrasse 12, Berne, Switzerland
P. N. Windmeijer, The Winand Staring Centre for Integrated Land, Soil and Water Research, Department of International Cooperation, PO Box 125, 6700 AC Wageningen, The Netherlands
E. Woltjer, Social Democratic Party (PvdA), PO Box 20118, 2500 EA Den Haag, The Netherlands
A. M. Yemelianov, President's Advisory Council, Moscow and Agricultural Economics Department, Moscow State University, Moscow, Russia

Preface

The past decade has been characterised by a number of important developments in relation to land use: an unprecedented pace of population growth, our increasing awareness of the fragility of the earth's resources, the impressive growth of information technology and, last but not least, a growing recognition of the importance of involving all parties—in particular the land users—in the process of decision-making. Land use planning is evolving into a continuous process of interaction and adjustment of goals and options. Today, more than ever, we should carefully plan the use of natural resources and the application of indigenous knowledge to select appropriate land use scenarios that meet divergent goals of society, local communities and individuals.

For many years, scientists and planners from 'Wageningen', that is to say the Agricultural University and the National Agricultural Research Institutes in the Netherlands, have played a clear role in developing concepts and methodologies of land evaluation and formulation of land use options. The 75th anniversary of the Wageningen Agricultural University presented an excellent opportunity to take stock of these approaches and experiences, to exchange views with other scientists, policy-makers and all others interested in the use of the land, including the users themselves, and to consider future directions for land use planning. For this occasion, an international inter-disciplinary conference entitled, 'The Future of the Land: Mobilising and Integrating Knowledge for Land Use Options' was held in Wageningen on 22–25 August 1993. The conference was attended by 250 participants working in 46 different nations.

In addition to the plenary and parallel group sessions, numerous poster sessions and software demonstrations that were permanently on display formed an integral part of this conference. The invited contributions to this conference form the basis of this book.

In order to facilitate comparison of approaches and to foster optimal exchange of experiences, this book is structured rather tightly around issues of land use planning methodology at different scales or levels of aggregation (supranational, national, regional and farm). These are highlighted in a dual fashion: through extensive case studies and through accompanying full papers and abstracts dealing with specific issues or other cases at the same scale. The abstracts have mainly a signalling function and they therefore do not contain references. These abstracts, added in alphabetic order, provide the reader with a quick overview of the state-of-the-art of the subject matter. This book does not focus on one specific climatic zone but aims at illustrating common features among agro-climatic and economic zones in the world.

A total of seven case studies representing four different scales of aggregation are presented. An essential feature of each case is that different perspectives on land use planning are presented, including those of land users, scientists and decision-makers. This approach illustrates the requirements and interests of different parties, which may serve as a basis for negotiation. Methodological aspects related to this objective are also treated.

Section 1 sets the stage for the case studies by providing key background information on recent developments in land use planning, new techniques and tools, tensions between levels of aggregation, the struggle for interdisciplinarity, environmental goals and challenges, and the capability of people and institutions.

Section 2 deals with land use planning at the supranational level and centres around the case, 'Possible changes in future land use in the European Community'. The aim of this case is to illustrate the methodology and results of a comprehensive study on options for land use in the European Community.

Section 3 deals with the national level and centres around the case, 'Land use planning and land development in Thailand'. This case aims to describe: (1) current land use planning and land development activities and procedures in Thailand and problems encountered in their implementation, (2) the role of land management research in land use planning and extension in Thailand, and (3) integrated methods to assess biophysical and economic consequences of the introduction of alternative land use practices.

Section 4, dealing with the regional level, describes three cases: (i) 'Sustainable land use planning in Costa Rica: A methodological case study at farm and regional level' is aimed at definition of current and potential agricultural systems at farm and regional level based on extensive land resource analysis and mapping as (spatial) options in a multiple goal programming model; (ii) 'Competing for limited resources: options for land use in the fifth region of Mali', is aimed at confrontation of policy objectives and technical possibilities; and (iii) 'Peat land reclamation areas in the north of the Netherlands' aims at giving an adequate description and analysis of agricultural change, regional development, the planning process and the consequences for farmers.

Section 5, dealing with the farm level, describes two case studies: (i) 'QFSA: a new method for farm level planning', aimed at the introduction of Quantitative Farming Systems Analysis (QFSA) as a tool to explore development options for small mixed farms taking into account the existence of various stakeholders, multiple goals and development scenarios; and (ii) 'Future land use planning and land use options in Eastern Europe', aimed at the consequences of changing landownership in Eastern Europe and at describing the introduction of a Dutch type of buying and selling cooperative in a Russian agricultural society.

Finally, in Section 6 an attempt is made to look forward and present lessons and challenges with respect to the future of the land. A plea is made to activate existing platforms and create new ones for negotiating land use options between planners and farmers as one of the means to overcome in future years the current 'silence of thc users'.

L. Stroosnijder

Acknowledgements

Many institutions and individuals have participated in the process that resulted in this book. It reflects the ideas brought together in 1993 during the anniversary conference 'The Future of the Land, Mobilising and Integrating Knowledge for Land Use Options' of the Wageningen Agricultural University (WAU), and all those instrumental in the success of that conference deserve our thanks.

The Wageningen Agricultural University in general initiated and hosted this important conference. The personnel of its Congress Office, headed by Dr J. L. Meulenbroek, was most useful in coordinating the activities.

The International Agricultural Centre (IAC) provided logistic assistance and the scientific meetings were held successfully at the Wageningen International Conference Centre (WICC), where many foreign guests found a temporary residence.

The Programme Committee, consisting of scientists reflecting the institutional complexity of the Wageningen knowledge centre, spent many days discussing the numerous issues arising from the complex subject of land use planning. The active participation of the International Advisory Committee greatly contributed in bringing together such a diverse and outstanding group.

The conference would not have materialised without the financial sponsoring of the WAU, the Directorate General for International Cooperation (DGIS), Euroconsult, Heidemij and Grontmij consultants, The National Agricultural Research Organisation (DLO-NL), the Royal Netherlands Academy of Sciences (KNAW) and John Wiley & Sons.

The final work of the Editorial Committee has been made possible only through the secretarial assistance of the WAU Congress Office and the Departments of Agronomy and Tropical Land & Water Management in general. Thc wholehearted devotion of Ir S. W. Duiker, who acted as secretary to the Editorial Committee and the Program Committee, deserves special mention.

The Editorial Committee
November 1993

INTRODUCTION

Imaginable Futures: A Contribution to Thinking about Land Use Planning

L. O. Fresco

Wageningen Agricultural University, Wageningen, The Netherlands

When Macbeth pondered his future land use plans, which in his case meant murdering King Duncan in an attempt to gain access to various things, including land, he gave a very apt summary of his planning process, in words well known to most of us:

> If it were done when 'tis done, then
> 'twere well it were done quickly (*Macbeth*, I. vii. 1).

We know, of course, that as a planning procedure, his was not the most appropriate. The least we can say about Macbeth's plans is that they lead to a whole series of unintended consequences in the future.

This conference about land use planning deals specifically with the unexpected and with the future. The contributions to this conference show overwhelmingly that land use planning is evolving from a centralised process executed by experts at considerable distance from the field to something quite different: an integrated, interactive process with due attention to its unexpected consequences.

SELECTED ASPECTS OF LAND USE AND LAND USE PLANNING

Current land use

Strange as it may seem, accurate data on actual land use and land use changes are not easily found. This applies both to the global and continental scales as well as to the national and regional ones. Global figures about very broad land use categories are known, and show, for example, that as a result of production increases the aggregated

The Future of the Land: Mobilising and Integrating Knowledge for Land Use Options
Edited by L. O. Fresco, L. Stroosnijder, J. Bouma and H. van Keulen.

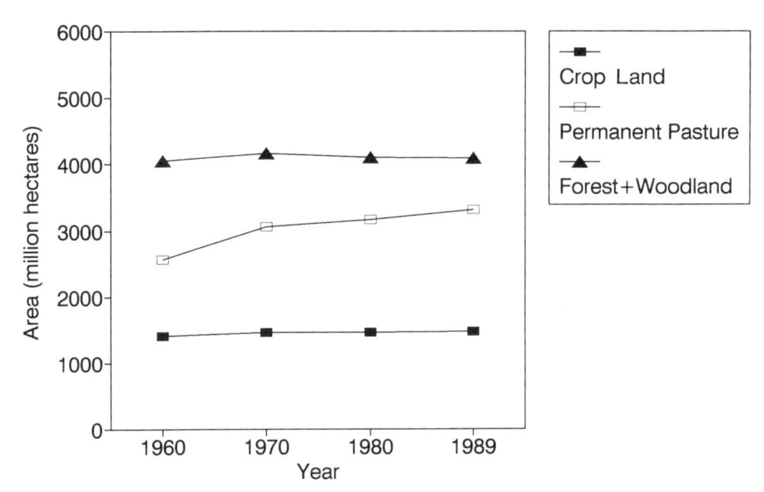

Figure I.1 Land use in the world, 1960–1989. Source: FAO production yearbooks

under crop land, forest and pastures, has remained stable during recent decades (Figure I.1). This is somewhat counterintuitive to the generally held view that the area under forest is rapidly declining, and that the cultivated area has increased considerably, at least in developing countries. This global trend does not reflect, obviously, the difference between primary, secondary or replanted forest, nor the results of a progressive takeover of forest land by crop land as a result of shortening fallows.

In fact, the figures show that our understanding of land use trends is severely curtailed by the idiosyncratic definitions of land use types adopted by various statistical agencies. A world-wide agreement on land use classification and its descriptors would be an important first step. Conceptual work on the unit of analysis of land use at different scales (for example, can land use be seen as a scale-neutral set of parameters?) is also required.

One thing that has become very clear as a result of our awareness of global processes is that many land use resources—be it CO_2, topsoil, water, biodiversity or labour—have become common resources that transcend national boundaries and do not belong to a single interest group. This is most obvious in the case of major catchment areas, such as the Amazon, that are shared by many nations. It implies that land use planning has truly international ramifications, even if the planning scale is regional or national and other goals or groups are not explicitly considered.

Forces driving land use changes

It goes without saying that the forces driving land use change are complex and act at various scales with differential rates of change. Among them, the most obvious are population growth and concomitant growth in demand for land use products, which differ considerably across the continents (Figure I.2). Infrastructural development, as a proxy for access to inputs and markets, is also geographically explicit, but many other socio-economic drivers are not. These are more difficult to include in GIS-based models.

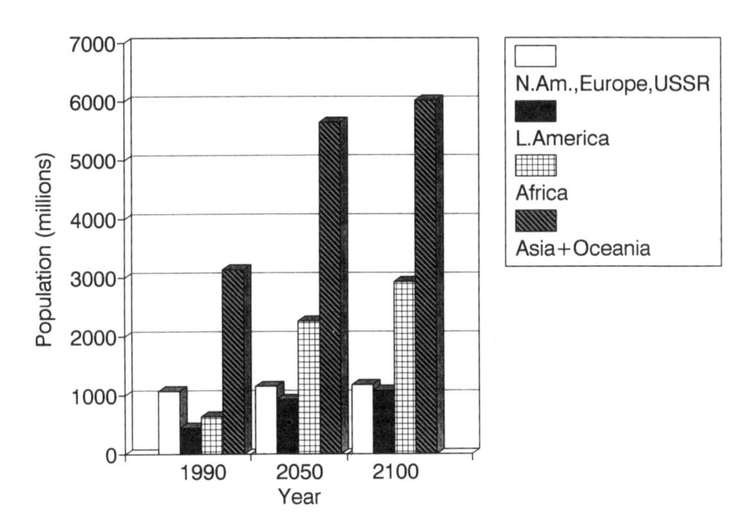

Figure I.2 Projected populations of major areas of the world, medium extension (UN projections)

We can speculate somewhat about the real important future changes, both on a global scale and regionally. In general terms, although the use of hydroponics and continental shelves and estuaries may gain in importance, the bulk of the worlds's carbohydrate production will remain land-based. Competition between agricultural and other land use types is likely to increase, in particular in the highly populated lowlands, leading to increased land use intensification and possibly concentration of agriculture in areas that are ecologically most favourable.

At present, we must be concerned also about the effects of current land use on the environment. Although many local aspects such as pollution and erosion are relatively well documented, it is still difficult to include these adequately in our land use planning models. The effects of past land use on present and future uses (for example, land degradation) are becoming so widespread (affecting nearly one-quarter of vegetated land in Central America; Figure I.3), that they cannot be ignored in descriptions of production techniques.

Land cover and land use

Land cover can be conceptualised as the layer of soils and biomass, in particular vegetation, that cover the land surface. About 90% of the land surface of the earth is covered by vegetation of some sort (10% is covered by ice or snow, and 1.5% by built-up areas; Table I.1). More than half of the world's land cover consists of cultivated land and pastures. Land use is the combined human action affecting land cover. From a global change perspective, land cover is the most important thing, and for the assessment of many aspects of sustainability—biodiversity, erosion, nutrient balances—understanding the linkage between cover and use is essential. Land use planning is man's systematic way of changing land cover. (The importance of land cover may be illustrated by the realisation that fossil fuels are nothing other than the remains of land cover from

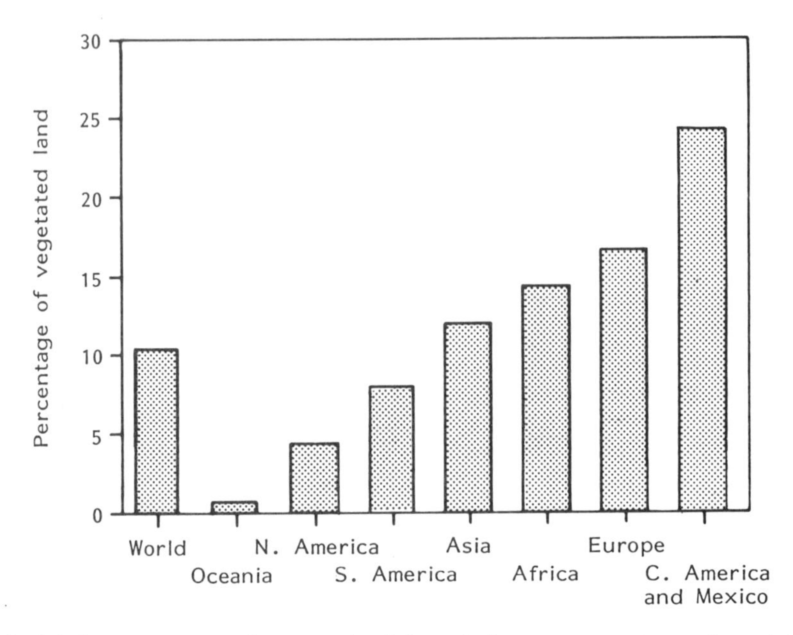

Figure I.3 Moderate, severe and extreme land degradation as a percentage of vegetated land, 1945–1990. Note: Data for European and Asian parts of the Soviet Union are included with Europe and Asia. Source: World Resources Institute (1992) (Reproduced by permission of the World Resources Institute)

a time on earth without humans.) Land cover has many functions, not only in terms of biogeochemical cycles, or food and shelter for humans, but also aesthetically, as landscapes.

Land cover is constantly changing, as a result not only of seasonal climate variations, but also of human action. The past centuries have been marked by massive land use conversion: the change from one category, natural ecosystems, generally forests and savannas, to agriculture and pastures. The rate of this conversion stepped up rapidly after 1945. As fewer and fewer areas of natural ecosystems are left, the future will be one of changes within the broad land use category of arable lands. In some regions, these may be reconverted to 'nature', when land is taken out of production, for example in the European community, and reverts back to what we tend to call 'nature'—often

Table I.1 Land cover categories

Covers	Area (km × 10^6)	Area (%)
Snow/ice	15	10
Cultivated	15	10
Pastures	65	44
Forest, etc.	45	30
Wetlands	6	4
Built-up areas	2	1.5

Source: D. Groetz (LUCC/IGBP/HDP CPPC, 1993).

an anthropogenic landscape of impoverished lands. But by far the most important land cover change appears to be the intensification of land use through better management of production factors.

Land use studies

Interdisciplinary research on land use is a relatively new field, although many aspects of land use, in particular agriculture, forestry and ecology, have been studied for many decades. Fascinating new subjects are emerging that allow us to integrate results from various disciplines into a better understanding of what is land use, what drives land use and how do land use changes lead to land cover changes. In particular, both historical and palaeoenvironmental studies focusing on the landscape are useful. The latter may help to understand the trajectory that a particular ecosystem is following and therefore indicate, for example, whether a particular rate of soil loss is typical of the Holocene or has been accelerated in recent times. Historical studies, on the other hand, may, for example, highlight the driving forces in different socio-economic and demographic conditions of land tenure.

The tools

During the last decade, the combined use of simulation models, expert systems, geographical information systems, various types of databases and multiple goal planning techniques have allowed us to formulate technical options for land use in a much more precise and varied way. Not as one-dimensional blue prints, but as scenarios for policy choices.

The presentation of scenarios is not a new technique; it is already well known in demography and economic planning. Scenarios are not to be confused with forecasts: they do not predict, but allow us to explore technical options based on explicit assumptions given a set of goals. More specifically, the cost of attainment of one goal can be expressed in terms of the reduction of the attainment of other goals, thus forcing policy-makers and land users to make explicit choices. In other words, land use models may help to make potential conflicts more visible.

THE CHALLENGES OF LAND USE PLANNING

Today's paradox is that, notwithstanding the great technological advances and our increased knowledge of the natural resource base, land use planning has not become easier and the challenges are perhaps greater than ever. We may attribute this to several factors.

The diversity of land users

Many goals, values and production techniques cannot be expressed adequately in quantitative terms, nor can they always be geo-referenced in a direct way. This applies to such diverse features as differences between individual households, or cultural differences between societies. Land users are far from homogeneous, even if they are all farmers: some may be full-time farmers, others depend on income outside agriculture, yet others have opted for specialised ecological farming or on-farm processing of farm products. Some may have been market-oriented farmers for generations, others may still depend, in part, on hunting and gathering. One of the challenging issues is, therefore, how to incorporate this local cultural diversity and understanding of the environment into the planning process. One word of caution is required, however, for those who tend to overemphasise so-called traditional or 'folk' knowledge indiscriminately: this kind of knowledge is rarely adequate in rapidly changing or environmentally degrading situations.

A crucial question, therefore, is to what extent both crop modelling and multiple goal planning techniques can really do justice to the diversity of land users, who use such a variety of crop species and animal races, and whose production techniques may vary so widely from one household to the next, even within a seemingly uniform agro-ecological and socio-economic environment. This problem can only be solved partly by extending the input–output matrix. We may need to use frequency distributions or probability functions, or stochastic models if the pattern of diversity between land users is unknown.

Diversity of goals in the planning process

All over the world, human societies have diversified tremendously in the last two centuries and nowhere can we find societies that act as monolithic blocks with unified objectives. On the contrary, as witnessed by many political conflicts in recent times, the rift between social and political groups seems to be widening and diverging objectives remain. Land use planning as a top down and static exercise has not only deserved its bad reputation, but it is also bound to fail, especially today. Indeed, perhaps the term 'planning' in itself should be avoided because of these past connotations.

We do not know very well how to deal with conflicting societal goals, although several case studies presented at this conference make such an attempt. Most models operate with a rather limited set of agricultural activities, and only one category of natural ecosystems, if any. As diversification of agricultural land use becomes more important, for many reasons, we must find ways to deal with this type of diversity as well, without unduly burdening the activity matrices in the models. New goals, in particular the goal of sustainable land use and the need to preserve the earth's natural resource base, require the building of a collective consensus within societies and between nations. This implies that planning can never be just a technical venture, but must involve a long-term participatory process which gives a voice to those groups that are seldom heard.

Many scientists at this conference work with interactive multiple goal planning, yet the advantages of this approach are often under-utilised. It is unclear how interactive the process really is, or how many examples there are of a truly participatory planning process. Procedures, in technical and organisational terms, to provide users' feedback into the models, seem to need much more development.

Future uncertainties

We are planning for an uncertain future. Perhaps uncertainty has always prevailed, but we are certainly more aware of it than before. Not only are the projections of population increase and economic development uncertain, our understanding of the processes involved in global change, and the magnitude of their impact is extremely limited. Yet these biophysical processes, both locally and globally, set the conditions for future options. In other words, many features of land use determinants, as well as their interactions, can only be guessed at today. This also applies to important ecological parameters such as biodiversity, for which there is as yet little consensus on definition or methods of measurement.

Model limitations

The ease of data processing, computation and mapping has greatly improved the degree of detail with which results can be presented. At the same time, however, there is increasing awareness of the inherent limitations of modelling. The quality of data sets, and the costs of obtaining these for the purpose of modelling, present a serious problem. An important question is, therefore, to what degree the accuracy of input data are reflected in the model's outputs, or, put differently, how much inaccuracy our models can tolerate without producing ridiculous results. In the same vein, questions may be raised about the sensitivity of models, not only to the quality of input data, but also to the fact that some routines or subsets of parameters are included in more detail than others and tend therefore, to influence model results disproportionally.

In land use, we are dealing with processes that are highly dynamic in themselves and result from the interaction of many parameters, each with their own time and spatial scales. Optimisation, a technique used in much of the work presented at this conference, is not without its difficulties, in particular on the conceptual side. It is the land use modeller who selects what is optimal, be it the balance of nutrient inputs and outputs at farm level, or the highest national production of biomass at the lowest energy costs. Possibly, mathematical procedures that allow us to retreat from optimisation to the generation (for example, through Monte Carlo techniques) of a set of acceptable outcomes may be worth exploring.

Validation of the procedures, and of the model as a whole, is quite often impossible, because of the dominance of context-dependent knowledge. Yet, as in any scientific endeavour, this must be a fundamental concern to all of us. This becomes all the more urgent due to the proliferation of models all over the world, and the lack of exchange and standardisation between modellers.

Even more important, we must never forget that models are, at best, tools that are useful only if applied with wisdom and commitment. We all know this, of course, but we must also bear the moral responsibility of watching closely the use of our models by governments and pressure groups. Models, especially if accompanied by glossy maps and displays in which the uncertainties are rendered invisible, are impressive but may be misleading tools. Unfortunately, it would be difficult to formulate a professional code to define the responsibilities of land use modellers and scenario builders, because of the inherent value judgment embedded in them. But nevertheless, perhaps we should try.

To conclude, land use planning is undergoing radical changes, as a result of three fundamental developments:

(1) new information technology,
(2) increased awareness of the need to apply scientific knowledge to safeguard the future of the earth and to distribute resources evenly, and
(3) the growing political wish to involve all parties in the definition and implementation of the best possible sets of land use and to refrain from overly centralised government interventions.

Unintended consequences of planning are, by definition, difficult if not impossible to foresee and they are inherent to any type of planning. There is always a danger that governments or land users, like Macbeth, have some equivalent of Lady Macbeth as a land use planning consultant, who, in response to Macbeth's doubts:

> If we should fail—

will reply:

> We fail!/ But screw your courage to the sticking place,/ And we'll not fail (*Macbeth*, I. vii. 59).

We know what came of that. Notwithstanding the critical attitude that one should constantly have about planning and planners, we are in dire need of some form of concerted action to decide about the future of the land. The mobilisation and integration of knowledge is not a sufficient condition for this, but is definitely a necessary one.

PART I

RECENT DEVELOPMENTS IN LAND USE PLANNING

CHAPTER 1

Recent Developments in Land Use Planning, with Special Reference to FAO

R. Brinkman

Land and Water Development Division, FAO, Rome, Italy

INTRODUCTION

The relative dominance of planning by national governments, intermediate organisational structures, individual farm families or land users has varied with the concentration of power and the need for coherence or synchrony in land use activities. Concentration of power made it possible, for example, to build run-of-the-river irrigation systems in arid or semi-arid areas, providing water to land hitherto beyond the reach of river water from annual floods, but in proportions under control of a single authority, rather than nature and the individual land users. In rainfed farming areas with grazed fallows, the decisions on timings of sowing and harvest are often made communally rather than by individuals, to minimise the need for control of livestock movements during the fallow periods.

In cases where land users have been on the land for generations and where the physical environment or the technical possibilities or the socio-economic conditions have not changed drastically, planning and management decisions made by these people tend to be near optimal within their total environment. Optimal is meant here in the sense that it is very difficult to find modifications of planning or management that would significantly improve the economic or social situation of the land users in the short or medium term within the limits of the technology known to them and available to them. Land users generally do much less well when faced with a different (new or changed) environment. People may have moved to find land or escape hazardous conditions; the physical environment may have changed (a multi-year drought, for example); traditional crops may have become subject to severe pest or disease pressure,

The Future of the Land: Mobilising and Integrating Knowledge for Land Use Options
Edited by L. O. Fresco, L. Stroosnijder, J. Bouma and H. van Keulen. Published in 1994 by John Wiley & Sons Ltd

or their price may have become uneconomic; the overall price relationships between produce and inputs may have shifted; or new inputs, tools, crops or varieties, or new management techniques may have become available. Probably the most important single change agent in many developing countries is the very rapid rise in population pressure.

Governments also have been doing less well in conditions of rapid change. Land use policies in many countries have been based on general estimates or assumptions on land qualities, or on the assumption that past trends in production can be extrapolated to the future, by means of a technology development factor 'explaining' the steady productivity increases, per unit of land and per person per day, that have characterised agriculture in most of the world over the last several decades. In a number of industrialised countries these policies, combined with protectionist objectives favouring existing agricultural production systems, have led to land and water degradation by pollution through the excessive applications of plant nutrients and agrochemicals. In a number of developing countries similar policies, combined with objectives favouring urban and industrial growth at the cost of returns to agriculture, have led to land degradation by processes such as water and wind erosion, plant nutrient depletion and salinisation. In conditions of rapid change, the informal, gradual accumulation of local experience as a basis for planning and management decisions becomes a too slow and expensive learning process. More formal and rapid methods of information gathering and interpretation are then needed to support decision-making with a broader and more systematic information base.

PLANNING POSSIBILITIES AND NEEDS FOR ANALYSIS, PLANNING AND NEGOTIATION

Where land is relatively abundant compared with the needs of the people dependent on it, land use can be planned informally or, if by formal methods, relatively simply, for example, by sequentially deciding on the best locations for different uses. With increasing pressure on the land, and increased competition by different users or classes of use (or different options for a single user), planning requires recognition of the fact that even an individual has multiple objectives, rather than a single objective such as income maximisation. These objectives can be made explicit to a greater or lesser extent. Optimisation, therefore, requires methods which assign weights and priorities to the different objectives. The availability of quantitative or semi-quantitative information about the objectives, as well as about the potentials of the land, is one of the preconditions for the application of such optimisation methods.

To date, semi-quantitative methods have been developed to estimate land productivity under different uses and with different combinations and intensities of inputs (labour, energy, nutrients, information, etc.), allowing a comparison of benefits with different inputs. Linear and nonlinear optimisation procedures are available to arrive at optimal land use patterns in space, once an objective function has been collapsed into a single objective (by translating certain objectives into constraints, and others into a common variable). Multi-objective evaluation methods, necessarily interactive, have become available for 'simple' problems, but are not yet designed to deal with spatially differentiated data.

So far, the concept of time—or better, change—has hardly surfaced in land use planning. The physical information base underlying decision-support systems, derived from primary information by agro-ecological zoning methods, is static: it reflects the situation at a given time (including any major land improvements), or an average over a period of several years or decades, or as in more recently developed examples, it describes a time-invariant probability distribution. Dynamic models have been developed for primary productivity or crop yield under specified management and environmental conditions, for individual growing seasons or years as well as for sequences of years, but these are normally limited to a site; not spatially articulated. Similarly, dynamic sectoral or intersectoral economic planning models have been developed, in which the agriculture sector, or a few agricultural subsectors, are considered as aggregates without spatial articulation.

Planning for optimal land use patterns in space under conditions of scarcity or competition, presupposes that the concept 'optimal' is defined or at least described in explicit operational terms. The descriptions will generally not be identical for different land users, or for government or commune compared with individual or farm family. The multiple objective function of an individual farm family may include, for example, food security of the family from its own land by maintenance of a variable stored surplus; a level of 'assured' income (achieved in most years, not interrupted by more than one year with lower income); a reasonably steady surplus of a readily traded or stored commodity, or other objectives. The multiple objective function of a central government, in so far as relevant for the country's land resources, might include a reasonably steady or increasing supply of exportable commodities; food self-sufficiency of the country as a whole, including its cities, or of specific provinces or regions; maintenance of certain areas of natural vegetation including its wildlife, as a basis for tourism; or physical security by an adequate density of population in specific parts of the country.

Sound policies start from the ground upwards in two ways. First, from the perceptions, needs and objectives of the land users (or the government) and, secondly, from systematic knowledge of the resources (natural environment, water and land potentials, degradation hazards, present infrastructure and inputs available, existing policy environment, people's expertise and the availability of technology).

Ideally, the land use plans of governments and their reflection in legal and institutional structures and in physical infrastructure, form a facilitating environment within which land use decisions by farm families, grazier communities and others can lead to near-optimal satisfaction of their objective functions. In practice, the situation is, and was, different in most times and places. Major early examples are the forced cultivation of certain cash crops by smallholders at the behest of various colonial governments. More recent examples are the centrally planned cotton cultivation in the Syr Darya and Amu Darya plains in the southern republics of the former Soviet Union, and efforts by several central governments to eradicate or control opium or coca cultivation. A more differentiated example is the different land use objectives of Indian tribes, gold miners, smallholder settlers, large-scale ranchers, and state and national governments in the Amazon region.

Next in the conceptualisation of decision-support systems is, therefore, a meta-objective: a consensus on land use, to be approached in three steps:

(1) identification of the degree to which the objective functions of the different actors in the land use planning process overlap and the ways in which they contrast or may give rise to conflict;
(2) land use optimalisations on the basis of the various objective functions of the different actors and analysis of the extent to which the different optimalisation runs lead to similar land use patterns for the area; and
(3) development and application of interactive methods to maximise the extent of consensus in a land use pattern.

The information used and developed in these steps can then form a common basis and a tool for arriving at a negotiated solution for the remaining differences.

DECISION-SUPPORT CONCEPTS IN FAO

The basic decision-support concepts are straightforward, but have taken some time to become clear. The general decision sequence (Figure 1.1) comprises:

(1) identification of goals or objectives;
(2) identification of (land use) options and their potentials; and
(3) selection of the mix of options that optimises attainment of the objectives.

The general sequence, and the decision framework derived from it, are independent of scale and of the nature of the objectives or of the options identified. Therefore, a decision support system built along these lines can be used at any level, for national or province level land use planning by a central authority, as in several applications to date, or for local extension support to planning and management decisions by farm families. A decision-support tool for land use planning can also be used to estimate what would be the probable consequences of central, high-level decisions changing the policy environment on land use decisions by large numbers of individuals. Such an application might have great benefits in the case of major decisions such as on changes in the Common Agricultural Policy. Tools are needed for each of the three steps mentioned.

While decision support will work at any level of detail, the basic data and the derived information underlying the system need to be at the appropriate scale and level of detail. The land-resources database could be at a 1:1 million scale or about 1 km resolution, for example, for a national land use plan in a medium-sized country, with land uses broadly defined in terms of produce and input levels (intensities). For meaningful support and advice within a village territory, land-resources data should be much more detailed, or semi-detailed with additional information supplied by the land users themselves, and land uses much more closely defined. Social and economic data should also refer to the specific environment and viewpoint of the user, for example, local market prices for village-level advice, and perhaps shadow prices for medium-term national planning.

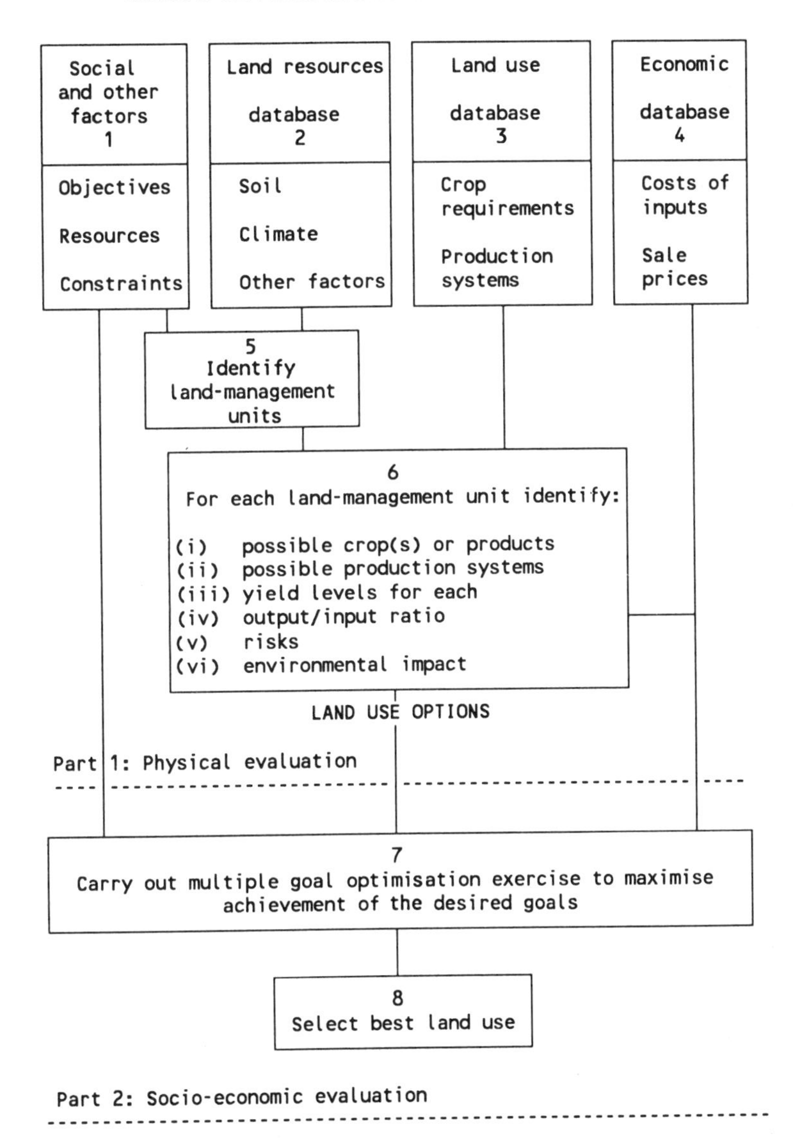

Figure 1.1 Simplified sequence for making decisions about the best use for land. Note: political factors are included under 'Social and other factors'; the desire to conform with the constraints and values of the social environment is included under 'Objectives'

COMPONENTS FOR A DECISION-SUPPORT SYSTEM

A number of tools and components are available, or being developed, that facilitate the assembly of a decision-support system for land use planning. It is most efficient *not* to attempt to build a monolithic system, but to create the possibility of using alternative modules for a given task, depending on the availability of data or on the needs of the user, and to encourage continual work on upgrading the least satisfactory modules.

FAO has published *Guidelines for Land-use Planning* (FAO, 1993), primarily intended for people engaged in making land use plans, or training to do so. The guidelines also provide an overview for administrators and decision-makers. These guidelines can be adapted to local conditions, and adjusted, often at the initiative of the land users, as the planning and implementation proceed.

Databases and agro-climatic interpretation

After publication in the 1970s of the FAO–Unesco Soil Map of the World, updated and more detailed soil-resources information has become available in many, but not all, countries. Much of this can be correlated into a common system of classification. The Revised Legend of the *Soil Map of the World* (FAO/Unesco/ISRIC, 1990) and the *Guidelines for Soil Description* (FAO, 1990) provide a common methodological base; a manual for third-level classification within the Revised Legend structure, a system for topsoil classification and a systematic descriptive guide for soil climate are in preparation. The FAO–ISRIC Soil Database (FAO, 1989) provides a common format for recording, retrieval and interpretation of point information on soil profiles, and is in use in several countries. The latest version of the SOTER Manual (UNEP/ISSS/ISRIC/FAO, 1993) provides a computer-compatible framework for national (and continental-scale) integrated soils and terrain databases that can be used in a GIS environment.

Agro-climatic databases compatible with agro-ecological zoning requirements can now be built on the basis of data extracted by various FAO programs, including METEO, from records in CLICOM format, the WMO standard currently in use in many countries. These data can be analysed in terms of means and probability distributions needed to identify and estimate risks and optimal response strategies for land users, as done by Brammer et al. (1988) in Bangladesh.

Goals and objectives

Several early tools to elicit the main goals and objectives of land users ranged from time-consuming (but solid) to ineffective to indiscriminate: some required the presence of a trained researcher in a village for a year or more; others merely provided the investigator with the answers that the respondents guessed the questioner would wish to hear; still others largely reflected the views of the dominant group. Recent improvements enable the land users, for example, to express themselves more freely, and also help them to visualise for themselves and convey to the researchers their main goals, needs and concerns.

The USDA Land Capability Classification (Klingebiel and Montgomery, 1961) and its several variants developed in different countries have a built-in priority sequence of uses that make them a simple land use planning tool—albeit an inflexible one. Other decision-support tools for land use planning developed in the 1950s and 1960s were restricted to one main kind of use, such as large-scale irrigation agriculture, and used a single, economic, criterion as the objective.

Land use options and their potentials

The *Framework for Land Evaluation* (FAO, 1976) provided a summary of the procedures to be followed for a systematic assessment of the potentials of the different kinds of land for a number of uses likely to be relevant in the area. Most of the other aspects mentioned (subsumed under 'social and economic context') were touched upon but the Framework did not provide systematic procedures to deal with them. Economic aspects were dealt with in later *Guidelines: Land Evaluation for Irrigated Agriculture*, for example (FAO, 1985). Several national applications are touched upon below.

During the 1980s, the methodology used for global agro-ecological zoning and the estimation of potential population-supporting capacities in the developing world (Higgins et al., 1982) was refined and adapted for use in national-level planning, using data on land resources and other information available for Kenya as a basis for the pilot study on *agro-ecological resources assessment for agricultural development* (FAO/IIASA, 1991, 1993). The method involves:

(1) formulation of land uses and their ecological requirements;
(2) compilation of a national land-resources and land use database;
(3) assessment of land productivity potentials under different uses; and
(4) development planning, including assessment of potential population supporting capacities and input requirements to address policy issues.

The crop productivity program, on the basis of land uses and their ecological requirements, presently covers 25 crop species (seven cereals, six grain legumes, three roots and tubers, nine cash crops), differentiated into 64 crop types and into 174 crop land utilisation types on the basis of management differences and input levels. For the livestock productivity program, information on 20 pasture and four fodder grasses and eight pasture and fodder legumes was combined with six livestock husbandry types. Cattle, goats and sheep were considered under pastoral and non-pastoral regimes; camels under a pastoral regime only; poultry and pigs under non-pastoral systems only. The fuelwood productivity program includes 31 species, of which 15 are N fixers. The soil erosion and productivity model quantifies implications of different land uses on topsoil loss and on its impact on land productivity under different soil conservation measures, leading to an estimation of soil conservation needs and costs. The photosynthetic and phenological crop requirements were matched against a climatic inventory for the country, including day length (of minimal importance in Kenya), thermal zone and length of the growing period, including its year-to-year variability. Edaphic crop requirements were matched with available soil characteristics considered relevant (soil unit, phase, slope, depth, texture, salinity, stoniness, etc.) in terms of optimum ranges and minimum and maximum values. For planning purposes, a range of other information items was included in the land-resources database: cash crop zones, forest zones, parkland zones, irrigation schemes, tsetse infestation areas, provinces and districts.

In the *Land Resources Appraisal of Bangladesh for Agricultural Development* (Brammer et al., 1988), flooding was systematically recorded and used in the appraisal for the different wetland uses (the various types of wetland rice and jute cultivation).

Although flood levels may vary by as much as a metre in different years, land levels in relation to mean flood level, in combination with estimated rate of rise of floodwater, are used by planners and farmers to estimate the potentials of the land for different crop types. Similar modifications of standard land evaluation procedures are needed in other lowland areas, such as the Mekong delta.

The set of programs in *CYSLAMB, Crop Yield and Land Assessment Model for Botswana* (De Wit et al., 1993), tuned to a semi-arid environment, can be used to estimate the yield of a particular crop-production system, well defined, on a specified land unit. It can be used with average weather data, but also with data for individual years, allowing the evaluation of year-to-year yield variability and risk assessment as part of land suitability estimation.

In CYSLAMB, yield estimates are made by:

- calculating radiation- (and temperature-) limited biomass yields for all possible planting decads in the growing season;
- identifying each planting opportunity from a moisture balance involving effective rainfall, bare soil or weed evapotranspiration, percolation and runoff;
- calculating moisture-limited yield from crop evapotranspiration through the crop growth period as limited by water balance, starting at each planting opportunity;
- reducing the moisture-limited yield for nutrient supply, drainage conditions, salinity or sodicity and high pH.

The system is completely modular, uses easily accessible databases on climate, soils, vegetation, crop requirements and production systems. The calculations were tested against a data set comprising sorghum, maize, cowpeas and groundnuts for the aspects of radiation, actual crop evapotranspiration, bare soil evaporation, water balance, moisture-limited yields and fertiliser effects.

Results are being used for different purposes, and at different scales of planning. In conjunction with socio-economic data for some well identified classes of farmers, a decision tree for crop-production practices is being built, covering options for crop production in different stages of the season depending on the resource base of the farmers. Also, guidelines and zoning maps for the extension of village and peri-urban areas are being prepared on the basis of objective criteria, including land suitability for the main crops and actual land use.

Arriving at optimal solutions

Computerised system for agricultural planning and policy analysis

A database for food and agricultural policy analysis and planning, with proven tools for policy analysts and planners, K2, is being built on the basis of CAPPA, the existing system for agricultural and population planning assistance and training (Verceuil et al., 1991). The new version will include aspects of environment and sustainability, and will deal with forestry and fisheries besides agriculture. It will have a facility for disaggregated analysis on the basis of a module for land resources utilisation and projection, as well as for analysis of livestock production. The user will be assisted at the time of making projections, by indicators to guide projection towards better consistency of the scenario, by adjustment procedures for sensitivity analysis of a scenario to a change in parameters, and by a linear programming facility in the production modules.

The land resources utilisation and projection module will have tools to analyse land resources and their potential sustainable use, and to compare this potential with actual or projected land use; to project land resources (degradation, investments for improvement) and to project changes in major types of land use (forest, pastures, arable cultivation, etc.). The module will include a spatial dimension, with the possibility of using a GIS. The methodology will draw on AEZ work by FAO's Land and Water Development Division and will be designed to suit three levels of data availability.

The ongoing work on land use planning in Mozambique is a recent example of a good balance, from the start, between social and physical aspects, and of a situation where the urgent need for the results of land use planning is forcing innovation, people's participation and strict limits on the degree of detail and perfection to be aimed for. There is relatively good climate and land resources information at a national level, complete soil map coverage at 1:250 000 scale for two provinces out of six and for parts of other provinces; a computerised system of land evaluation linking METEO (Schalk, 1990), the *FAO-ISRIC soil database* (FAO, 1989), and a system of defined land/climate units for the country. Some of this information still requires checking for accuracy and the database program and codes need to be translated into Portuguese. Besides this information, land cover is being mapped; two districts have been completed so far. Also, social surveys have been started and are planned to be conducted in each district with a view to identifying the present socio-economic communities, finding out how they operate and what their resource management customs are.

Land tenure issues in the country are complicated by policies enforced before and after Independence, and by the fact that customary tenure and inheritance systems differ from one community to another. Also, the inflow of people into the more secure areas and abandonment of land elsewhere, as well as the reverse flow that has now started, have further complicated the present occupation of the land. Therefore, the actual communities now existing need to be identified, a system of communal management of common resources recreated, and land allocated for cultivation and other purposes. The communities now identified will become the *de facto* lowest level of local government, and it is recognised that they will have different resource allocation and management systems.

Different government agencies and non-governmental organisations are involved, and there are major communication, correlation and coordination tasks ahead, both among these and with the people in the villages and districts. For districts for which social and physical information is available or being assembled, the land use planning group is to draft a simple land use and management plan by broad land types based on, among other things, landforms, soils, present land cover (crop sequences or vegetation) as well as on the (more detailed) knowledge of the local population. After full discussion and modification by the local community, this will then be formally adopted as the basis for future development. The management area and its boundaries will be recognised by the Government as an administrative unit. This will be followed by emphasis on the development of an effective agricultural extension programme and of infrastructure including roads and markets.

CONCLUSIONS

Several semi-quantitative methodologies are now available for agro-ecological zoning on the basis of more or less complete data sets. Because they were designed for country-specific natural and socio-economic conditions and tuned to data sets available or potentially available, most are monolithic programs rather than modular constructs. This restricts their immediate utility even in rather similar environments elsewhere.

A clear structure of well-defined modules for agro-ecological zoning procedures should be set up and disseminated to facilitate upgrading and modification. A general agreement on module boundaries will be needed, with exact definition of those interfaces, so that applicable parts of different systems can be matched into a new structure for a new environment. FAO's programs ECOCROP-1 (FAO, 1994), providing a general crop selection procedure for environments, and ECOCROP-2 (in preparation), providing a quantitative match between crop and environment, form part of the ongoing effort to create a set of independent, but well-matched modules for interpretation and decision support.

Linear programming methods should be more generally used to explore the different land use patterns emerging as optimal under different combinations of objectives and constraints.

Interactive multi-objective methods should be adapted to accept spatially differentiated data sets, or at least, several distinct kinds of land with defined extents and potentials for various uses.

Planning tools allowing dynamic modelling of the economy, or of a sector, on the basis of different choices of values for initial conditions and rates or other parameters, should similarly be adapted to allow the introduction of several distinct kinds of land with their specific production functions, rather than the agriculture sector as a whole or complete subsectors without spatial differentiation.

A plan based on imperfect information but having the agreement and consensus of the people and the government is better than a 'perfect' plan prepared without the people.

Applications should be developed that enable groups of land users to explore and identify different land use options and management alternatives for their own land.

A structure and procedure of negotiation, founded on a common information base on land and water resources and their potentials, is needed to assist governments and groups of land users to come to a maximum degree of agreement on land use decisions and thus to aim for the creation of optimal economic and physical environments for agricultural development within the government's wider objectives. Such agreement is probably approached most efficiently if at the outset, the goals and objectives of the different land users and of government are set out explicitly.

REFERENCES

Brammer, H., Antoine, J., Kassam, A. H. and Van Velthuizen, H. T., 1988. Land resources appraisal of Bangladesh for agricultural development: Report 4, Hydroclimatic resources: Vol. I, Climatic resources inventory. BGD/81/035 Technical Report 4.I. FAO, Rome.

De Wit, P. V., Tersteeg, J. L. and Radcliffe, D. J., 1993. Crop Yield Simulation and Land Assessment Model for Botswana (CYSLAMB), Part I: Theory and validation. TCP/BOT/0053 Field Document 2. FAO/Government of Botswana. FAO, Rome.

FAO, 1976. *A Framework for Land Evaluation*. Soils Bulletin 32, FAO, Rome.

FAO, 1985. *Guidelines: Land Evaluation for Irrigated Agriculture*. Soils Bulletin 55, FAO, Rome.

FAO, 1989. *FAO-ISRIC Soil Database*. World Soil Resources Report 64, FAO, Rome.

FAO, 1990. *Guidelines for Soil Description*, 3rd edn. (revised). FAO, Rome.

FAO, 1993. *Guidelines for Land-use Planning*. FAO Development Series 1, FAO, Rome.

FAO, 1994. *ECOCROP1, the adaptability level of the FAO crop environmental requirements database*, version 1.0. Brochure and two-diskette set, FAO, Rome.

FAO/IIASA, 1991. *Agro-ecological land resources assessment for agricultural development planning: a case study of Kenya. Resources data base and land productivity*. World Soil Resources Report 71 (main report) and 71/1 to 71/8 (technical annexes), FAO, Rome.

FAO/IIASA, 1993. *Agro-ecological Assessments for National Planning: The Example of Kenya*. Soils Bulletin 67, FAO, Rome.

FAO/Unesco/ISRIC, 1990. *Revised legend, FAO-Unesco Soil Map of the World*, reprinted with corrections. World Soil Resources Report 60, FAO, Rome.

Higgins, G. M., Kassam, A. H., Naiken, L., Fischer, G. and Shah, M. M., 1982. *Potential population supporting capacities of lands in the developing world*. Technical Report of project INT/75/P13, Land resources for populations of the future, undertaken for UNFPA by FAO/IIASA, FAO, Rome.

Klingebiel, A. A. and Montgomery, P. H., 1961. *Land-Capability Classification*. Agricultural Handbook 210, Soil Conservation Service, US Government Printing Service, Washington, DC.

Schalk, B. 1990. METEO: a climate database system for agricultural use. System description and diskette. AG:BOT/85/011 Technical Paper 5. FAO/UNDP/Republic of Botswana, Gaborone.

UNEP/ISSS/ISRIC/FAO, 1993. *Global and National Soils and Terrain Digital Data Bases (SOTER)*. Procedures manual, World Soil Resources Report 74, FAO, Rome.

Verceuil, J., Bretton, J. and Marcoux, A. 1991. *CAPPA Manual: Computerised Systems for Agricultural and Population Planning Assistance and Training*. Training Materials for Agricultural Planning 22, FAO, Rome.

CHAPTER 2

New Techniques and Tools

J. Bouma[a] and K. J. Beek[b]

[a]Wageningen Agricultural University, Wageningen, The Netherlands.
[b]ITC, Enschede, The Netherlands

INTRODUCTION

The study of the land is being revolutionised by the introduction of many new techniques that allow continuous observation of land features and monitoring of land conditions, including development of crops and natural vegetations. Application of remote sensing techniques can provide information on conditions of the land surface as a function of time. Aside from observation and monitoring techniques, natural processes can increasingly well be characterised by simulation modelling, allowing predictions of future conditions of the land to be a function of different management scenarios. Many models are available, ranging from qualitative expert systems to detailed process-oriented models. Their use is facilitated by fast computers which are widely available, even in remote areas.

Such simulation techniques follow different procedures for different objects of study at different scales, characterising conditions at observation points: fields and ever larger areas, ranging from counties to the complete world, require different data and models. However, studying the land implies more than characterisation of its dynamic physical, chemical and biological properties and the associated development of its vegetation or agricultural crops. Consideration of alternative forms of land use implies also the need to express the effect of social and economic factors on land use. For this purpose, interactive multiple goal programming techniques have been developed to allow objective comparisons of land use alternatives.

The challenge at present is to channel these modern opportunities for research into a systems approach that recognises the interrelationships of the many factors involved in land use and that is focused on clear and relevant objectives rather than on the techniques themselves. In this chapter, technical developments are discussed in terms of observation and monitoring techniques to be applied at different scales. Also, simulation models are reviewed and the need to manipulate many data in geographical

The Future of the Land: Mobilising and Integrating Knowledge for Land Use Options
Edited by L. O. Fresco, L. Stroosnijder, J. Bouma and H. van Keulen.

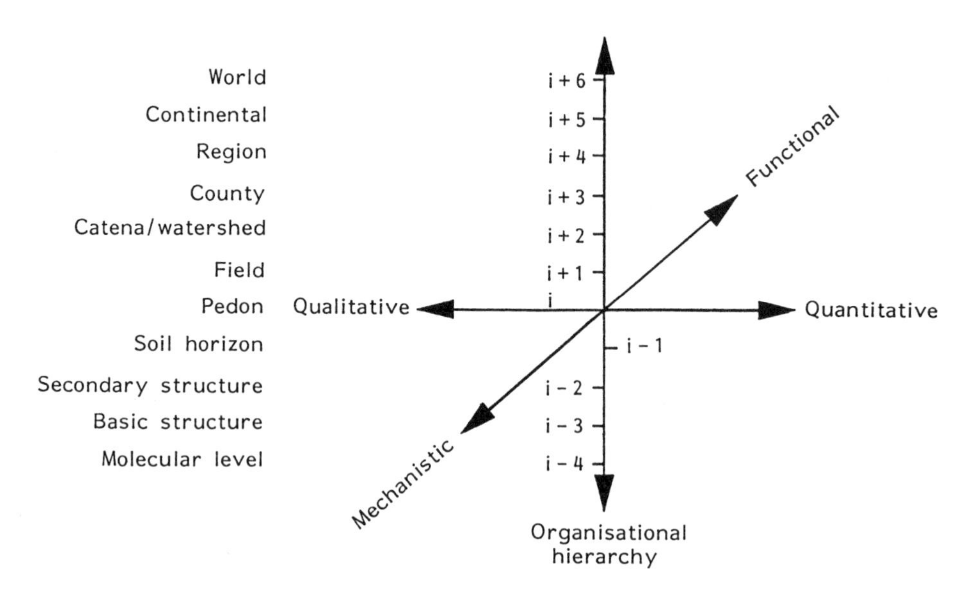

Figure 2.1 Schematic diagram illustrating types of modelling and different levels of hierarchy (after Hoosbeek and Bryant, 1992)

information systems is emphasised. Planning procedures to derive alternative land use scenarios are briefly reviewed as an example of applying socio-economic conditions to models that are focusing on solute movement and plant growth, based on physico-chemical processes. A final plea is made to always involve expert knowledge for the purpose of guiding studies on the land: even the most detailed simulation model is only a sketchy pitiful representation of a highly complex, and unique, reality.

In discussing techniques and tools, we follow the classification of organisational hierarchy as proposed by Hoosbeek and Bryant (1992). In discussing models of pedogenesis, they distinguish different levels of hierarchy, starting at level i for the pedon, the individual soil. Higher hierarchial levels are catenas (i + 1); watersheds (i + 2); etc., while lower levels are horizons (i – 1); soil structure (i – 2), etc. (see Figure 2.1).

MONITORING AND OBSERVATION TECHNIQUES

Development of electronics and information technology has resulted in many new techniques that allow reliable continuous monitoring of subsurface physical conditions *in situ*. Pressure heads of the soil water (level i – 1) can be measured with transducer tensiometry and water contents with time domain reflectometry (TDR) (Topp et al., 1980). At i – 2 level, non-destructive CT (computer tomography) and NMI (nuclear magnetic imaging) techniques have been used to characterise pore patterns in soil and flow processes of water (Heijs et al., 1994). Various radiation techniques have been used successfully to determine *in situ* the chemical and mineralogical composition of soil materials at i – 3 level (Jongmans et al., 1994). On a broader scale, electromagnetic measurements of apparent soil electrical conductivity and ground-penetrating radar have been used successfully to establish water-tables and contrasting soil horizons at levels i + 1 and higher (e.g. Collins et al., 1989; Brus et al., 1992). A variety of remote sensing techniques have been used to characterise land conditions at the i + 1 and higher levels.

The most familiar operational systems that are currently being used in natural resources in inventory and monitoring are the Landsat TM and MSS multi-spectral imaging systems, the SPOT HRV scanner system, and the TIROS-N/NOAA scanning system. All the earth-viewing satellites have made important contributions to various geoscientific applications, and in the future will also provide a means for monitoring terrestrial and aquatic ecosystems, or for operational assessment of natural resources. Several shortcomings are still recognised in the existing systems, which are hoped to be overcome in the future. In order to be able to study the earth as a unified system, future instrument platforms will carry extensive sets of instruments rather than one unique sensor system. Both NASA and ESA committed themselves to launching so-called 'earth observation polar platforms' (EOPP), carrying multi-spectral imaging devices with a lower spatial resolution, but a highly improved spectral resolution (after Hill, 1991).

Although already in use for over 30 years, radar remote sensing is often portrayed as a new technique with enormous potential for providing data required by natural resource specialists. The ERS-1 mission is the forerunner of a new generation of space missions for the 1990s employing radar sensors for the study of the environment. The ERS-1 mission is designed to provide essential data for addressing a wide range of primary environmental problems (Churchill and Sieber, 1991). During the last 10 years, many efforts have been made to use different kinds of video imagery for environmental monitoring. The range of (airborne) video imagery includes normal colour, near infrared, thermal infrared and the recently developed colour infrared (CIR video imagery). The applications include studies on inventory and control of fallow land, mineral occurrence, pollution, crop damage, crop yield, evaporation measurements, etc. (Braam, 1993, Verhoeven, 1993). Much research has been done on detecting and assessing variables relating to crop land, rangeland, tree crops and soils. In addition, videographic research and its applications have been conducted in forestry, water quality, wetlands, land cover inventory and urban land use (Mausel et al., 1992).

As well as the prospects of future systems, developments in the field of remote sensing are taking place not so much in equipment-related terms, but in the development of methodologies for field applications. In the framework of the Tropenbos project in Colombia, work is being carried out on the development of a methodology for the detection, monitoring and modelling of changes in land cover and land use. The monitoring system is largely based on the use of ERS-1 images, because of their higher availability (Bijker, 1992). The work concentrates on the combination of radar and multi-spectral scanner data, together with information from other sources and modelling of vegetation (re)growth to come to an optimal result in mapping and monitoring. Although several studies have been devoted to this topic in recent years, research tends to concentrate increasingly on the integration of remotely sensed data with data derived from other sources, such as topographic data, digital terrain models, and auxiliary thematic information. In the framework of a BCRS-project, Delft Hydraulics, ITC and IWACO are jointly developing a remote-sensing-based information system for monitoring and simulation of erosion and flash-flood impacts of deforestation and reforestation, as a supporting tool in planning for water resources and environmental development (Van Lieshout and Kempers, 1993). The project develops an operational system in which remotely sensed data are integrated with spatial data from other sources and a variety of models for the analysis of hydrologic conditions and erosion hazards.

In many cases it has been shown that it is still impossible to beat a human being in the interpretation and classification of a remote sensing image. Pure digital analysis frequently shows errors that have to be corrected by visual interpretation of textural features; the success of visual image interpretation serves to emphasise the potential for further improving the accuracy of digital image analysis, by including information other than spectral data. Given the nature of visual interpretation, the use of image texture in classification schemes seems appropriate, and a lot of effort is being put into the development of procedures for textural analysis. The inclusion of textural information in a classification can considerably improve the automated identification of those features whose spectral characteristics exhibit spatial heterogeneity at the image pixel scale (Ducros-Gambart and Castallu-Etchegorry, 1990; Trotter, 1991).

New techniques discussed here have provided a large body of data indicating the inadequacy of many of our previous concepts which were based on less data and more subjective interpretations of reality. For example, mapping units of soil surveys are often more heterogeneous than expected when observed with remote sensing techniques. Flow of solutes in undisturbed field soils is much more heterogeneous than is assumed by models based on homogeneous porous-media flow. Thus, the application of new techniques and tools makes it necessary to update our concepts.

SIMULATION MODELLING

In their paper, Hoosbeek and Bryant (1992) distinguish different types of models. In terms of 'the degree of computation', qualitative and quantitative models are distinguished, while 'complexity of structure' allows the distinction between functional and mechanistic models (Figure 2.1). Quantitative mechanistic models require much data and are usually applied at the i + 1 level or lower. Qualitative functional models often take the form of expert systems which are useful to broadly characterise land suitabilities in large areas of land (Rossiter and Van Wambeke, 1993). Increased emphasis on environmental quality in relation to agricultural production, requires use of quantitative models. Crop production can be calculated as well as fluxes of soil solutes which can cause groundwater pollution when extending beyond the root zone. Such models allow quantitative expressions, balancing agricultural production and environmental interests, thus addressing crucial issues of modern agricultural production systems.

Preferably, calculations are to be made for point data while expressions for areas of land can subsequently be obtained by interpolation procedures, such as kriging. Use of disjunctive kriging allows the generation of expressions in terms of probabilities of exceedance, which is a realistic representation of spatial variability (e.g. Finke, 1993). A mix of quantitative models and expert systems has proved to be particularly effective, using expert knowledge for problems where simulations are not feasible and combining this knowledge with simulations for aspects where simulations are feasible (e.g. Bouma et al., 1993). Models are, of course, only as good as the quality of the basic data that go in. A major concern now is use of detailed quantitative and mechanistic models for conditions where inadequate data are available. Such use is scientifically unjustified. Use of pedotransferfunctions, which relate existing soil survey data to parameters needed for simulation, is recommended to overcome, at least partly, lack of measured data (e.g. Bouma, 1989; 1992).

In addition to simulating solute movement and plant growth, models are also being used increasingly for planning purposes in comparing different alternative land use scenarios as a function of a series of economic and social boundary conditions. A study by the Netherlands Scientific Council for Government Policy (1992) (see Chapter 9) demonstrates that many factors which determine land use can be substituted when making an economic analysis defining land use alternatives. However, the factor 'land' does not allow such substitution. Each land unit has a set of given properties that can, to a certain extent, be manipulated by management. However, this flexibility is limited and desirable land use will remain a strong function of permanent soil properties. This has become painfully clear in many areas where marginal lands have been developed for agriculture with unsatisfactory results.

Huizing (1992) describes the application of interactive multiple goal linear programming (IMGLP) as a tool that facilitates the assessment of alternative plans in land use planning. IMGLP is a method that helps to make decisions on land use issues in situations where multiple, conflicting land use goals are involved.

GEOGRAPHICAL INFORMATION SYSTEMS (GISs)

The impressive growth of digital information technology such as GISs and remote sensing, application models, databases, spreadsheets, etc., has made a set of valuable tools and techniques available for land use planning. These tools have largely developed in isolation from each other. At best, GISs are being linked to other software packages such as databases and spreadsheets or statistical analysis packages, but mostly on an *ad hoc* basis. The integration of remotely sensed data as an information source for geographical information systems has advanced more rapidly, but has not yet been realised to its full potential and a truly operational integration has not yet been reached.

One of the main tasks that lies ahead is the true integration of all these tools into intelligent geographical information systems. There are several possibilities for structuring such a system, choosing either the GIS or an expert system as the main focus (Burrough, 1992; Loran and Rabou, 1992; Lam and Swayne, 1993). System components may include the GIS, a database, spreadsheets, expert systems, geostatistical packages, library of applications models, etc. Despite the fact that a wealth of information is available on how to use the individual components of a system (e.g. GIS, models, etc.) most users do not know how to arrive at the best possible results. An intelligent geographical information system would have to include a component which is able to guide a user in such a way that appropriate models and analysis procedures can be selected and minimum data requirements are indicated.

The ease with which information layers, originating from various sources, can be combined in current GIS to produce new output tends to veil the fact that the absolute accuracy of this new product may be well below the desired standard of reliability.

The influence of spatial variation and the propagation of errors in overlaying and other analysis procedures are important in this respect, but are often conveniently forgotten. An intelligent geographical information system would be able to analyse the magnitude of errors, indicate when the border of acceptable accuracy and reliability is crossed, and indicate options to come to better results (Burrough, 1992).

CONCLUSIONS

(1) Recent decades have seen an almost explosive development of new techniques and tools allowing a dynamic and continuous characterisation of physical, chemical and biological soil conditions in space and time. These techniques involve in-situ monitoring and measurements above the land by various sensing techniques using hand-held equipment, aeroplanes and satellites. Data can be stored and manipulated in geographic information systems. Tools have largely been developed in isolation and an operational integration has not yet been achieved. Yet these new techniques offer new possibilities to observe actual soil behaviour, and data obtained have made it necessary to re-evaluate many of our concepts originating from earlier periods.

(2) Simulation modelling of crop production and solute movement in soil allows quantitative characterisation of the effects of a large series of alternative land use scenarios in space and time which can be visualised by geo-referencing using GIS technology. Modelling is crucial to answer 'what if...' questions that can never be answered by monitoring current conditions. A combination of modelling with interactive multiple goal linear programming allows selection of optimal scenarios to be defined in close interaction with users, be it farmers or planners. Modern methods tend to focus on defining options rather than clear-cut solutions, as users become more sophisticated and knowledgable.

(3) Use of complex simulation models is only justified when adequate data are available and when proper validation procedures are followed. Scientists have a clear responsibility here. Pedotransferfunctions are used to predict model parameters from existing land data, while expert systems are useful to screen and restrict the number of simulations to be made and by checking intermediate results as a means of continuous quality control. Interpretations of remote sensing images also require expert screening during the interpretation process.

REFERENCES

Bijker, W., 1992. Outline of a land cover monitoring system for the San Jose del Guaviare area, based on remote sensing. Draft paper for internal use, ITC-Tropenbos Colombia Project 18.

Bouma, J., 1989. Using soil survey data for quantitative land evaluation. *Advances in Soil Science* **9**(31): 225–239. Springer Verlag, New York.

Bouma, J., 1992. Effect of soil structure, tillage and aggregation upon soil hydraulic properties. In: Wagenet, R. S., Baveye, P. and Stewart, B. A. (eds), *Advances in Soil Science, Interacting Processes in Soil Science*, pp. 1–37. Lewis Publishers, Am Arber, USA.

Bouma, J., Wopereis, M. C. S., Wösten, W. J. H. M. and Stein, A., 1993. Soil data for crop–soil models. In: *Systems Approaches for Agricultural Development*, pp. 207–220. Proceedings of an International Symposium, Kluwer, Dordrecht.

Braam, B., 1993. CIR Video: an operational rool for environmental monitoring. In: *Proceedings International Symposium Operationalization of Remote Sensing*, vol. 7. ITC, Enschede.

Brus, D. J., Knotters, M., Van Dooremolen, W. A., Van Kernebeek, P. and Van Seeters, R. J. M., 1992. The use of electromagnetic measurements of apparent soil electrical conductivity to predict the boulder clay depth. *Geoderma* **55**: 79–93.

Burrough, P. A., 1992. Development of intelligent geographical information systems. *International Journal of Geographical Information Systems* **6**: 1–11.

Churchill, P. N. and Sieber, A. J., 1991. The current status of ERS-1 and the role of radar remote sensing for the management of natural resources in developing countries. In: Belward, A. S. and Valenzuela, C. R. (eds), *Remote Sensing and Geographical Information Systems for Resources Management in Developing Countries. Euro Courses, Remote Sensing*, Vol. 1. Kluwer Academic, Dordrecht.

Collins, M. E., Doolittle, J. A. and Rourke, R. V., 1989. Mapping depth to bedrock in a glaciated landscape with ground-penetrating radar. *Soil Science of America Journal* **53**: 1806–1812.

Ducros-Gambart, D. and Castallu-Etchegorry, J. P., 1990. Layered, textural, supervised classification land cover mapping with SPOT in Indonesia. *Asian-Pacific Remote Sensing Journal* **2**: 9–15.

Finke, P. A., 1993. Field scale variability of soil structure and its impact on crop growth and nitrate leaching in the analysis of fertilizing scenarios. *Geoderma* **60**: 89–109.

Heijs, A. W. J., De Lange, J., Schoute, J. F. Th. and Bouma, J., 1994. Computed tomography as a soil structure analysis tool. *Geoderma* (in press).

Hill, J., 1991. Remote sensing systems: sensors and platforms. In: Belward, A. S. and Valenzuela, C. R. (eds), *Remote Sensing and Geographical Information Systems for Resources Management in Developing Countries. Euro Courses, Remote Sensing*, Vol. 1. Kluwer Academic, Dordrecht.

Hoosbeek, M. R. and Bryant, R. B., 1992. Towards the quantitative modeling of pedogenesis—A review. *Geoderma* **55**: 183–211.

Huizing, H., 1992. Interactive multiple goal analysis for land use planning. In: *Proceedings of a Workshop on GIS and Remote Sensing for Natural Resource Management by ILWIS*, 25–27 November 1992, Bangkok. Dept. of Land Development, Min. of Agriculture and Cooperatives, Thailand, and International Institute of Aerospace Survey and Earth Sciences, The Netherlands.

Jongmans, A. G. J., Van Oort, F., Buurman, P. and Jannet, A. M., 1994. Micromorphology and submicroscopy of isotopic and anisotopic Al/Si coatings in a Quaternary Allier terrace, France. In: Ringrose-Vose, A. (ed.), *Proceedings 9th International Working Meeting on Soil Micromorphology*, Townsville, Australia (in press). Elsevier, Amsterdam.

Lam, D. C. L. and Swayne, D. A., 1993. An expert system approach of integrating hydrological database, models and GIS: application of the RAISON system. In: Kovar, K. and Nahtnebel, H. P. (eds), *Proceedings of an International Conference on the Application of Geographic Information Systems in Hydrology and Water Management*. ITC, Enschede, Netherlands.

Loran, T. M. and Rabou, A. E. B. M., 1992. Development of a flexible and problem-oriented geographic information system. Paper presented at the International Conference on Geography in the ASEAN Region, 31 August–3 September, Yogyakarta, Indonesia.

Mausel, P. W., Everitt, J. H., Escobar, D. E. and King, D. J., 1992. Airborne videography: current status and future perspectives. *Programmetric Engineering and Remote Sensing* **58**: 1189–1195.

Netherlands Scientific Council for Government Policy, 1992. Ground for Choices. Four perspectives for the rural areas in the European Community. Report to the Government No. 42. Staatsdrukkerij, Den Haag, The Netherlands.

Rossiter, D. G. and Van Wambeke, A. R., 1993. *ALES: Automated Land Evaluation System. Version 3 User's Manual*. Soils, Crops and Atmospheric Sciences, Teaching series No. 2, Dept. of SCAS, Cornell University, Ithaca, NY.

Topp, G. C., Davis, T. L. and Annan, A. P., 1980. Electromagnetic determination of soil water content in coaxial transmission lines. *Water Resources Research* **16**: 574–582.

Trotter, C. M., 1991. Remotely-sensed data as an information source for geographical information systems in natural resources management: a review. *International Journal of Geographical Information Systems* **5**: 225–239.

Van Lieshout, A. M. and Kempers, R., 1993. Remote Sensing based Information System for River Basins—Demonstration Package. ITC-WL-IWACO, Enschede, The Netherlands.
Verhoeven, R. W. M. E., 1993. Airborne Colour Infra-red (CIR) CCD-Videography as a remote sensing tool. In: *Proceedings International Symposium Operationalization of Remote Sensing*, Vol. 3. ITC, Enschede, The Netherlands.

CHAPTER 3

Tension between Aggregation Levels

R. Rabbinge and M. K. van Ittersum

Wageningen Agricultural University, Wageningen, The Netherlands

INTRODUCTION

Land use studies comprise contributions of various disciplines and take place at several levels of aggregation. Tension between aggregation levels and also between disciplines occurs frequently and is partly due to misunderstanding, improper definitions and the absence of well-defined aims of a study. In this contribution definitions of concepts are given and different objectives of land use studies are discussed in an analysis of real and apparent conflicts between disciplines and aggregation levels.

All land use studies consider systems, at a high aggregation level. Crops, or cropping systems, are building blocks in land use studies for farming systems or systems at regional level. Systems are limited parts of reality with well-defined boundaries. The boundaries are selected on the basis of the objectives of the study.

In land use studies, agricultural and other land use systems have to be well-defined. The definition comprises three sets of classification criteria: time, space and the influence of man. The last criterium requires an appropriate description of objectives of a study. Roughly three types of land use studies with different objectives may be distinguished: (i) descriptive and comparative studies, (ii) explorative studies, and (iii) planning studies.

In descriptive, comparative land use studies, the functioning of the system (e.g. the farm household or a region) is investigated. By analysing the various descriptions of the system it is possible to explain the current situation and to gain insight in its limitations. By means of descriptive, comparative system analysis it may be possible to tell something about the near future. In this type of study the influence of man is a very important driving variable in the system analysis.

Another group of land use studies aims at exploring possibilities and potentials for a particular farm or area in the long run. This can be done from a biophysical and technical point of view or in such a way that socio-economic factors are also involved. In studies meant to investigate the biophysical potentials of a particular area, the way

The Future of the Land: Mobilising and Integrating Knowledge for Land Use Options
Edited by L. O. Fresco, L. Stroosnijder, J. Bouma and H. van Keulen.

man manages the system at present is excluded. The potentials in land use of the area are dictated by the soil, the climate and the characteristics of a crop. In explorative studies, concepts like best technical means and best ecological means are used rather than best practical means. In such studies the potentials cannot be translated into consequences for the farming system, or in the day-to-day management of a cropping system or crop. In explorative studies in which socio-economic factors are also taken into account, assumptions on farm management have to be introduced. These studies explore which land use changes can take place, taking into account socio-economic factors also. Management decisions are very often determined by socio-economic objectives and constraints rather than production–ecological possibilities. Land use studies that are meant to identify and explore technical possibilities and limitations are usually relatively narrow and need mainly biophysical knowledge and insight, whereas explorative land use studies in which socio-economic objectives and constraints are included need a much wider interdisciplinary approach.

After certain land use options for the future have been chosen, studies for planning and management become important. The question of how the land use options that have been chosen can actually be achieved is crucial. Policy instruments play an important role in it. Predictive models at various aggregation levels may be very useful. They may help in strategic and tactical planning. At various levels of aggregation other models are needed.

It is vital for any study to identify the appropriate level of aggregation, i.e. the level that corresponds with the objectives of the evaluation. Studies on possibilities at farm level have a different character to those at a regional level. It is important to choose the appropriate aggregation level in relation to the objective of the study and to be explicit about the aggregation level of the study.

In this chapter, production-ecological and socio-economic aggregation levels are discussed, and possible reasons for tension and conflicts between aggregation levels and disciplines are mentioned. An interface between the farm and regional level and various disciplines is given. Finally, possible rules and recommendations to prevent tensions and conflicts are suggested. It may be useful to screen the various case studies on land use against these guidelines.

AGGREGATION LEVELS IN PRODUCTION-ECOLOGICAL TERMS

The basis of all primary and secondary production in agro-ecosystems is the photosynthesis of plants. Individual leaves intercept the radiation of the sun and use its energy for the production of sugars by means of a reaction between carbon dioxide and water. The underlying diffusion of CO_2 and photochemical and biochemical processes can be quantified, thus enabling us to compute the sugar yield from the number of CO_2 and H_2O molecules involved. The sugars are used for the construction of all types of structural components. This construction requires energy that is obtained through combustion of sugar—the so-called 'growth respiration'. Together with the photosynthesis and the growth respiration, the maintenance respiration closes the C-balance at leaf level. Insight in the physical, chemical and physiological processes

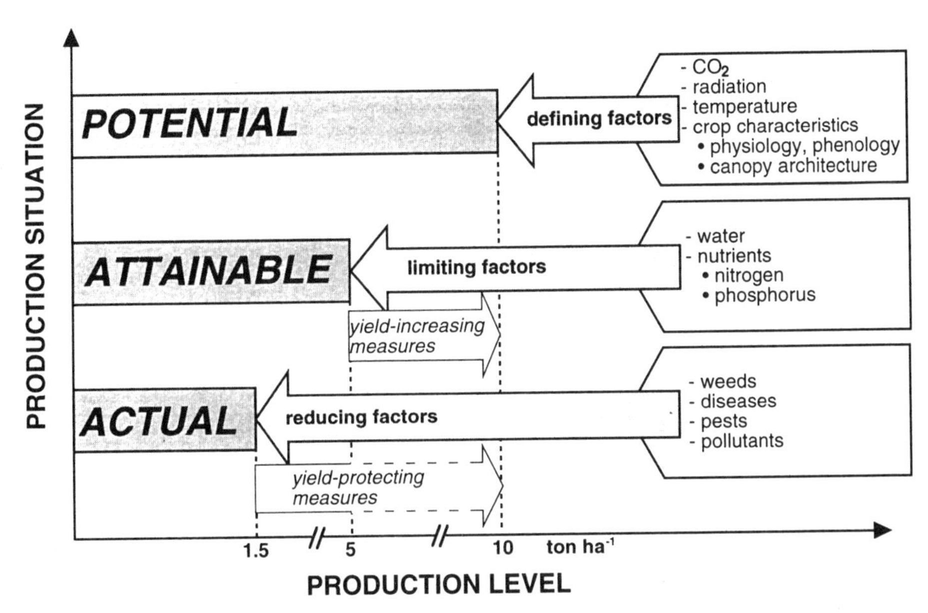

Figure 3.1 Schematic overview of different types of production factors and their corresponding production levels (Reproduced from Rabbinge, 1993, by permission of the Ciba Foundation)

involved in photosynthesis of leaves and respiration forms the basis of production-ecology.

Integration of leaf photosynthesis to crop level (De Wit, 1965) enables the quantification of crop performance under various circumstances. Dynamic simulation models are used for that purpose. At the crop level, various growth factors can be distinguished; the growth-defining, limiting and reducing factors (Figure 3.1). From a plant's point of view, the potential production is dictated by the growth-defining factors climate and characteristics of the crop. The attainable and actual production are determined by production-limiting and production-reducing factors.

A crop is part of a cropping system, and a cropping system is part of land use on the farm and in the region. In land use studies at the farm or regional level, production-ecological concepts and insights may be used to *explore* the system. The biophysical potentials of land units within a system can be investigated using crop growth simulation models. The production-ecological concepts and insights are used in this case only for the exploration of land use potentials and not for prediction, explanation and management support. The latter objectives would require completely different approaches and models.

In a system analysis, the level of detail of each of the underlying levels is dictated by the questions posed at the higher level. The more accuracy or the more quantitative aspects needed at the higher level, the more detail at the underlying level is necessary. In dynamic simulation studies, e.g. crop-growth simulation studies, this is made explicit by using the concept of aggregation level and the spatio-temporal characteristics, scale and time coefficients. Many production-ecological studies, e.g. leaf photosynthesis or

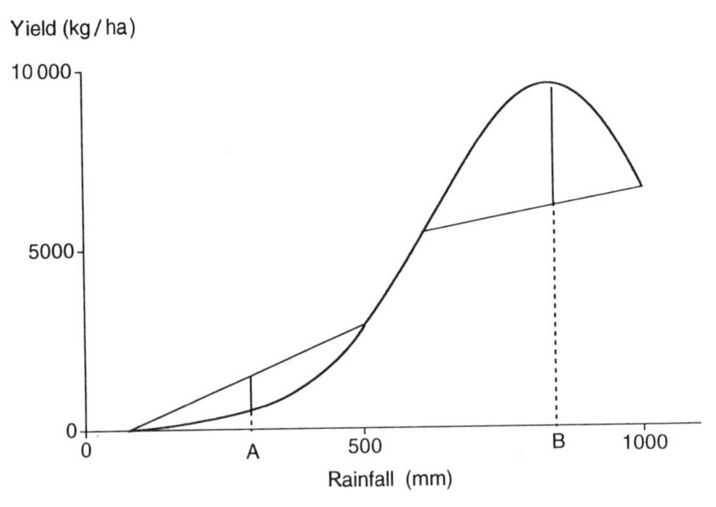

Figure 3.2 The influence of averaging rainfall on its calculated yield response. The yield is underestimated by averaging in the lower rainfall region (A) and overestimated in the higher rainfall region (B). Source: De Wit and Van Keulen (1987) (Reproduced by permission of Elsevier Science Publishers BV)

crop growth analysis, aim at understanding and explanation, and in these studies a detailed crop-growth simulation model is required. However, it does not make much sense to work with a detailed crop-growth simulator in explorative land use studies. The results of detailed crop-growth simulators are very plot-specific and cannot be used for the objective of quantifying (quesstimate) various yield levels in different land evaluation units. Land use studies need crop-growth simulators of a low level of complexity. They should be adequate for the determination of a first estimate of potential and attainable yield for homogeneous units on the soil and climate map. Time coefficients are then normally big and the land units for which yield is estimated are of a considerable size.

Aggregating information from a lower to a higher aggregating level is dangerous when homogeneity at the lower level is absent. Two examples of possible consequences of aggregation in land use studies are given in Figures 3.2 and 3.3. The first concerns heterogeneity in rainfall, which affects the results of a quantitative land evaluation (Figure 3.2). Because of the curvilinear relation between rainfall and yield, the yields are underestimated in the lower rainfall region and overestimated in the higher rainfall region by using average rainfall figures. The second example shows a possible consequence of aggregating land units or sub-regions in a land use study using linear programming (Figure 3.3). By aggregating heterogeneous land evaluation units or sub-regions and averaging the corresponding input–output data of the different units or sub-regions, the extremes in the original input–output data level off. When these aggregated data are used in an LP model the results of an optimisation may be less extreme than when non-aggregated data are used. Suppose a particular region consists of four land evaluation units of 1000 ha each. The wheat yields with a certain production technique in these regions are supposed to be 8, 6, 4 and 2 ton/ha for units 1, 2, 3 and 4, respectively. The objective of the optimisation in this example is to minimise the area in the region

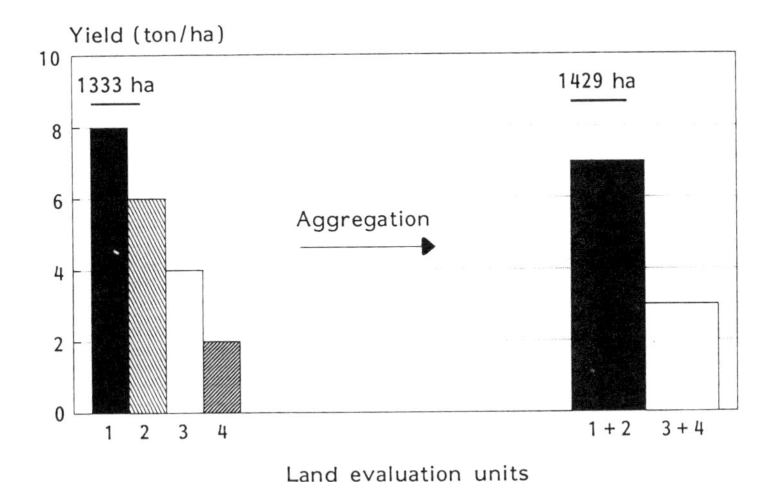

Figure 3.3 The effect of aggregation of four land evaluation units of 1000 ha each to two units of 2000 ha each on the minimum agricultural area required to produce 10 000 ton wheat in the region (4000 ha total). The average wheat yields in units 1, 2, 3 and 4 are 8, 6, 4 and 2 ton/ha, respectively. The average yields in the aggregated units 1+2 and 3+4 are 7 and 3 ton/ha, respectively. The minimum area in the non-aggregated problem is 1333 ha, wheras after aggregation the minimum area is 1429 ha. Source: unpublished data of R. J. Hijmans and M. K. van Ittersum

required to produce 10 000 ton wheat. In Figure 3.3 the effect of aggregating units 1 and 2 and units 3 and 4 is shown. Without aggregation, the minimum area is 1333 ha (1000 ha in unit 1 with 8 ton/ha and 333 ha in unit 2 with 6 ton/ha), whereas after aggregation the minimum area is 1429 ha (1429 ha in unit 1+2 with an average yield of 7 ton/ha). In the aggregated model the minimum agricultural area is larger than in the non-aggregated model.

Heterogeneity in time and space should be handled with care. In land use studies the credo 'first calculate, then average' is valid. Heterogeneity and curvilinearity in input relations should be retained as long as possible and their consequences should be included in the evaluation studies.

AGGREGATION LEVELS IN SOCIO-ECONOMIC TERMS

Many land use studies in socio-economic terms describe the functioning of farm households at micro-level. By analysing the various descriptions at this level it is possible to explore options for change and to get a better insight into limitations. These limitations may be biophysical but are often institutional. They comprise limitations due to shortage of labour, capital, knowledge, infrastructure or policy. Solutions to eliminate such limitations may be described when the functioning of the farm systems is better understood. Lower levels of aggregation than the farm are usually not considered in socio-economic terms. The farm is the basic unit.

Studies on a regional level may be based on the aggregation of studies on a farm level. However, that creates many difficulties as there is variation between individual farms, and limitations at regional level may differ from those at farm level. The 'bottom-up' approach has, therefore, its limitations; aggregating farm level to regional level will lead to ambiguous results. On the other hand, it is very hard to draw conclusions from regional studies about decisions to be made at the farm level. An iterative approach between the regional and the farm level seems a more apt approach, but before substantiating that for explorative studies, we will deal with the tension between the different types of land use studies and disciplines.

TENSION AND CONFLICTS BETWEEN AGGREGATION LEVELS AND DISCIPLINES

The presence of various aggregation levels and many disciplines in land use studies may easily lead to misunderstanding, polarisation and finally absence of any communication. The conflicts and tension often arise from differences in objectives or unclear objectives of a study.

Many studies are concerned with a static description of the present situation and do not take into account the dynamics of systems. On the basis of static observations of present land use, conclusions on the possibilities in the future are possible only up to a certain extent. In the short term, insight into possibilities for change can be given but it is virtually impossible to explore long-term options. In those cases, the future is an extrapolation of past and present, and discontinuities in trends are absent. The future is regarded as restricted by the past and shows no unexpected possibilities. In many socio-economic studies for the short term this type of extrapolation makes sense. Deterministic, descriptive studies are then sufficient.

New results are possible when the past is not used as a measure for the future but when political or societal desires are combined with technical possibilities. This means explorative instead of predictive studies, which use plausibility and predictive value as the criteria for measuring the quality. In explorative technical studies plausibility is not important but consistency, completeness and scientific soundness of the technical possibilities are the most important criteria for measuring the quality of long-term studies. In explorative studies for land use the relations are not described but based on insight into the input–output relations. In studies that describe the present situation, correlations between various variables and characteristics are used. Causal relations based on an understanding of basic processes are then absent. For explorative studies, a good understanding of the input–output relations is necessary. This may lead to the definition of techniques that are not yet widely used in practice. The feasibility of various options is not based on their relation to the present status, but on the biophysical and technical limitations and possibilities that determine the potential.

Often, in land use studies the behaviour of actors is incorporated and seen as an integral part of land use. In descriptive analytic studies this approach is necessary as actors form part of the way the present situation may be explained from developments in the past. However, in explorative land use studies behaviour of actors should be explicit and choices should be transparent. This may help in the judgement of *a priori* assumptions, expectations and objectives.

Production-ecological studies are in most cases explorative and deterministic, using explanatory models that are integrated in multiple goal explorative models.

Socio-economic studies are in most cases predictive, using descriptive input–output relations based on the past.

POSSIBLE INTERFACE BETWEEN VARIOUS AGGREGATION LEVELS AND DISCIPLINES

In explorative land use studies at the regional level, technical information about land use is confronted with different objective functions in an interactive multiple-goal linear programming model. The technical information can be derived from crop-growth simulation models, literature and expert knowledge, and the objective functions can be distilled from the different policy views in the region. The time horizon (e.g. 25 years) of these studies is far enough away to limit its effect. In this way, different land use options can be generated which represent the different policy views in the system; they demonstrate the extremes in land use from a technical point of view for the long run.

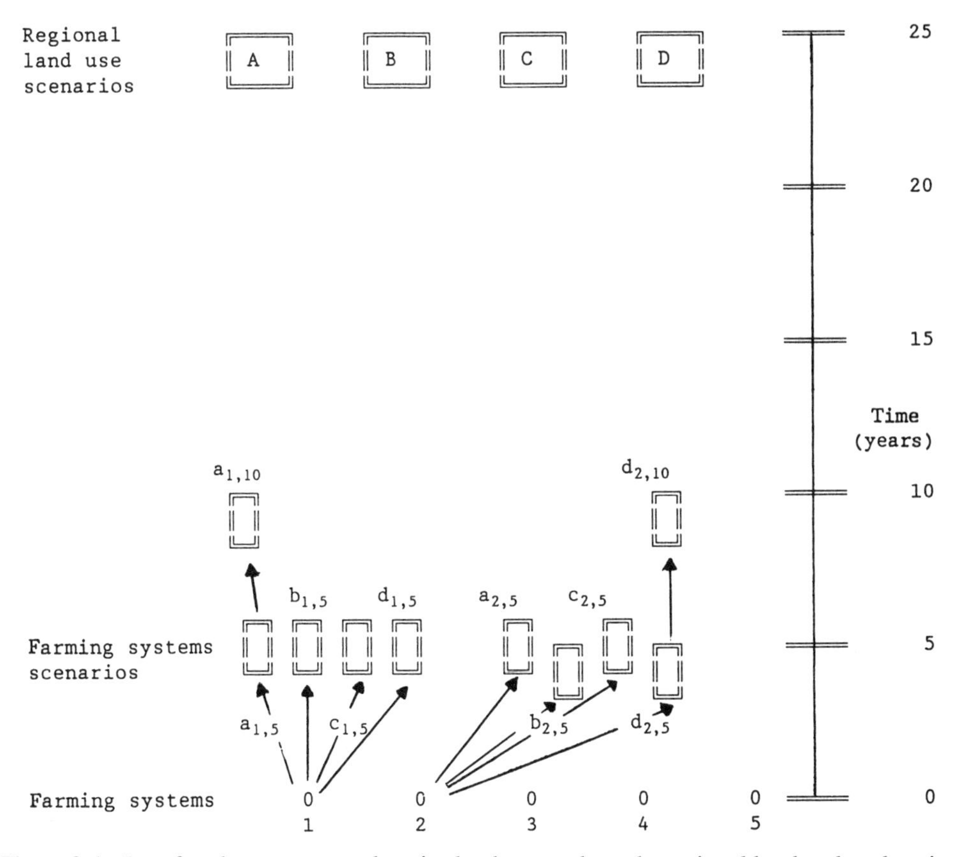

Figure 3.4 Interface between an explorative land use study at the regional level and explorative land use studies at the farming system level within the region. A, B, C and D are regional scenarios representing different policy views. $a_{1,5}$: Scenario a (priority of different objectives similar to those in Scenario A for the region) for farming system 1 in year 5; $d_{2,10}$: Scenario d (priority of different objectives similar to those in Scenario D for the region) for farming system 2 in year 10

However, these studies do not show the consequences for the individual farms within the region. The relation with the actual situation and short-term options is absent but may be achieved with a procedure explained in Figure 3.4.

The regional land use options set the scene for more detailed studies at farm level. This can be done in an iterative way. Different farming systems in the region can be distinguished. For each of these farming systems explorative studies can be carried out with much shorter time steps (e.g. 5 years) by confronting technical information about land use on the farm with the objectives of the farmer and those of the region in LP models. Different scenarios for each of the farming systems can be generated by putting different weights to the objectives of the region. For the first time interval, the land use activities that are offered to the linear programming model are more closely related to the current land use activities, whereas for later intervals they are more closely related to possible future land use activities. From these explorative farm studies an answer to the question of whether individual farms (farming systems) may be capable of reaching the land use patterns found in the regional study might be obtained. Often structural changes in size and structure are necessary. Exploration with studies at farm level may help to gain insight into the way transfer can take place. These studies should demonstrate whether institutional, socio-economical or cultural factors limit the attainable changes in the near future. Such studies require detailed economic and social analysis of the present situation. The connection with explorative studies for the long term may take place from both sides, from the present and from the potential future. When necessary the regional studies may be made more dynamic by the introduction of various time horizons (Spharim et al., 1992).

Both in the regional and farm studies the definition of the boundaries of a system, the elements of the system and the influences of the environment are very important. At the farm level the 'farm gate' is the appropriate boundary; however, it should be clear whether income generated outside the farm but used for investment in the farm is taken into account. At a regional level, for instance, it is important to be explicit about possible imports and exports of products; in other words, to distinguish between policies aiming at self-sufficiency for agricultural products or those aiming at free market and free trade.

GUIDELINES

To prevent misunderstanding and lack of communication between disciplines and aggregation levels, some guidelines have been formulated:

(1) Describe the objectives of land use studies explicitly. The objectives of the study determine the size of the system, its boundaries and the environment. The objectives may vary from an investigation of the short-term perspectives of an individual farm to an exploration of possible land use in a region.
(2) Define the system and its boundaries in time, space and influence of man. Systems as a limited part of reality are not a construct but are quantifiable and identifiable phenomena. Models are simplified representations of systems.
(3) Describe the next lower level and next higher level of aggregation. The definition

of the system and boundaries enables a clear description in general terms of the next lower and next higher levels of aggregation. The relation between aggregation levels can be identified in that case. It is impossible to consider at once (for example, in one model) more than three aggregation levels. It will lead to unreliable results or to unjustified conclusions.

(4) Identify the external influences and constraints. Their influence in terms of driving force (e.g. as an effect on the demand for agricultural products in the system) and constraints that dictate the ultimate limits (technical limits, e.g. maximum yields or water availability, and normative limits, e.g. a minimum employment rate) should be defined.

(5) Determine the internal variables (activities) related to land use, their interaction and their relation with the environment. The objectives and size of the system dictate these variables. A minimal number of variables is as a preferable strategy.

(6) Make explicit the necessary technical information and the various policy issues. In regional land use studies an indication of various techniques and their organisation is sufficient, whereas studies on a household level require much more detail in the way of labour organisation, income generation, etc. Depending on the objectives and limits of the system, other technical information (e.g. consumption level, imports and exports) is needed.

(7) If explanation is the aim, distinguish clearly between levels. Systems behaviour is explained from the underlying process level. Quantification of explicit relations at a process level form the backbone of the explanation and understanding of systems behaviour.

(8) If prediction is the aim, be sure of the reliability of the models. Models that are used for predictions should be validated and their sensitivity for changes in inputs and input relations should be tested. Their robustness or fragility should be quantified and considered in the predictions.

(9) If exploration is the aim, do not pretend to predict. Often explorative studies are interpreted as predictions. If plausibility and not consistency or technical possibility is considered as a criterion for the value of an explorative study, this may lead to the wrong type of discussion.

(10) If decision-making is the aim, determine exactly the appropriate decision variables. Decision-making, be it strategic, tactical or operational, requires proper identification of the decision variables. Description of the ultimate decision variables and consequences of change should be quantified. In this way, decision-making is supported by land use studies.

(11) Aggregate or average as late as possible. Aggregation or averaging input data may lead to the wrong results. 'First compute/calculate and then average' should be the credo. Another order leads to the wrong results in cases of curvilinear input relations or in cases of much variation due to stochasticity of input data.

(12) Never disaggregate in order to derive guidelines for management decisions at a lower aggregation level. The relation between micro, meso and macro level in socio-economic studies is a critical one. The same holds for aggregation levels in production-ecological studies. It is dangerous to draw conclusions from studies on the meso or macro level for individual situations at micro levels. The study at the

meso or macro level shows the ultimate consequences of the choices of policy-makers at that level. They do not indicate what decisions have to be made at the micro level.

The guidelines and suggestions described above may be used as a checklist in the evaluation of case studies on land use. Awareness of these guidelines may increase the quality of these studies and indicate what may be expected and for what purpose the studies may be used.

REFERENCES

De Wit, C. T., 1965. *Photosynthesis of Leaf Canopies*. Agricultural Research Report no. 663. Centre for Agricultural Publications and Documentation, Wageningen, The Netherlands.
De Wit, C. T. and Van Keulen, H., 1987. Modelling production of field crops and its requirements. *Geoderma* **40**: 253–265.
Spharim, I., Spharim, R. and De Wit, C. T., 1992. Modelling agricultural development strategy. In: Alberda, T., Van Keulen, H., Seligman, N. G. and De Wit, C. T. (eds), *Food from Dry Lands*, pp. 159–192. Kluwer Academic, Dordrecht.

CHAPTER 4

Resource Use Analysis in Agriculture: A Struggle for Interdisciplinarity

C.T. de Wit

Abersonlaan 31, 6703 GE Wageningen, The Netherlands

INTRODUCTION

Referring to the law of diminishing returns, it is often taken for granted that the large increases in crop yield since 1945 in the industrialised world and since 1970 in a large part of the developing world require a much more than proportional use of resources. This would imply diminishing returns, when yields per hectare over time are correlated with, for instance, nitrogen use per hectare. However, it appears (De Wit, 1991, 1992a,b) that the returns to increased fertiliser application are at least equally high in the upper range as in the lower range. This does not mean that the famous law of diminishing returns does not hold, but that any decreasing return has been compensated by increased efficiency as a consequence of other technical changes in the production process.

Neither in agronomic research nor in agro-economic research has much work been done on the relation between resource use and technological change. As a consequence, policy measures may be taken that, despite good intentions, contribute little to the efficient use of resources and the control of pollution, they may be even counter-productive. This justifies a further analysis of production principles, with special emphasis on the efficiency of resource use.

PRODUCTION FUNCTIONS

Much agronomic research before the Second World War was directed towards the search for laws governing the relation between the input of so-called production factors and the yield of crops. This research on production functions has been replaced by research on the physical, chemical and biological processes that govern the growth of crops.

The Future of the Land: Mobilising and Integrating Knowledge for Land Use Options
Edited by L. O. Fresco, L. Stroosnijder, J. Bouma and H. van Keulen.

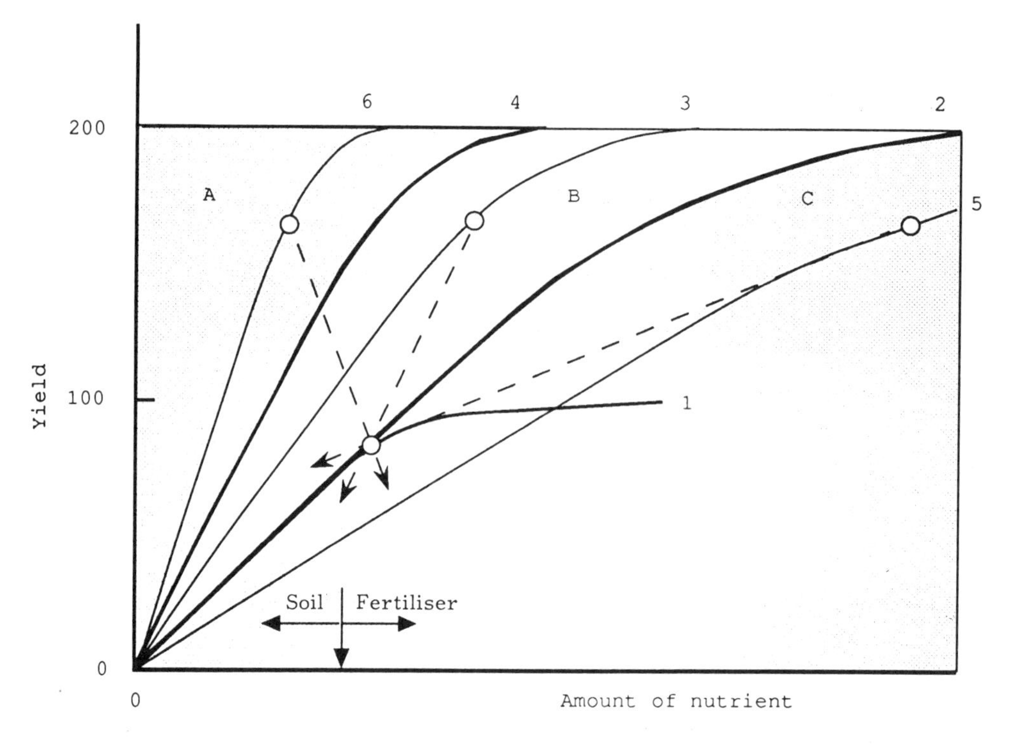

Figure 4.1 Schematic presentation of three classical yield response laws. Curves 1 and 2: Von Liebig's law of the minimum; curves 1 and 3: Liebscher's law of the optimum; curves 1 and 4: Mitscherlich's law of constant activity

However, the interest of agro-economists in production functions remained, so that gradually a situation has developed in which economists ask questions that cannot be answered by agronomists and agronomists give answers to questions not asked by economists.

Perhaps that stalemate can be broken by resuming where the agronomic discussion was left off some 50 years ago. For this purpose three production laws formulated in the century following the introduction of industrial fertiliser are schematically presented in Figure 4.1. Here the amount of a nutrient in the soil is given along the horizontal axis and the yield along the vertical axis. As is usual in this type of analysis, the available amount of nutrient in the soil and from sources like rain are expressed as an equivalent amount of nutrient applied as fertilizer. Both amounts are arbitrarily distinguished here by the vertical arrow at the horizontal axis. All curves reflect the well-known phenomenon of decreasing marginal returns with increasing amounts of nutrients. In the region of liberal supply, two maximum yield levels are distinguished: a basic level at 100 and an enhanced level at 200. The difference is due to the availability of other production factors, like water, radiation or other nutrients.

The law of the minimum, attributed to Von Liebig (1840, 1855), implies that the yield of a crop is proportional to the availability of the production factor that is most limiting.

This law is schematically presented in Figure 4.1 by curves 1 and 2 with a maximum of 100 and 200, respectively. Characteristic for this law is that curves with different maximum yields have the same initial slope and thus coincide in the region where the nutrient presented along the horizontal axis is yield-limiting.

Liebscher (1895), however, did not observe such constant initial slopes in many experiments with the nutrients N, P and K. It appeared instead that the production factor that was in minimum supply contributed more to yield, the closer other production factors were to their technical optimum. This law of the optimum of Liebscher is presented by curves 1 and 3 in Figure 4.1. The difference with the law of the minimum is that the initial slope of the curves increases with increasing maximum yields. According to Liebscher, this law holds either when maximum yields are higher due to increasing supply of other nutrients, or to increased availability of water, better soil tillage, better weather or better pH of the soil.

As a special case of Liebscher's law, Mitscherlich (1924) assumed that the amount of nutrient in soil and fertiliser needed to reach a certain fraction of the maximum yield for a given production situation, is independent of the level of this maximum and the production factors that determine it. This heroic assumption of 'constant activity' is schematically presented in Figure 4.1 by curves 1 and 4. In this special case the initial slope of the curve is proportional to its maximum yield.

Van Der Paauw (1938) used these three functions for the classification of yield responses, as in Figure 4.1. He distinguished three regions of response, A, B and C, separated by the production functions of Mitscherlich (curve 4) and Von Liebig (curve 2). In region C the increase in yield is less than according to the law of Von Liebig: curve 5 is an example. The broken line that joins points of the same relative yield on curves 1 and 5 intersects with the horizontal axis to the left of the origin. Accordingly, the need for the nutrient that is in minimum supply increases both in absolute and relative terms under improved growing conditions. The efficiency of nutrient use or the nutrient productivity (defined as the slope of the relation between yield and nutrient in soil and fertiliser) is therefore lower, the higher the maximum. Such dismal responses are soil-chemically and plant-physiologically unlikely and examples have not been found.

In region A the responses are more favourable than according to the law of Mitscherlich: curve 6 is an example. Here, the slope of the broken line joining points with the same relative yield is now reversed, so that the need for the nutrient decreases both in relative and absolute terms under improved growing conditions. A most striking example of such a benign response (Van Der Paauw, 1938; De Wit, 1992a,b) is that crops growing under otherwise better conditions can stand a lower pH much better and therefore need less lime.

Region B at last is the domain of the optimum law of Liebscher. Here, the broken line that joins points of the same relative yield on curves 1 and 3 intersects with the horizontal axis to the right of the origin. This reflects that indeed the absolute need for a nutrient that is in minimum supply increases under improved growing conditions, but the relative need decreases: more nutrient is needed when expressed per unit of area (kg/ha), but less when expressed per unit yield (kg/kg). The marginal return at a given nutrient application increases therefore with increasing maximum yield, but the increase is less than proportional. Van der Paauw (1938) found, as did Liebscher 40 years earlier,

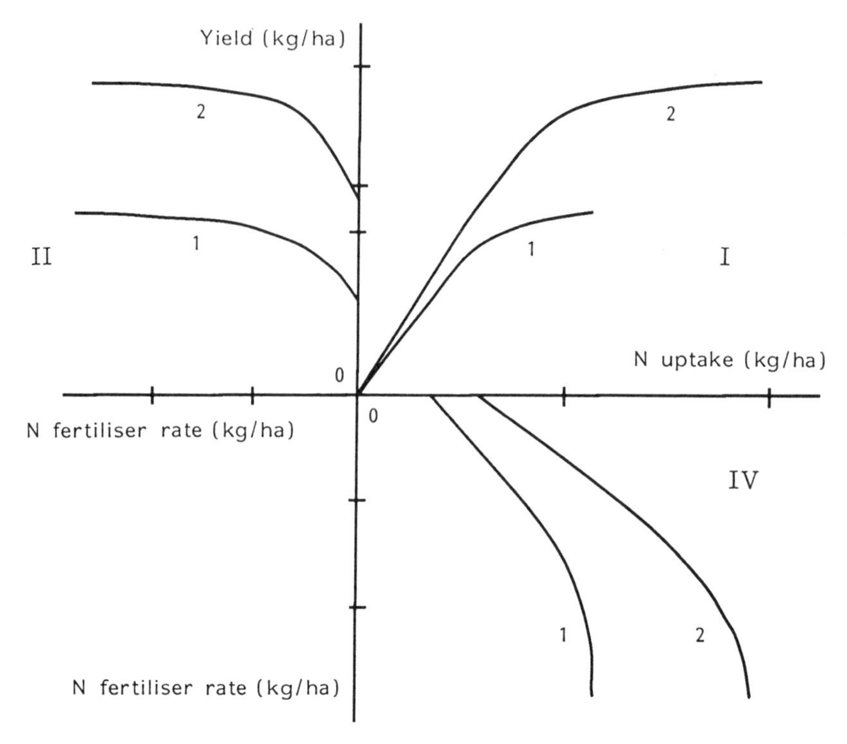

Figure 4.2 Three-quadrant diagram (De Wit, 1953) in which the relation between fertiliser rate and seed yield (quadrant II) is split up into the relation between fertiliser rate and uptake in seed and straw (quadrant IV) and between this uptake and seed yield (quadrant I)

that for experiments with more than one nutrient, by far the most production functions are located in region B, sometimes over the border in region A, but never in region C.

FURTHER ANALYSES

De Wit (1991, 1992b) analysed, on the one hand, the yield response of grain species to nitrogen as affected by the use of other fertilisers, choice of variety, supply of water, and control of diseases and, on the other hand, the effect of improved growing conditions on the efficiency of water use and the incidence of pests, diseases and weeds. Therefore, it suffices here to summarise the results.

By using above-ground nitrogen uptake as an intermediate between fertiliser rate and yield (Figure 4.2), the relation between both (quadrant II) could be split up into the relation between fertiliser rate and total uptake in seed and straw (quadrant IV) and the relation between this total uptake and yield (quadrant I). It was then found that both relations often respond to improved growing conditions according to Liebscher, as illustrated by the position of curves 2 with respect to curves 1. The improvement of the initial slope of the relation between uptake and yield is for grain species mainly due to the use of new varieties with a higher grain/straw ratio. The slope of the relation between uptake and fertiliser rate is referred to as the recovery of the fertiliser. This recovery often increases under improved growing conditions because the root system is more extensive and longer active and losses are reduced. For the same reason and due to improved mineralisation (Middelkoop et al., 1993), the uptake from the unfertilised soil may also be higher with improved growing conditions.

Improved growing conditions also lead to an appreciably higher efficiency of water use due to an extended soil cover and a more extensive and active root system, resulting in reduced losses by soil surface evaporation and leaching. The amount of water transpired per unit dry matter produced is far less influenced by growing conditions. In dry regions, water use exhibits a threshold value before the crop grows at all, so that water use efficiency is at first practically zero, but increases consistently with further improvement in water supply.

In more humid regions, water supply as such may be guaranteed, so that amelioration measures are directed towards optimising the temporal and spatial availability of water and air during the whole growth period and across the whole field. A fundamental difference exists with the optimisation of inputs, e.g. fertiliser. The inputs are freely available on the market, so that their prices are independent of their use. However, the closer the situation is approached where water and air are in optimal supply throughout time and space, the more the control of water supply becomes an infrastructural problem that requires more expensive solutions. Investments in better management then become a public concern and the ultimate yield level reflects the attitude of society to such investments.

Under improved growing conditions many crops become more susceptible to obligate, parasitic fungal leaf diseases and insect pests, such as mildew, rusts, aphids and plant hoppers, and require intensified control at high yields. Crops are, however, at the same time less sensitive for parasitic and often soil-borne diseases that attack plants when they are weak and under stress. Examples are *Fusaria* and *Verticillium* species and nematodes: under favourable growing conditions, yield reduction due to these diseases is relatively less severe and their control requires less efforts.

For weeds, crop 'look alikes' exist, with development and requirements similar to the crop they are associated with. Examples are red rice and *Echinochloa* species in rice. They may not even be recognised at transplanting and are in general more competitive under favourable growing conditions, especially if modern, short and less leafy varieties are used. The effort to control such species increases therefore with increasing yields. On the other hand, under less favourable conditions, some weed species always exist that are more tolerant and perform relatively better than the crop, whatever the cause of its sub-optimal growth. Thus a crop under sub-optimal conditions is always invaded by weeds that are adapted to the situation and that claim their share of resources. The control of such 'weeds of opportunity' requires less effort under more favourable conditions.

Consequently, there are some pests, diseases and weeds that require less effort to control under favourable growing conditions and others that require more effort. Overall, the use of biocides in kilograms of active ingredient per hectare has a large fixed component, so that their productivity expressed in kilogram yield per kilogram active material is likely to increase with increasing yields. This is confirmed by a statistical comparison of four regions in Europe: in the new Flevo Polders in the Netherlands, by far the most active material is used per unit surface, but at the same time productivity of its use is the highest (Jansma and Van Keulen, 1992).

Apart from the use of optional yield-increasing and yield-protecting production factors, certain activities are conditional for agriculture at any yield level and require the input of labour, capital and energy. Their requirements are partly area-related (e.g. plowing, harrowing and sowing) and partly yield-related (e.g. transport and drying of harvested products). Harvest activities themselves occupy intermediate positions. All in all, at increasing levels of production the use of labour, capital and energy increases per unit surface, but decreases per unit product.

Hence, variable production factors and more or less fixed activities have to be distinguished. For variable production factors the law of Liebscher has general validity, so that with increasing yields their need expressed per unit surface may increase, but expressed per unit product it decreases. Or formulated otherwise, the marginal productivity of resources that are limiting increases with improvement in growing conditions. Marginal productivities do not exist for fixed activities, but their productivity increases with increasing yields by definition.

RETURNS TO SCALE AND SUBSTITUTION

Instead of increasing the supply of one production factor, the supply of a number of production factors may be increased concurrently, such as nitrogen, phosphorus and potassium in a composite fertiliser. It is often assumed by agricultural economists that in such situations the law of diminishing marginal returns is also valid (Dillon and Anderson, 1990), which is then referred to as the phenomenon of decreasing returns to scale of the production per unit area. However, this cannot be the case if the optimum law of Liebscher has general validity.

To prove this, two production factors p_1 and p_2 are considered. According to the law of constant activity of Mitscherlich and at low input levels, yield is proportional to the input of each of the production factors when varied on its own. Hence, it is proportional to the product $p_1 \times p_2$ and increases initially in a quadratic fashion when both are varied concurrently. This increase levels off again at high inputs, as shown in Figure 4.3 for a situation where the yield response to both production factors is supposed to be identical. The combined response is therefore S-shaped and thus the more pronounced, the larger the number of production factors involved.

According to the law of the minimum of Von Liebig, one or the other production factor is limiting, depending on the ratio in which both are applied. Hence, the yield response to combined application remains linear until a limit, dictated by a third production factor, is reached. The optimum law of Liebscher assumes an intermediate position: at increased supply of two production factors the initial yield increase is less than quadratic but more than proportional. Hence, across the whole yield range, there are first increasing returns to scale which gradually change into decreasing returns to scale when the maximum yield is approached, and the S-shape is more pronounced the larger the number of production factors involved and the more the law of Liebscher approaches that of Mitscherlich.

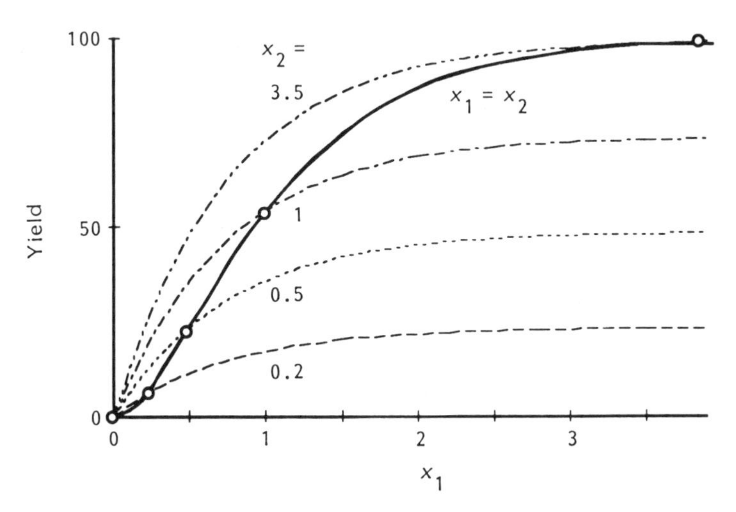

Figure 4.3 Returns to scale ($x_1 = x_2$), as constructed for a situation where the yield responses for two inputs are the same

Many production factors have unique physiological functions: solar radiation cannot substitute for lack of water and nitrogen not for lack of phosphorus. Such absence of substitutability is reflected in the law of the minimum of Von Liebig. However, according to the law of Liebscher, there is always a possibility for partial substitution, as in Figure 4.1, where in the yield range of 0 to 100 the same yield can be attained with less of the nutrient, when growing conditions are more favourable. Therefore, agricultural economists (Dillon and Anderson, 1990) like to use flexible mathematical expressions for production functions:

$$Y = F(X_1, X_2, \ldots)$$

with the only restriction that they are so smooth that they can be differentiated twice and inputs and outputs are homogeneous, as for instance yield for Y and nitrogen and phosphorus fertiliser rates for X_1 and X_2. The economic optimum is then defined as that combination of inputs that maximises the difference:

$$W = p_y Y - (p_1 X_1 + p_2 X_2)$$

which is found by solving:

$$(\mathrm{d}Y/\mathrm{d}X_1)\,p_y = p_1 \qquad \text{and} \qquad (\mathrm{d}Y/\mathrm{d}X_2)\,p_y = p_2$$

with p_y being the price of the yield, and p_1 and p_2 the prices of the fertilisers. As the position of the maximum and hence the fertiliser rates depend on the price ratios, there is no technical optimal of applying fertilisers.

Nitrogen (X_1) and phosphorus (X_2) in such equations are substitutable in the range where the same yield increase can be obtained by applying either only P or only N.

However, if phosphorus is applied, not only yield increases, but the nitrogen reserve in the soil decreases due to increased uptake in the harvested material. Similarly, if nitrogen is applied, the stock of phosphorus in the soil decreases. There are more changes, but this suffices to show that there is not only one homogeneous output (i.e. the yield), but also at least one inhomogeneous output (i.e. a change in soil fertility). Hence, simple mathematics to determine optimal fertiliser rates are not applicable. Also the results are not very meaningful, because situations where only one nutrient is applied are always unsustainable. Similarly, optimal yields are only by coincidence sustainable, because agro-ecological sustainability is defined in natural science terms only, and optimal yields are also defined in economic terms.

If the reasonable demand of agro-ecological sustainability is imposed, yields can in general only be maintained by fertilising in such a way that the uptake of one nutrient is matched by the uptake of others. Moreover, if yield is improved by, for instance, improved water supply, better varieties or better disease control, increased nutrient uptake is required. Consequently, agronomic production factors are more complementary than substitutable in sustainable agriculture. This does not imply that the law of Von Liebig holds and that there are no positive returns to scale, but that input combinations are only sustainable in a restricted range. Other combinations overuse some inputs and are exhaustive for other inputs. The relevant question is then, not what are the marginal returns to increased fertiliser application under otherwise constant conditions, but what fertiliser rates are needed to realise a given target yield in such a way that the fertility of the soil is brought to or maintained at its corresponding equilibrium level.

To answer such questions, elementary dynamic simulation models are available for nitrogen and phosphorus (Wolf et al., 1987, 1989) that require as main input the results of fertiliser experiments, where in addition to yield, uptake of the nutrient under consideration is also determined. An example of the results is given in Figure 4.4. The target uptake of nitrogen is given as an *independent* variable along the horizontal axis and the nitrogen fertilisation that is necessary to sustain this uptake as a *dependent* variable along the vertical axis. Compared with Figures 4.1 and 4.2, the axes in Figure 4.4 are switched. Higher target uptakes also require other improved growing conditions, but these are taken for granted here and are not considered further. Taking wheat as an example, an uptake of 25 kg of N in straw and seed suffices to produce 1000 kg of seed under any growing conditions, without falling into the traps of luxury consumption and under-fertilisation (De Wit, 1992a,b). Further, the law of Liebscher is reflected in the assumption that the recovery of inorganic fertiliser is 0.3 for target uptakes below 15 kg N/ha and increases linearly to 0.6 for target uptakes of above 150 kg N/ha. The recovery of nitrogen from other sources is set at two-thirds of these values and the background supply by, for instance, rain is 60 kg N/ha per year. This latter value has been taken so high for illustrative purposes.

The bold curve 2 in Figure 4.4 presents the equilibrium fertiliser rates needed to sustain the target uptakes. Transient processes are accounted for in the models, but are not considered further here, apart from observing that such equilibrium values are, for all practical purposes, reached in a period of 5–10 years. The marginal productivity of the N fertiliser is defined in this context as the increase in target uptake per unit increase

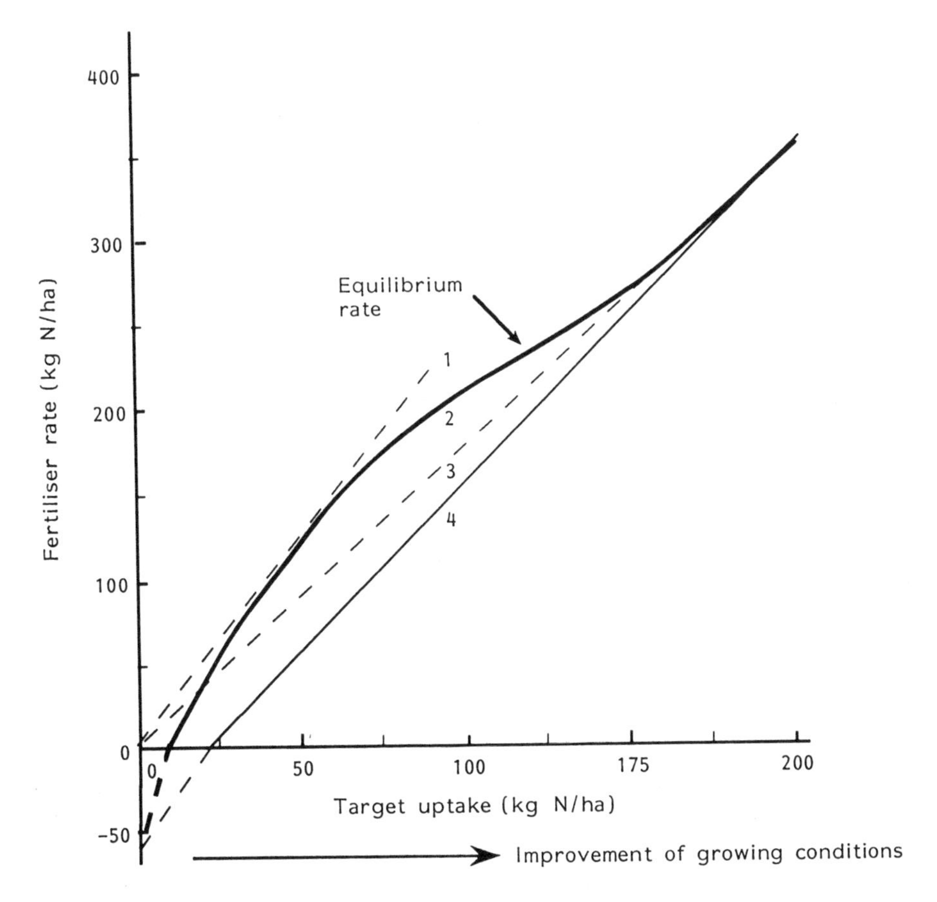

Figure 4.4 Relation between target nitrogen uptake and the nitrogen fertilisation that is necessary in the equilibrium situation to sustain this uptake in case of concurrent improvement of other growing conditions

in equilibrium fertiliser rate and its productivity as the ratio of target uptake and fertiliser rate. The marginal productivity increases over a wide range of target uptakes. The productivity, however, is at its lowest at a target uptake of about 50 kg N/ha, where curve 1 touches curve 2. At higher target uptakes, the productivity increases until a maximum is reached at the point where curve 3 touches curve 2. At lower target uptakes the productivity increases rapidly to infinite because the background supply allows sustainable uptakes without any fertiliser at all. At less illustrative but more realistic background supplies of about 10 kg N/ha per year, yields are so low that to make cultivation of a crop worthwhile, it is always necessary to apply inorganic or organic fertiliser, to grow green manures or to use fallow periods. The point of contact of curves 1 and 2 is then so close to the origin that the productivity of the fertiliser increases with increasing target uptakes over practically the whole range.

Curve 2 reflects equilibrium situations, so that the amount of nitrogen that is not taken up by the crop is lost by leaching and denitrification. Since at lower target uptakes, the amount of fertiliser needed for their realisation is relatively larger, the losses are also relatively larger. For response functions according to Von Liebig, recovery of fertiliser N does not increase with increased target uptake. Curve 2 is then a straight line and the relative losses are independent of the target uptake. This conclusion of constant or decreasing relative losses with increasing absolute losses seems contradictory to the common experience in nitrogen fertiliser experiments. However, these represent a completely different situation where the fertiliser rate is varied under otherwise the same conditions. It is then self-evident that finally the crop is over-fertilised, so that both absolute and relative losses increase with increasing rates of fertilisation. Undue emphasis on this trivial result diverts attention from the more sensible question of what fertiliser rates are needed to reach target uptakes as determined by the availability of other production factors and what are then the associated losses.

Curve 4 is approached if nitrogen is in minimum supply, while the other production factors are maintained at levels necessary for high target uptakes and thus sustain high recoveries of nitrogen. Then, fertiliser productivity at lower rates exceeds that for curve 2 and the distance between both curves could be considered the maximum substitution possibility between nitrogen fertilisation and all other variable production factors. However, curve 4 implies under-fertilisation with nitrogen or over-fertilisation with other nutrients and does not reflect a sustainable situation.

Nevertheless, many substitution possibilities remain in sustainable agriculture, but much more so at the management level than at the agronomic level. Many activities that are conditional for agriculture at any yield level, like plowing and seedbed preparation, can be executed with much labour and little capital and energy, or the other way round. Weed control is always necessary, but can be done with ecological (i.e. dense planting) or mechanical means or by herbicides. And if an ecotax on nitrogen is introduced, a more efficient application method may be developed, but the uptake that is necessary to achieve the target yield will remain the same.

SOME RAMIFICATIONS

In agriculture, one agronomic measure to improve growing conditions leads to others in a heuristic process of trial and error and based on limited and sometimes flawed knowledge of the production system. The analysis suggests that the law of Liebscher has general validity, so that this heuristic process occurs in an environment where returns to scale of yield follow an S-shaped curve. Therefore, it will be rewarding to examine a wide yield range for increasing returns to aggregate supplies of two or more production factors. For this purpose one should systematically consider whether each resource is used in such a way that other resources are used most efficiently.

Increasing returns to aggregate supplies reflect that agriculture is more difficult the more control is restricted, as is the case with a sub-optimal supply of resources. This not only leads to low yields per unit area but also to inefficient use of all other resources. In other words, the increase in production per unit area and in the efficiency of resource use are closely interlinked. Comparison of food grain production in the Punjab in the pre-green revolution period of 1962–1965 and the post-green revolution period of

1970–1973 indeed confirms that both yield and the productivity of all inputs increased (Bhalla et al., 1984).

In agro-ecologically sustainable situations, the reserves of nitrogen and minerals in the soil build up to their equilibrium level for the yield that is aimed at and the damage by pests, diseases and weeds does not systematically increase with time. As shown, the substitution possibilities at the agronomic level are then considerably less than in the usual field experiments. It is then less complicated, as mentioned above, to identify the minimal production factors and activities necessary to achieve a given target yield than to determine production functions that give yield as a function of all possible input combinations, disregarding their sustainability.

Activities like seedbed preparation, sowing and harvesting hardly require more effort at increasing target yield. This holds also for the application of micronutrients, for the use of lime to maintain an acceptable pH of the soil and for maintenance-breeding. Expressed in terms of active ingredient per hectare, biocides have a considerable fixed component, so that their needs increase relatively little with increasing target yield. Fertiliser requirements increase with increasing target yield, but above some minimum yield level the efficiency of their use increases as well. Water use efficiency increases also, but to create an optimal water supply and uniform growing conditions at the field level may require amelioration measures that demand a more than proportional effort with increasing target yield.

Formulated in economic terms, the productivity of the traditional fixed activities increases with increasing yields, while the marginal productivity of more variable yield-increasing and yield-protecting production factors remains high because of their complementarity. This implies that within a rather wide range, their input level is less dependent on their costs and their reward is less dependent on their scarcity, than for independent production factors (Van Dijk and Verkaik, 1989). Accordingly, within a wide margin, relative prices have little influence on their optimal mix (De Veer et al., 1992). As said before, substitution possibilities do exist but mainly at the management level and then mostly between labour, capital and energy.

THE FRONTIER OF MINIMUM COSTS OF PRODUCTION

So far in this discussion, prices have played a minor role and the considerations could be largely restricted to the technical level. This is different for the farmer, who requires acceptable financial returns for his/her work, without jeopardising the continuity of his/her enterprise. To elucidate the link between technical and economic considerations, the relation between production and costs over the whole range from extensive to intensive farming, is schematically presented in Figure 4.5, as inspired by the work of Holt (1987, 1988). The production target, expressed in ECU/ha (European currency unit) for a full rotation is given along the horizontal axis, with higher production targets representing more intensive production systems. The curve in the graph is the frontier of minimum production costs. It represents the costs that have to be met to reach the production target on the horizontal axis in an agro-ecologically sustainable way. Many possibilities exist to do worse, but none to do better. Sustainability implies here that the quality and fertility of the soil, the damage level of pests, diseases and weeds, and the capital

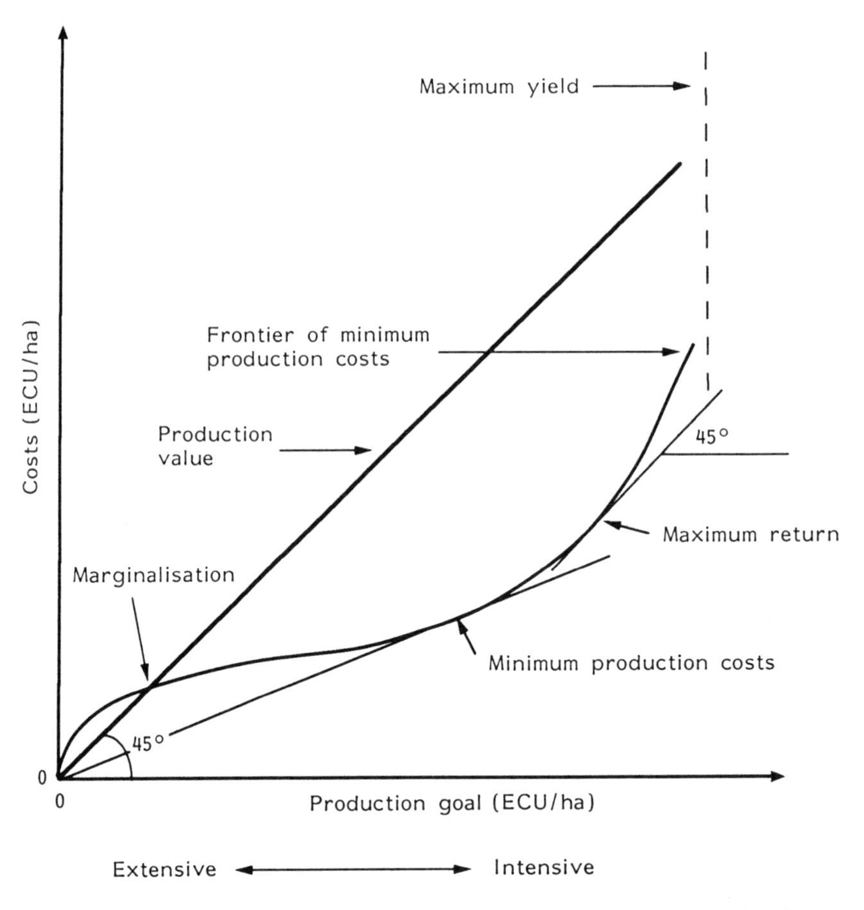

Figure 4.5 Minimum cost of production in dependence of production targets of farming systems. Source: Holt (1988)

stock do not deteriorate systematically in the course of time and is therefore not restricted to a certain production level.

The inverted S-shape of the frontier (Figure 4.5) reflects the results of the preceding analyses. In this presentation, the amelioration level is fixed, but all other costs, including those of capital goods, are considered variable. Further amelioration reduces the minimum costs to reach a certain target, and shifts the maximum attainable production to the right. The 45° line in the graph represents the gross production value when the production target is attained. Net return is the difference between gross production value and costs of production and is by definition equal to the entrepreneural reward plus the soil rent. It is assumed that prices are independent of the production target, to avoid unnecessary complexity. This implies that all labour is supposed to be hired on an hourly basis or that the farm size is adapted to the availability of labour in the farm household.

It requires a major effort to determine such a frontier of minimum production costs across the whole range from extensive to intensive farming, in spite of the limited substitution possibilities at the agronomic level. Initial attempts should, therefore, be directed to production systems with a simple crop rotation as, for instance, only grain or only maize and soybeans and for the present level of mechanisation. But even the discussion on the basis of a schematic presentation remains enlightening.

Three characteristic points may be distinguished in the graph (Figure 4.5). The lowest costs per unit product are represented by the point of contact of the line through the origin and the frontier of minimum costs. This is also the point of highest ECU productivity. The increase in production and in the costs of production are the same at the point of contact of the minimum cost frontier and a line parallel to the gross production value. Here net return is at its maximum. This point is always to the right of the point of highest productivity if, as in this example, a range exists where net return is positive. Finally, the intersection of the line of gross production and the frontier of minimum production costs represents the point below which farming does not pay. It is emphasised that the point of highest efficiency in financial terms is not the point of highest efficiency of resource use. For some resources the latter may require more extensive production targets and for others more intensive production targets. This only shows up in a more detailed presentation of results.

It appears from agro-ecological surveys that in the greater part of the European Community (EC) a substantial gap exists between current production levels and the levels that could be reached at the present amelioration level, and that current production would be possible with a much more efficient use of resources (WRR, 1992). Farms are, therefore, not clustered around the point of maximum return in Figure 4.5, but scattered above the frontier of minimum costs and, hopefully, below the gross production line. There are many reasons for this, such as attachment to the established way of running the farm, lack of knowledge of the production situation and alternatives, imperfectly functioning markets for credit and for products and production resources not widely used in the region, and the time needed to acquire and apply new knowledge. Given this yield gap and the options to improve the productivity of the soil and other resources, it is in most regions of the EC economically attractive for the farmer to aim at higher net returns by setting higher production targets. The rate of intensification is then not so much dependent on prices, but on public and private activities of research and extension.

A continuing increase in production volume without much increase in demand leads sooner or later to reduced prices of agricultural products, even in the EC where agricultural policy struggles to control prices and production volumes. Such a reduction at otherwise the same price ratios may be presented in Figure 4.5 by a decrease in the slope of the line for the gross production value: the production target on the horizontal axis is then no longer formulated in current prices but in prices in some base year. With such a decreasing slope of the line for gross production, the point of marginalisation moves upward and the point of maximum return downward along the frontier of minimum production costs. The point of highest productivity is, however, independent of the price of agricultural products and remains in place. Hence, in a case where price reduction continues, a situation will be reached where the three points coincide. Agriculture becomes uneconomical if product prices decrease still further. The point of highest productivity may thus also be referred to as the vanishing point of agriculture.

Research that is aimed at continuity of agriculture should, therefore, not be centred around the more elusive point of maximum return, but around the more robust point of maximum productivity.

Well- and less-endowed regions in the EC are distinguished by the position of the frontier of minimum production costs: in less-endowed regions the maximum is lower, and the minimum costs—although to some extent compensated by lower wages—are higher. With decreasing prices the vanishing point of agriculture is, therefore, reached earlier the less endowed the region is. After some years of exhaustive exploitation and hardship, the land is abandoned or at best used for extensive grazing. Since the vanishing point of agriculture coincides with the point of highest ECU productivity, such changes do not occur along a pathway of gradual extensification. In regions of the EC where agriculture remains an economically attractive enterprise, yields continue to increase, so that there are in fact only two alternatives: completely terminating agricultural activities or continuing intensification (De Veer et al., 1992).

Preventing pollution is best served by the efficient use of external resources. This is also achieved by concentrating arable farming in well-endowed regions. However, there are a few caveats to consider. Although the use of resources per unit product is lower, such a concentration of agriculture also leads to a concentration of pollution, so that environmental standards may be threatened in regions where agriculture is concentrated. Moreover, the concentration of agriculture would make it impossible to mine large soil surfaces for plant nutrients as is done in extensive forms of agriculture and to exploit the possibilities of reducing the effects of pollution by dilution.

There are equity arguments for support of farmers in less-endowed regions and ecological and environmental reasons to maintain forms of agricultural land use in these regions. World-wide, geopolitical and environmental arguments exist to produce the food where the mouths are. Such goals will place heavy demands on the political process, because it requires not only national and international solidarity, but invariably leads to a less efficient use of production resources and unnecessarily exposes soils to the risks of agricultural use.

ACKNOWLEDGEMENTS

Constructive comments on this version and an earlier Dutch version of the article (De Wit, 1992b) by Dr A. J. Oskam, Dr F. Veeneklaas and Dr J. de Veer and collaborators of the Centre for Agro-Biological Research (CABO) and the Department of Theoretical Production Ecology are gratefully acknowledged. Dr H. van Keulen also reviewed and edited the manuscript very thoughtfully.

REFERENCES

Bhalla, G. S., Alagh, Y. K. and Sharma, R. K., 1984. Food grain growth—A distinctive study. Center for the study of regional development, Jawaharlal Nehru University New Delhi. Mimeo cited through Y. Mundlak (1992).

De Veer J., Mansholt, S. L., Van Dijk, G. and Veerman, C. P., 1992. Naar een vaste grondslag in het Europese landbouwbeleid. *Spil* **3**: 18–22.

De Wit, C. T, 1953. A physical theory on placement of fertilizers. Verslagen Landbouwkundige Onderzoekingen 59.4, Pudoc, Wageningen.
De Wit, C. T., 1991. On the efficiency of resource use in agriculture. In: Böhm, W. (ed), *Ziele und Wege der Forschung im Pflanzenbau*, pp. 29–54. Triade-Verlag, Göttingen, Germany.
De Wit, C. T., 1992a. Resource use efficiency in agriculture. *Agricultural Systems* **40**: 125–151.
De Wit, C. T., 1992b. Over het efficiente gebruik van hulpbronnen in de landbouw. *Spil* **5**: 40–52.
Dillon, J. L. and Anderson, J. R., 1990. *The Analyses of Response in Crop and Livestock Production*, 3rd edn. Pergamon Press, London.
Holt, D., 1987. Agricultural production systems research. In: Proceedings and minutes of the 36th annual meeting of the Agricultural Research Institutes (Rockville Pike, Bethesda), 7–9 October, Washington, DC.
Holt, D., 1988. Agricultural production systems research. *National Forum* **68**: 14–18.
Jansma, E. and Van Keulen, H., 1992. Gewasbescherming nu en in de toekomst. *Spil* **3**: 29–35.
Liebscher, G., 1895. Untersuchungen uber die Bestimmung des Düngerbedürfnisses der Ackerböden und Kulturpflanzen. *Journal für Landwirtschaft* **43**: 49–125.
Middelkoop, N., Ketelaars, J. J. M. H. and Van Der Meer, H. G., 1993. Produktie en emissiefuncties voor het gebruik van stikstof bij de teelt van gras ten behoeve van de rundveehouderij. In: Verkennende studies over input–output relaties, pp. 9–52. NRLO rapport nr. 93/9. NRLO, Den Haag, The Netherlands.
Mitscherlich, E. A., 1924. *Die Bestimmung des Düngerbedürfnisses des Bodens*. Paul Parey, Berlin.
Mundlak, Y., 1992. *Agricultural Productivity and Economic Policies: Concepts and Measurements*. OECD Development Centre Studies Series, OECD, Paris.
Van Der Paauw, F., 1938. Over de samenhang tussen groeifactoren en opbrengst, en de principes die dit verband bepalen. *Landbouwkundig Tijdschrift* **51**: 1–36.
Van Dijk, G. and Verkaik, A. P., 1989. Resource allocation for animal health research. *Netherlands Journal of Agricultural Science* **37**: 301–310.
Von Liebig, J., 1840. *Organic chemistry in its application to agriculture and physiology*. Playfair, London.
Von Liebig, J., 1855. *Die Grundsätze der Agricultur-Chemie mit Rücksicht auf die in England angestellten Untersuchungen*, 1st edn. Vieweg und Sohn, Braunschweig, Germany.
Wolf, J., De Wit, C. T., Jansen, B. H. and Lathwell, D. J., 1987. Modeling long-term crop response to fertilizer phoshorus. I—The model. *Agronomic Journal* **79**: 445–451.
Wolf, J., De Wit, C. T. and Van Keulen, H., 1989. Modeling long term crop response to fertilizer and soil nitrogen. I—Model description and application. *Plant and Soil* **120**: 11–22.
WRR, 1992. Ground for choices: Four perspectives for the rural areas in the European Community. Report to the Government of the Scientific Council for Government Policy. Sdu uitgeverij, Den Haag, The Netherlands.

CHAPTER 5

Land Use Options and Environmental Goals

J. H. J. Spiertz,[a] H. van Keulen[a] and B. J. A. van der Pouw[b]

[a]Research Institute for Agrobiology and Soil Fertility, Wageningen, The Netherlands.
[b]The Winand Staring Centre for Integrated Land, Soil and Water Research, Wageningen, The Netherlands

INTRODUCTION

In the Netherlands, crop and animal production have been intensified in terms of the use of energy, water, fertilisers, pesticides and concentrates. The rate of intensification depended on the availability and costs of inputs and new agricultural technology (for example, high yielding cultivars and harvesting machinery) and on local, regional or international markets. Subsidies and price support regulations, as a consequence of the Common Agricultural Policy within the EC, aiming at parity income for farmers have resulted in growing surpluses. Prices of major commodities (cereals, pulses and oil seeds) have been reduced. Already the area under cereals has been reduced and a substantial part has been included in the set-aside scheme. This illustrates the importance of economic boundary conditions on land use options.

Environmental restrictions will also play an increasingly important role in the future development of farming systems and land use. The general objective of the agricultural policy in the Netherlands is to promote competitive, safe and sustainable agricultural development. Aiming at clean air, clean water and clean soil, regulations have been formulated for:

- reducing the use of chemical crop protection agents by 50% in the year 2000 compared to 1990;
- reducing ammonia emissions from animal manure by 70% relative to the level in 1980. Ammonia depositions strongly contribute to nitrogen emission and acidification in forests and nature areas. According to current technical and economic views a 50% reduction is possible;

The Future of the Land: Mobilising and Integrating Knowledge for Land Use Options
Edited by L. O. Fresco, L. Stroosnijder, J. Bouma and H. van Keulen.

- reducing nitrate leaching in such a way that the EC-norm of 50 mg nitrate per litre of groundwater will be met;
- preventing further accumulation of phosphorous in soils that are saturated and where the risk of leaching of mobile phosphorous is imminent.

On a regional scale, more specific and strict regulations are imposed to protect water quality and to maintain nature and landscape. To implement these policies, the general criteria should be 'translated' into location-specific and production-system-specific criteria. The most suitable aggregation levels for implementing policy instruments are the farm and the plot level. At these levels decisions are made about the amount and timing of inputs. For evaluation of the effectiveness of policy instruments, monitoring of water, air and soil quality at regional level would be most appropriate. At this level the time-scale plays an important role, because the processes involved are dynamic and complex. Crop response to weather and soil conditions has a diurnal and a seasonal cycle, whereas nitrate leaching to deep soil layers may take place within a time-scale varying from some days to many years. Also variation in soil properties affects the yield stability of crops and, therefore, the water and nutrient use efficiency and hence, the economic returns of the farmer.

At the moment, sustainable development is the guiding principle for solving environmental problems (Meerman et al., 1992). In the industrial countries, the growing concern about food quality and environment has led to new initiatives in agriculture, like organic farming, integrated farming (Wijnands and Vereijken, 1992) and sustainable farming (Vereijken, 1992). Progress has been made in developing alternative farming systems; however, quantitative evaluation of the extent to which various environmental goals are attained is still difficult.

In the Netherlands, land use options in the future will be largely determined by legislation on environmental restrictions for farming. Hence, legislation has already been introduced for the use of organic manure and pesticides. In the near future further restrictions are expected for water use and quality of ground and surface water.

This chapter discusses some opportunities and limitations of mobilising knowledge for the analysis of the interactions among land use patterns and environmental impact, with special reference to nitrogen and phosphorous.

CONSEQUENCES OF ENVIRONMENTAL RESTRICTIONS

Plot level

Research at the plot level is needed to increase insight in the soil–crop–weather relationships for the development of integrated crop management systems. To minimise losses of nutrients to the environment, the supply of nutrients both from natural sources and from amendments (fertiliser and manure) should coincide as closely as possible with the demands of the crop. However, a poor quantitative understanding of the biotic and abiotic processes that play a role in the supply of nutrients from organic sources and the losses by leaching and denitrification hampers the fine-tuning in nutrient management.

In the Netherlands, current nitrogen fertiliser recommendations for arable crops, aiming at predicting the economically optimum application rate, are based on the availability of soil mineral nitrogen (N-min) in early spring. However, the large error associated with assessment of the optimum doses results in response curves with a rather wide optimum range. The recommendations according to the N-min method are too crude for application in specific situations. Soil nitrogen mineralisation rates during the growing season and expected yield levels, which have a profound effect on the fertiliser requirement of a crop, may vary considerably from field to field. Hence, farmers often apply higher rates of nitrogen fertiliser than recommended, aiming at maximisation of crop yields, with the associated risks of losses to the environment. To minimise these risks, recommendations should be based on methods that explicitly take into account nitrogen mineralisation and expected yield level. The most refined method of nitrogen fertiliser recommendation is the use of a simulation model that predicts daily crop nitrogen requirement and nitrogen supply to the crop from various pools in the course of the growing period.

A recent study by the Nitrogen Committee (Goossensen and Meeuwissen, 1990) concluded that the existing simulation models did not accurately predict nitrogen emissions to the environment under a range of soil conditions and farming practices. Therefore, it was recommended that nitrogen leaching be limited by restricting nitrogen inputs in agricultural production systems to such an extent that the expected soil nitrate reserves in autumn do not exceed 70 kg N/ha in the soil layer to 1 m depth. This recommendation was mainly based on empirical relationships between soil nitrogen reserves and leaching losses during the winter period. It was also concluded that nitrogen emissions to the environment, both in the form of nitrate leaching to the groundwater and volatilisation of ammonia and nitrous oxides to the atmosphere, mainly originated from animal manure.

The consequences of environmental restrictions on the performance of individual crops and on soil characteristics at the plot level may be used as a basis for aggregation at higher levels. As an example, some results referring to sandy soils in the Netherlands, predominantly used for forage production with grass and maize as the major crops, are discussed. The dairy farming systems of the recent past are characterised by inefficient utilisation of mineral nutrients from forage and concentrates, manure and chemical fertilisers, leading to high losses of these elements to the environment (Van Der Meer and Van Uum-Van Lohuyzen, 1986; Reijerink & Breeuwsma, 1992). In efforts to avert this development, the Dutch Government has introduced legislation, regulating the quantities of organic manure and especially the inputs of phosphorous and nitrogen allowed in the system and their timing.

Analysis of the effects of such measures (Aarts and Van Keulen, 1990; Aarts et al., 1990) shows that if irrigation is prohibited to protect the groundwater resources, so that crops have to rely on the moisture supplying capacity of the soil, production will be seriously reduced. However, the effects are variable, depending on the physical conditions of the soil. Addition of the restriction of the nitrogen load of the groundwater to 50 mg/litre NO_3 further restricts production, because of a temporary nitrogen shortage during crop growth. In reality, the situation is even more complex as even 'individual fields' are not homogeneous in terms of physical properties. Van Noordwijk and

Wadman (1992) show that 'environmentally acceptable' production levels, associated with a restricted level of residual N in the soil at crop harvest to avoid excessive nitrate leaching, are lower in heterogeneous fields. Knowledge at plot level has to be integrated to derive recommendations applicable at higher levels of aggregation, such as farm and regional level.

Farm level

To operationalise knowledge of the underlying processes related to environmental consequences of the agricultural production process, analyses at farm level are required, because aggregation from the plot level is not directly possible. An example of such an analysis is the work of Van Der Meer (1985) on nitrogen utilisation on dairy farms in the Netherlands, comparing intensive experimental farms with an extensive commercial farm. The results show that at the extensive farm the 'utilisation efficiency' of nitrogen, expressed as the ratio of output in products and input, is 0.34, compared to only 0.16 in the intensive experimental farms. The difference between input and output, which at the intensive farms is much higher in absolute terms (Table 5.1), is to a large extent lost to the environment, either as nitrate to groundwater, or as ammonia, nitrogen gas or nitrous oxides to the atmosphere. Our understanding of the underlying processes is still insufficient to accurately quantify the distribution among the various terms. Therefore, policy measures derived from such data leave a wide margin of uncertainty.

An alternative analysis at the farm level has been worked out by Aarts and Grashoff (1993) to analyse the effects of nitrogen and phosphorous limitations on the production possibilities of dairy farms on sandy soils in the Netherlands. Starting from a dairy farm of 25 ha, they quantified the restrictions for nitrogen leaching and phosphorous accumulation on the possibilities for milk production, under the condition that all animal manure produced should be used on the farm. Their results show that on soils with a moisture supplying capacity of 125 mm (a rather drought-susceptible soil) the optimum

Table 5.1 Nitrogen balance (kg/ha/yr) at farm level of 14 intensive experimental farms and an extensive commercial farm in the Netherlands for the period May 1975–April 1976

	Experimental farms	Extensive farm
Nitrogen inputs		
Chemical fertiliser	383	—
Clover	—	65
Concentrate	127	24
Deposition	23	23
Total	533	112
Nitrogen outputs		
Milk	72	31
Liveweight gain	12	7
Total	84	38
Difference (input – output)	449	49
Utilisation efficiency	0.16	0.34

Source: Van Der Meer (1983). Reproduced by permission of The British Grassland Society.

distribution of land use is 60% grass and 40% maize, allowing an annual milk production of 14 200 kg/ha. The exact land use distribution and the associated potential milk production depend on the physical properties of the soil. These results indicate again that actual land use, under strict environmental conditions, is to a large extent dictated by the physical properties of the soils available at the farm.

Hence, it appears that in addition to the uncertainty originating from spatial and temporal variability, uncertainty associated with 'scaling up' from plot level to farm level adds to the overall margins (Fresco and Kroonenberg, 1992). At the farm level, we also encounter the conflict between environmental goals and socio-economic goals, as the farmer should be able to attain parity income, while respecting the environmental goals set by society or policy-makers.

Regional level

Analysis of land use options at a regional level requires knowledge of the abiotic system, especially soil and groundwater conditions. The essential procedure in dealing with the variable soil–groundwater continuum is schematisation. In the analysis, the area under consideration is subdivided into a number of units, representing typical combinations of soil and groundwater conditions. Schematisation may start already at the farm level, as a farm usually does not have uniform soil and groundwater conditions.

The role of regional soil and groundwater heterogeneity in schematisation depends on the aim of the research: if its aim is assessment of agricultural production levels, schematisation starts from the soils; if its aim is directed to, for instance, water management, it may start with a hydrological subdivision, e.g. catchments.

The core of the procedure in these studies is the combination of regional information on soil heterogeneity, usually in a geographical information system (GIS) and qualitative and/or quantitative models to assess crop production potentials. The combination of a GIS and models has developed into a powerful tool for regional studies. Application of a GIS allows the combination of various spatial data on, for example, soils and land use into units for which parameters are subsequently derived, using models. Reijerink and Breeuwsma (1992) used this technique to calculate the area of phosphorous-saturated soils in regions with a manure surplus. Kroes et al. (1990) followed a similar approach to analyse the effect of six fertiliser application scenarios on the N and P load of surface waters. In that study, part of the policy analysis of water management for the Netherlands, schematisation was based on 77 districts, representing catchments. They were subdivided, finally, into 500 units, taking into account soil type, drainage conditions, altitude, land use and availability of sprinkler irrigation. For each unit, the simulation model ANIMO (Agricultural Nitrogen Model) was applied to calculate the N and P load from the rural areas to the surface water (Rijtema et al., 1993).

Application of models in regional studies presents two major problems: lack of information on model parameters, and lack of spatial detail. The latter can be illustrated by the way soil units in such studies are characterised, i.e. by a single 'representative' soil profile. Ideally, statistically reliable information on the variability within these units should be available.

TRANSLATION OF KNOWLEDGE FOR POLICY SUPPORT

Application of knowledge for analysing the effects of environmental goals on land use options requires analysis of limitations and possibilities to meet various (often conflicting) goals.

Such an approach can be followed at the regional level, starting from a policy statement that describes a set of desirable environmental and/or ecological goals; for example, a certain surface water quality for a given area. Analysis of the hydrological or ecohydrological system of the area, combined with soil vulnerability research, should result in indications of 'acceptable' nutrient loads to the system. These should subsequently be translated into land use options, taking into account that a given load may be realised by different combinations of land use. Such an approach was followed by Van Walsum (1992) in a study of water management for the 'Groote Peel', a bog-forming area in the Netherlands. The combined use of simulation and optimisation yielded scenarios for water management indicating opportunities for reaching a compromise between nature conservation and agriculture.

Such an analysis can also be performed using multi-criteria optimisation methods. Application of such methods requires quantitative formulation of the production techniques ('land use types') available, in terms of their inputs and outputs. The outputs should include both the desired outputs (production, both in economic product and crop residues) and 'side effects', such as contribution to nitrate leaching, ammonia volatilisation, accumulation of phosphorous, etc. Results of simulation models and (expert) knowledge on production systems can be combined to generate the required technical coefficients.

An optimisation method that has shown promise in this respect is the interactive multiple-goal linear programming technique (IMGLP; De Wit et al., 1988; Van Keulen, 1990). In cooperation with relevant interest groups, such as decision-makers, targets (objectives) can be defined and the possibilities of attaining such targets can be explored, as well as the exchange values among different objectives. Problems still exist with respect to incorporation of social objectives, methods for evaluation of the results of such explorations and translation of the desired developments into policy instruments.

The method may be used to aid decision-makers in exploring the consequences of various policy options and in clarifying the consequences of choices. Of course, the margins of uncertainty that have been indicated earlier, and that are partly reflected in the technical coefficients of the various production activities, directly influence the results of the optimisation procedure.

CONCLUSIONS

Considerable knowledge is available on various aspects of agricultural production systems, both in terms of their potentials and in terms of their environmental impact. Methods and tools have been (and are being) developed for operationalisation of that knowledge and translation into decision support systems and policy recommendations for land use planning. Uncertainty often exists with respect to the quantitative effects due to temporal and spatial heterogeneity. Methods for (des)aggregation of processes at plot level are still under development.

The application of multi-criteria optimisation methods requires sufficient knowledge of planners and decision-makers at all relevant levels. Decision support systems or a combination of a GIS with mechanistic models should be developed for planners and decision-makers at various levels of aggregation, i.e. regional, local and farm level. The farmer of the 'future' may need a GIS of his/her own farm, combined with a decision support system that allows analysis of farm strategy to attain parity income, while practising environmentally sound production techniques.

An urgent need exists for further development of these tools, and for a more systematic appraisal of the ways in which the use of existing knowledge can be 'optimised'.

REFERENCES

Aarts, H. F. M. and Grashoff, C., 1993. Voederproduktie op droogtegevoelige zandgronden bij beregeningsverboden. In: Van Keulen, H. and Penning De Vries, F. W. T. (eds), *Watervoorziening en gewasproduktie (Water supply and crop production)*, pp. 83–101. Agrobiologische Thema's 8, DLO-Centre for Agrobiological Research (CABO-DLO), Wageningen.

Aarts, H. F. M. and Van Keulen, H., 1990. De praktische gevolgen van verscherpte milieu-eisen voor de weide- en voederbouw op zandgrond: een theoretische benadering. CABO-verslag 139, CABO-DLO, Wageningen.

Aarts, H. F. M., Biewenga, E. E. and Van Keulen, H., 1990. Dairy farming systems based on efficient nutrient management. *Netherlands Journal of Agricicultural Science* **40**: 285–299.

De Wit, C. T., Van Keulen, H., Seligman, N. G. and Spharim, I., 1988. Application of interactive multiple goal programming techniques for analysis and planning of regional agricultural development. *Agricultural Systems* **26**: 211–230.

Fresco, L. O. and Kroonenberg, S. B., 1992. Time and spatial scales in ecological sustainability. *Land Use Policy* **9**: 155–168.

Goossensen, F. and Meeuwissen, P. C. M., 1990. *The Nitrogen Report*. Report of the Nitrogen Committee, DLO/IKC, Ede/Wageningen, (in Dutch).

Kroes, J. G., Roest, C. W. J, Rijtema, P. E. and Locht, L. J., 1990. De invloed van enige bemestingsscenario's op de afvoer van stikstof en fosfor naar het oppervlaktewater in Nederland. SC-DLO Report 55, SC-DLO, Wageningen.

Meerman, F., Van De Ven, G. W. J., Van Keulen, H. and De Ponti, O. B. M., 1992. Sustainable crop production and protection. A discussion paper. Ministry of Agriculture, Nature Management and Fisheries, The Hague, The Netherlands.

NRC, 1989. *Alternative Agriculture*. National Academy of Sciences, Washington, DC.

Reijerink, J. G. A. and Breeuwsma, A., 1992. Ruimtelijk beeld van de fosfaat-verzadiging in mestoverschotgebieden. SC-DLO Report 222, SC-DLO, Wageningen.

Rijtema, P. E., Roest, C. W. J. and Kroes, J. G., 1993. Formulation of the nitrogen and phosphate behaviour in agricultural soils, the ANIMO model. SC-DLO Report 49 (in prep.). SC-DLO, Wageningen.

Van Der Meer, H. G., 1983. Effective use of nitrogen on grassland farms, pp. 61–68. *Proceedings of the 9th General Meeting of the European Grassland Federation*, Reading, British Grassland Society Occasional Symposium No. 14.

Van Der Meer, H. G., 1985. Benutting van stikstof op weidebedrijven. *Landbouwkundig Tijdschrift* **97**: 25–28.

Van Der Meer, H. G. and Van Uum-Van Lohuyzen, M. G., 1986. The relationship between inputs and outputs of nitrogen in intensive grassland systems. In: Van Der Meer, H. G., Ryden, J. C. and Ennik, G. C. (eds), *Nitrogen fluxes in Intensive Grassland Systems*, pp. 1–18. Martinus Nijhoff, Dordrecht.

Van Keulen, H., 1990. A multiple goal linear programming basis for analysing agricultural research and development. In: Rabbinge, R., Goudriaan, J., Van Keulen, H., Penning De Vries, F. W. T. and Van Laar, H. H. (eds), *Theoretical Production Ecology: Reflections and Prospects*, pp. 265–276. Simulation Monograph No. 34. Pudoc, Wageningen.

Van Noordwijk, M. and Wadman, W. P., 1992. Effects of spatial variability of nitrogen supply on environmentally acceptable nitrogen fertilizer application rates to arable crops. *Netherlands Journal of Agricultural Science* **40**: 51–72.

Van Walsum, P. E. V., 1992. Water management in the Groote Peel bog reserve and surrounding agricultural area; simulation and optimization. SC-DLO Report 49, SC-DLO, Wageningen.

Vereijken, P., 1992. A methodic way to more sustainable farming systems. *Netherlands Journal of Agricultural Science* **40**: 209–223.

Wijnands, F. G. and Vereijken, P., 1992. Region-wise development of prototypes of integratd arable farming and outdoor horticulture. *Netherlands Journal of Agricultural Science* **40**: 225–238.

CHAPTER 6

Challenges to the Biophysical and Human Resource Base

C. H. Bonte-Friedheim[a] and A. H. Kassam[b]

[a]ISNAR, The Hague, The Netherlands. [b]The Secretariat of the Technical Advisory Committee to the CGIAR, Research and Technology Development Division, FAO, Rome, Italy

INTRODUCTION

The primary and perennial function of land in human societies, beyond providing space for settlement and recreation, for industry, for communications, and for conservation and as a source of biodiversity, is to form a biophysical resource base for the production of biological products including foodstuffs, fodders and feeds and raw materials for people, livestock and industry. Fresh water is a major product of land whose role in its capture, storage and conveyance forms part of the hydrological cycle. Land is a source of minerals, fossil and renewable energy, and other natural resources; land serves as a source of wealth and power, and together with labour, constitutes the two original economic factors of production.

Land as a constituent of the biosphere, and the geosphere is an integral part of the processes of transformation of energy and matter including waste. Through these processes, the earth maintains spatial and temporal equilibrium. The land–sea configurations, together with the geometry of the solar system, determine the circulation of the air masses and ocean currents, and thus the climate and weather patterns. At the planetary level, the earth and its atmosphere form a closed system in terms of the transformation of matter; it forms a semi-closed system in terms of its energy balance.

In the context of sustainable development, land should be managed to satisfy the growing and changing human needs while maintaining or enhancing the quality of the environment and conserving natural resources (TAC, 1988). Here, the word ‘development’ is taken to mean a change process in which land resources, as well as other resources which may become available at acceptable cost as development proceeds, are made more productive of goods and services which the people, and their governments, require (Bunting et al., 1986).

The Future of the Land: Mobilising and Integrating Knowledge for Land Use Options
Edited by L. O. Fresco, L. Stroosnijder, J. Bouma and H. van Keulen.

A major consequence of adopting the concept of sustainable development as an imperative, is that a different kind of relationship must now be sought between nature and man—a relationship in which land, as an integral component of nature, is understood, exploited for different purposes and cared for with a view to enhancing its quality and potential over generations.

We examine four classes of challenges to the biophysical and human resource base which must be addressed in most developing nations to facilitate the advancement of sustainable development in their agriculture and the rural sector. These challenges are posed by: (1) the demand challenges; (2) the supply challenges; (3) the knowledge and technological challenges; and (4) the cultural, institutional and policy challenges.

DEMAND CHALLENGES

The demand challenges are mainly imposed by population size and per capita consumption of biological products from the land.

At the turn of this century, the world population was 1.7 billion (10^9). From 1900 to 1930, the population increased by 25%, from 1930 to 1960 by 50%, and from 1960 to 1990 by 75% to 5.3 billion (Table 6.1). According to the UN median-variant population projection, the world's population will be around 6.2 billion by 2000, 8.5 billion by 2025 and 10.0 billion by 2050. The world is expected to reach a stable or stationary population of 11.5 billion by 2110, when 10.2 billion people (or 88% of the total) will be in the developing world. The various regions of the world will reach stable population levels in different years, ranging from about 2030 for Europe and 2060 for North America to 2100 for Asia and 2110 for Africa and the Middle East. Proportionately, the greatest

Table 6.1 Populations 1990, expected in 2000 and 2050, and estimated stationary (millions of persons)

	Population			
	1990	2000	2050	Stationary
World	5 297 (100)	6 204 (100)	10 035 (100)	11 514 (100)
Developed world	1 251 (24)	1 267 (20)	1 371 (14)	1 328 (12)
Developing world	4 046 (76)	4 937 (80)	8 664 (86)	10 186 (88)
Africa	642 (12)	870 (14)	2 275 (23)	3 049 (26)
Middle East	165 (3)	271 (4)	655 (7)	854 (7)
Asia	2 791 (53)	3 266 (53)	4 915 (49)	5 407 (47)
Latin America	448 (8)	530 (9)	819 (8)	876 (8)

Figures in parentheses are percentages of world totals.
Source: World Bank.

increases are expected to occur in Africa and the Middle East (over fourfold) and in Latin America and Asia (twofold, but from a much larger base in Asia).

Effective demand for goods and services can only exist when people have money to purchase them. Effective demand is a function of income, and is therefore inversely related to poverty. The national market for biological products from land is created primarily by the effective demand of the urban and peri-urban population; and if this population is not only proportionately small but also poor, the market is inevitably limited (Bunting, 1992a).

For this reason, a national policy which aims to hold the population on the land, or to 'base development on agriculture' alone, does not succeed unless there are both sufficient rural resources and an international market for the surplus which is also sufficiently large and reliable. This condition cannot always be guaranteed, and so we find in practice that development in agriculture is most likely to succeed if the national economy, as a whole, is both growing and becoming more diverse. Rural and non-rural development, in practice, are two sides of the same coin: the one does not progress without the other (Bunting, 1992a).

If we project the current trend in the growth of effective demand for cereals per capita in the developing countries, then by 2050 it will reach roughly the level in Japan today (350 kg per capita), and by 2110, roughly the level in the European Community today (450 kg per capita). Consequently, the 8.7 billion people in the developing world in 2050 will need a harvest of about 3 billion tonnes, and the 10.2 billion people in 2110 will need a harvest of about 4.6 billion tonnes. For the developing world as a whole, these output demands are 2.9 and 4.4 times greater than now (1.03 billion tonnes). The regional distribution of the increase in total output demand varies from three times in Asia to over 14 times in Africa (Table 6.2). Assuming that the future share of non-cereal food products remains similar to what it is at present (i.e. 35%), then a total harvest of about 4.6 and 7.1 billion tonnes in grain equivalent would be needed in 2050 and 2110 respectively.

For agricultural development to succeed, or at least to be sustained, it has to meet its capital, operating and maintenance costs. It can only do this, in general, if it generates surplus products of appropriate quality for sale at prices acceptable to both producers and consumers (or users) in national and international markets (Bunting, 1992a).

SUPPLY CHALLENGES

On the input side of the production equation, additional resources of land, water, nutrients, energy, etc., must be obtained. On the production side, more efficient and environmentally friendly systems with high total factor productivities must be available, including improved genotypes of plants and animals, and efficient nutrient and pest management techniques. On the output side, an efficient output delivery system must exist to meet the effective demand for agricultural and rural products from the non-rural population.

The most important overall supply-side challenge is to prepare for the future demand in the developing world, by helping producers of biological commodities everywhere to manage in sustainable ways the natural resources and the environments they use. This

Table 6.2 Land area, output and yields of cereals in 1990; land area of non-cereal crops and total cropped land area in 1990; land, output and yields of cereals and non-cereal crops and total cropped land area required in 2050 and in 2110 at per capita cereal consumption of 350 kg and 450 kg respectively

		Cereals					
Region	Year	Land area (10^6 ha)	Output (10^6 t)	Yield (t/ha)	Cereal to non-cereal land area ratio	Non-cereal land area (10^6 ha)	Total land area (10^6 ha)
Africa	1990	77	95	1.2	0.42	105	182
	2050	265	796	3.0		366	631
	2110	274	1372	5.0		378	653
Middle East	1990	28	41	1.5	0.51	26	54
	2050	65	229	3.5		63	128[a]
	2110	77	384	5.0		74	151[a]
Asia	1990	273	790	2.9	0.71	111	384
	2050	344	1720	5.0		140	484
	2110	487	2433	5.0		198	685
Latin America	1990	48	102	2.1	0.32	104	152
	2050	72	287	4.0		153	225
	2110	79	394	5.0		167	246
Developing world	1990	426	1028	2.4	0.55	346	772
	2050	746	3032	4.1		722	1468
	2110	917	4583	5.0		817	1735

[a]Total area required is larger than total potentially cultivable suitable rainfed and irrigable land.

will call for effective land use policies at national, regional and international levels in order to maximise the complementarities in the ecological land potentials between nations and between regions (FAO, 1982; Bunting, 1992a).

Nutrition and food supply

Although in some developing nations in Africa and Asia, levels of dietary intake of energy, protein or both do not yet seem fully adequate for normal health, growth and activity, the average situation in the developing nations of the world has been, in general, improving steadily. Between 1965 and 1990, the number of countries that met daily per capita nutritional requirements doubled from 25 to 50. The numbers of all people of all ages with a BMR (basal metabolic rate) below 1.54, over the period 1975–1990, decreased from 970 million to 786 million, corresponding to a decrease in the proportion of malnourished population from 33% to 20%. Over the same period, the proportion of malnourished children under the age of five years decreased from 42% (168 million) to 34% (184⅝million), so that the number of children who were not seriously malnourished increased from 232 to 356 million (ACC/SCN, 1992).

During the past half-century, as the population of the world grew from about 2.3 billion in the mid-1930s to its present level of about 5.4 billion, the output of cereals, the principle staples of mankind, increased from about 655 million to around 1.94 billion tonnes in 1989–1991. The output per capita increased over the same period from 300 to 380 kg per year. In the developed countries output increased from 355 to 900 million tonnes, and output per capita from 450 to 750 kg per year. In the developing countries output increased from 300 to 1040 million tonnes, and output per capita from 220 to 255 kg per year, although in sub-Saharan Africa output per capita did not increase. Indeed, in spite of food crises and shortages at particular times and in particular places, diets in most areas and in normal times are evidently adequate to support the large rates of population growth in many developing countries (Bunting, 1992a).

The record of yield improvement and output growth is also impressive. FAO data show that, in the developed world as a whole, these increases in cereal output have come from gains in yield, from an average of 1.2 t/ha in the mid-1930s to 3.2 t/ha in the early 1990s, on a more or less constant harvested area of about 280 million hectares. In the developing world, marked gains in output were at first due to increases in harvested area (including multiple cropping), but since the early 1960s increase in yield has become the dominant component in these countries also, with the exception of the countries in sub-Saharan Africa. Since the mid-1930s, cereal yields in developing countries have increased from 1.1 to 2.4 t/ha in the early 1990s on a harvested area which has increased from 265 to about 425 million hectares. Indeed, food production in the developing world grew by 3.1% per year during the past 25 years, faster than population growth. Of this growth, 75% was absorbed by population growth, the rest was additional demand from higher incomes leading to increased levels of dietary intake of energy and protein from animal sources.

Also, despite the debt crisis, and the poor performance in sub-Saharan Africa during the 1980s, the economies of developing countries as a group during the past 25 years grew at an average annual rate of 4.8%, significantly higher than the growth rate achieved by the developed world over the same period (Cesal and Rossmiller, 1989).

Supply of suitable cultivable land

A question that often arises is whether the land resource base in the developing world is large enough to meet the needs for biological products of future populations. The FAO studies of agro-ecological zones confirm that suitable land for expansion exists, particularly in Africa and Latin America, and in parts of Asia (FAO, 1978–1981; Dudal et al., 1982; Dudal, 1987). However, land resources are very unevenly distributed both between and within countries.

The total potentially suitable cultivable rainfed and irrigated land in the world is estimated at more than 3536 million hectares, of which 2526 million hectares (or 71%) are located in the developing world (Table 6.3). Latin America and Africa have the largest share of the world's potentially suitable cultivable land resources; 27% and 24% respectively. Asia has 18% and the Middle East 2%. Globally, 41% of the potentially cultivable land is currently under arable and permanent crop cultivation. In the developing world as a whole, 31% of the potentially suitable cultivable land is currently under use. The regional values for Latin America and Africa are 16% and 22% respectively, and for Asia and the Middle East, 56% and 70% respectively. About two-thirds of the unused portion of the potentially suitable cultivable land is located outside the humid forest zones—in the woodland and grassland savanna zones. Additionally, there are more than

Table 6.3 Potentially cultivable land

	Developed world	Developing world					World total
		Africa	Middle East	Asia	Latin America	Total	
Potentially cultivable suitable rainfed land[a]							
extent (10^6 ha)	877	802[d]	48	424[e]	894	2168	3045
% of land area	16	27	7	21	44	28	23
%of world potential	29	26	2	14	29	71	100
Currently (1990) cultivated rainfed land as % of potentially cultivable rainfed land	69	21	83	59	15	28	40
Potentially irrigable land[b]							
extent (10^6 ha)	133[c]	19[d]	29	266[e]	44	358	491
% of land area	2	<1	4	13	2	5	4
% of world total	27	4	6	54	9	73	100
Currently (1990) irrigated land as % of potentially irrigable land	48	58	48	50	36	48	48
Potentially cultivable suitable rainfed and irrigated land							
extent (10^6 ha)	1010	821	77	690	938	2526	3536
% of land area	18	28	10	31	46	32	26
% of world total	29	24	2	18	27	71	100
Currently (1990) cultivated rainfed and irrigated land as % of potentially cultivable land	66	22	70	56	16	31	41
Persons per hectare of potentially suitable cultivable land							
1990	1.2	0.8	2.1	4.3	0.5	2.0	1.52
2050	1.4	2.8	8.5	7.7	0.9	3.5	2.88
2110	1.4	3.7	11.1	8.4	0.9	4.1	3.30
Potentially cultivable suitable land per capita (ha)							
1990	0.81	1.28	0.47	0.23	2.09	0.61	0.66
2050	0.74	0.36	0.12	0.13	1.14	0.29	0.35
2110	0.76	0.27	0.09	0.12	1.07	0.24	0.30
Marginally suitable potentially cultivable rainfed land (10^6 ha)	295[c]	231	16	323[f]	162	732	1027

[a]Very suitable and suitable land; [b]irrigable land with short distance water transport; [c]author's estimates; [d]includes only currently cultivated rainfed and irrigated lands for South Africa; [e]includes only currently cultivated rainfed and irrigated lands for China and Korea; [f]includes an estimate for China and Korea of 97 million ha.

Source: FAO (1978–81), FAO (1982, 1984).

700 million hectares of marginally suitable, potentially cultivable, rainfed land in the developing world, and some of this land could be upgraded through land improvement.

The total area of rainfed and irrigated cultivated land under arable and permanent crops in the world in 1990 was 1444 million hectares of which 772 million hectares (or 54%) were in the developing world (Table 6.4). The total area of rainfed cultivated land of 1207 million hectares is about equally divided between the developed and developing world. The rise in population has encouraged investments in irrigation. Land under irrigation doubled between 1900 and 1950, and has increased more than 2.5 times since then, to reach a global total of some 240 million hectares. About 73% of the total irrigated land is located in the developing world, with 56% in Asia. In the Middle East and Asia, 26% and 34% of their respective cultivated areas are under irrigation. The population pressure on cultivated land in 1990 was 5.2 persons per ha in the developing world compared with 1.9 person per ha in the developed world, corresponding to cultivated land area per capita of 0.19 and 0.54 ha respectively.

Let us assume that, by 2050, average cereal yields could be raised to 3 and 3.5 t/ha in Africa and the Middle East respectively (i.e. corresponding to what has been achieved in Asia and Latin America over the past 50 years), and to 4 and 5 t/ha in Latin America and Asia respectively (i.e. continuing with the established trend). The cultivable land area needed to harvest the required output of cereals would be 265 million hectares in Africa, 65 million hectares in the Middle East, 72 million hectares in Latin America and 344 million hectares in Asia, a total of 746 million hectares (Table 6.2). This corresponds to the total cultivable land area required by 2050 of 1468 million hectares, assuming that the regional cereal to non-cereal (including permanent crops) land requirement ratios remain as they are now (0.55 average for the developing world; Table 6.2). This is less than the potentially available cultivable rainfed and irrigated land of 2478 million hectares in the developing world (Table 6.3). Only in the Middle East region is the total cultivable land required larger than the potentially available cultivable land. Similarly, the cultivable land required to meet the developing world's cereal demand in 2110 at an average yield of 5 t/ha (i.e. continuing with established trends in all regions) is 917 million hectares, corresponding to a total cultivable land requirement of 1735 million hectares, which is 69% of the total available suitable cultivable land. The net area of land required in reality would be smaller because of multiple cropping.

The challenge related to the supply of agricultural land is therefore not necessarily due to any direct absolute shortage in the supply of suitable land, but in developing and operationalising national and supranational land use policies that could deal rationally with the multiple role of land and its uneven distribution, particularly between nations and between regions.

Potential population supporting capacity

The FAO study on potential population supporting capacity in the developing world (FAO, 1982, 1984) suggests that at a feasible level of inputs, the land resources of the developing world, taken as whole, are able potentially to support their expected populations (Table 6.5). The study took into account the effect, on the population supporting capacity, of the likely expansion of irrigation up to year 2000 only. The

Table 6.4 Land use and population in 1990

	Developed world	Developing world					World total
		Africa	Middle East	Asia	Latin America	Total	
Land area							
extent (10^6 ha)	5 623	2 964	738	2 048	2 018	7 768	13 392
% of world total	42	22	6	15	15	15	100
Population (1990)							
no. of people (10^6)	1 251	642	165	2 791	448	4 046	5 297
% of world total	24	12	3	53	9	76	100
Currently cultivated rainfed and irrigated land (1990)							
extent (10^6 ha)	672	182	54	384	152	772	1 444
% of world total	46	13	4	27	11	54	100
Currently cultivated rainfed land (1990)							
extent (10^6 ha)	608	171	40	252	136	599	1 207
% of world total	50	14	3	21	11	50	100
% of currently cultivated rainfed and irrigated land	90	94	74	66	89	78	84
Currently irrigated land (1990)							
extent (10^6 ha)	64	11	14	132	16	173	237
% of world total	27	5	6	56	7	73	100
% of currently cultivated rainfed and irrigated land	10	6	26	34	11	22	16
Persons per hetare of cultivated rainfed and irrigated land (1990)	1.9	3.5	3.1	7.3	2.9	5.2	3.7
Cultivated rainfed and irrigated land per	0.54	0.28	0.33	0.14	0.34	0.19	0.27

Source: FAO Production Yearbook (1991).

Table 6.5 Potential population supporting capacities of the developing regions at low, intermediate and high levels of inputs (billions of persons)

	Low inputs	Intermediate inputs	High inputs
Africa	1.25 (20)[b]	4.49 (5)	12.87 (<5)
Middle East	0.18 (89)	0.24 (66)	0.32 (49)
Asia[a]	4.13 (70)	6.87 (45)	10.35 (34)
Latin America	1.71 (22)	5.85 (6)	13.67 (3)
Developing world	7.27 (51)	17.45 (33)	37.21 (13)

[a]Includes an estimate for China and Korea of 1.67, 2.51 and 4.02 billions of persons at low, intermediate and high inputs.
[b]Figures in parentheses show the contribution of irrigation potential to total potential as a percentage.
Source: FAO (1982).

potential for the enhancement of the land resource base through land improvement and specialised farming techniques was not taken into consideration.

The most important result of the study is that even at the intermediate level of inputs, the land resource base in the developing world taken together is potentially capable of supplying the needs of over 17 billion people, and at a modestly high level of inputs, over 37 billion people. All regions taken as wholes, with the exception of the Middle East, are each potentially capable of meeting their estimated needs at intermediate levels of inputs.

The FAO study indicates that in some nations population growth will sooner or later outrun the potential population supporting capacity. There may also be difficulties in the more densely populated parts of Asia. Fortunately, many of the potential deficit regions are adjacent to, or not too far from, more richly endowed regions which are potentially able to produce surpluses (FAO, 1982; Shah et al., 1985; Dudal, 1987; Bunting, 1992a). Where these ecologically and demographically complementary regions lie within the bounds of individual nations, it has generally proved possible, in the past, to manage the foreseeable difficulties. In other cases, transfers will have to be made, as they are now, between countries and between regions of the earth, but on a larger scale. To realise and pay for these exchanges is more a management and engineering task than an agricultural or applied biological one, and depends on the income level of people.

Nutrients, energy and water

Larger yields and extra outputs required in the future cannot be obtained without applying substantial amounts of external inputs. This applies particularly to nitrogen and other plant nutrients, energy and water.

Nutrients

The current production of about one billion tonnes of cereals in the developing world contains some 15 million tonnes of nitrogen and about three million tonnes of phosphorus. The 4.6 billion tonnes of cereals that will be needed in the developing world in 2110 will contain about 70 million tonnes of nitrogen and about 14 million tonnes of phosphorus. Some crops also require large amounts of potassium. The return of organic matter to the soil in crop residues and manures can replace some of the nutrients removed by crops and animals, but in the amount that can be returned to the soil under any system of agriculture there are insufficient quantities of nutrients for total replacement. The deficiency must be made good by application of mineral fertilisers, particularly as more and more of the produce is sold off the farm. While some extra supplies in the case of nitrogen can come from biological fixation, there is as yet no practical prospect of increasing supplies of nutrients, through biological means, on a sufficient scale. Although potential supplies of plant nutrients exceed future needs, the extra amounts of fertiliser nitrogen, phosphorus and other nutrients will have to be produced and delivered at an affordable price, and used by producers with maximum efficiency.

Energy

At present, the world consumes some 350 exajoules of energy per year, of which 330 exajoules (94%) is in the form of primary commercial energy. The approximate percentages of current use of these primary market and traditional energy resources are: oil 36%, coal 26%, gas 18%, renewables 8%, traditional fuels 6% and nuclear 6%. By 2100, consumption is expected to rise to around 1100 exajoules, mainly in the form of primary commercial energy (Gouse et al., 1992).

All energy sources—coal, oil and gas, as well as nuclear and renewables, particularly solar—will be needed in the coming century to support world development. For large-scale power generation, solar-thermal energy has immediate promise, and photo-voltaic energy has medium-term promise (Anderson and Ahmed, 1993). The potential contribution from biomass appears limited. Given the known sources of energy (excluding geothermal, wind and tidal), there does not appear to be a real energy resource problem for at least the next 50 years. But thereafter, acceptable nuclear- and solar-based technologies as part of integrated systems involving other forms of energy, would be necessary to provide solutions (Gouse et al., 1992). However, because fossil fuel may have other uses, its conservation may be beneficial.

Water

It is estimated that the volume of fresh water annually renewed through the hydrological cycle, involving sea, atmosphere and land, is sufficient on a global basis to meet the material needs of many times the existing world population (TAC, 1988). Despite such apparent abundance, shortage of water is already a growing problem in much of the drier areas of the world, particularly in the Middle East, and in parts of Asia. Over 40% of the world's population lives in shared drainage basins, and water is now recognised as a strategic resource of political significance. For nations whose surface water supplies originate from outside their national borders, international agreements and land use policies will be required to maximise water capture in the watersheds and to ensure equitable distribution.

Irrigated lands currently contribute more than a third of the global food supplies, and in Asia and the Middle East about half (Sarma, 1986; Brown, 1987). Sustainable management of irrigated lands is a growing challenge, particularly in the intensively cultivated systems in Asia where productivity gains in crops such as rice and wheat must not only be sustained but also increased further to meet future needs. In Asia, waterlogging and salinisation pose a particular threat to future performance. In sub-Saharan Africa both production techniques and management must become increasingly efficient to justify past and future investments in irrigation (Dudal, 1992).

KNOWLEDGE AND TECHNOLOGY CHALLENGES

To produce a larger surplus of biological products for the market, at a smaller unit cost, producers will need new, improved production technologies and knowledge. Such knowledge and technologies are the products of research.

The ultimate responsibility for this task lies with national agricultural research systems (NARSs), which encompass all those institutions in the public and non-governmental sectors, including universities and NGOs, that are potentially capable of contributing to research related to the development of agriculture.

Despite the many difficulties, it is noteworthy that the average size of the public-sector national research system in developing countries as a group, measured in number of researchers, has more than doubled since the late 1960s. However, too many national research systems, particularly in Africa, still have limited capacity to undertake anything but adaptive research on a few commodities (Hayward, 1987; Pardey and Roseboom, 1991). Also, most NARSs lack the expertise and resources to deal with research on natural resources management and its integration with productivity research.

At the international level, the CGIAR centres are engaged in strategic research and fill gaps that cannot be filled by national systems. They also provide linkages with advanced research institutions in the developed world. The global research community does not presently have an effective paradigm for research on natural resources management, and there are no models for achieving sustainable increases in agricultural production.

The level of external inputs used by farmers is a key factor affecting the land resource base, because both underuse or overuse have detrimental effects (TAC, 1992). Farming systems in which farmers use few or no external inputs but plant crop annually eventually deplete soil nutrient reserves and reduce vegetation cover, thereby exposing the soil to erosion and degradation. Many cropping areas of Africa have been affected in this way. For intermediate levels of output, research is needed to develop integrated nutrient supply systems based on a balanced mix of external inputs, organic manure, biological nitrogen fixation and efficient cycling of nutrients. To sustain high levels of production, high levels of external inputs are needed but may result in pollution problems.

Strategies for the improvement of both intensive and extensive production systems would need to focus more directly on providing institutional and technical support for improved soil fertility, pest and disease control, and water management. The inevitable increase in fertiliser application, which will be the main source of future agricultural growth and food security, must be balanced by efforts to improve efficiency of fertiliser use and to maximise the contribution from organic sources. Such efforts are needed to promote sustainability of the land resource base and to bring productive practices within the reach of the farmer.

CULTURAL, INSTITUTIONAL AND POLICY CHALLENGES

Development is about material and social change in which old or traditional ways are combined with new ways. Cultural, institutional and policy factors are therefore inextricably linked to questions of pace and direction of the desired changes and how they should be brought about.

In each nation, an enabling cultural, institutional and policy environment for sustainable development must be created and maintained so that: the public and private sectors can maximise the opportunities available for agricultural and economic growth; so that enterprise, competition and efficiency can flourish; and so that the rule of law is respected and promoted.

Cultural environment can influence what changes will be acceptable, in what form, and at what rate. Most peoples of the world, if they so choose, can respond positively to the challenges of change provided their nations adopt features such as the existence of market economy; the absence of rigid, doctrinal orthodoxy; the freedom to inquire, to dispute, to experiment; a belief in the possibility of improvement; a concern for the practical; a rationalism that defies ideologies, dogmas and folklore; and the promotion of professional meritocracy in the management of all national institutions (Kennedy, 1993). Education, more than any other social instrument, is a key factor in hastening the pace of cultural change and development.

Central to the effective management of national agricultural and rural development is the system of public institutions set up by governments, and the professionals who work in them. The institutions must have the right kinds of people and contribute to the formation and execution of policy for national development at three levels: central (national), intermediate (regions and districts) and local (Shah et al., 1985; Kassam et al., 1990; Bunting, 1992b; FAO, 1993).

Centrally, at the level of the nation, institutional capacity is required to produce the strategic long-term national land use development and management plans to facilitate integrated policy decisions, legislation, administrative actions and budgeting. At the intermediate level in regions and districts, institutional capacity is required to formulate more specific and detailed programmes based on the national strategies and programmes. At the local levels, the institutional capacity must be able to provide the field services of different ministries and departments for the different sectors or commodities (Bunting, 1992b). Consequently, at the national level, geographically referenced databases of information relating to natural resources, land use and land potentials, continuously kept up-to-date, are essential for the formulation and execution of policy for sustainable development in agriculture and the rural sector (Kassam et al., 1982; 1990; Brammer et al., 1988; FAO, 1993). Few nations have such databases.

Equally important is the institutional capacity to address questions of transnational concerns such as: (i) which set of neighbouring countries may constitute a natural and logistical cooperative unit for trade, food and economic security and development of renewable natural resources; and (ii) what levels of international assistance and cooperation will be needed to promote a certain level of regional agricultural and rural development?

Development takes place within individual nations; only the government of the nation has the authority to determine how and at what rate the resources of the nation are used to advance the economy of the nation and to improve the material and social condition of its citizens. However, wherever advances have occurred in farming practices and output, they were linked to other developmental changes, national and international, which created the enabling environment in which producers and those who purchase, process and deliver rural surpluses to consumers and other end users would reasonably expect to make profits from their investments (Bunting, 1992a).

Policy challenges operate at several levels. At the national level, they surround and affect all else, including the initiatives of rural communities and individual families at the local level. At the intermediate level, policy challenges relate to the need to reconcile national policy goals with local opportunities and enterprise to ensure that both long-term and short-term needs are addressed, and all actors become involved. Policy challenges also operate at the supranational level, and can be expected to become more acute and critical in the future as more and more nations attempt to reconcile national priorities with regional and global priorities. Also, factors at the international level affect trade in agricultural surpluses between nations. A major policy challenge to the developing nations relates to the need to expand export markets within the developing world, and to maximise complementarities between nations and between regions in meeting future needs (FAO, 1982; Bunting, 1992a).

The average statistics for the provision of food and other biological products from land, though encouraging, conceal the variations in distribution in both space and time which lead to disasters and tragedies. The biological and environmental causes of these variations can often be removed or ameliorated by appropriate technical actions. But the social and political causes are more intractable; for example, unequal control of land and other resources, wars of all kinds, repressive and exploitative social and political structures, and incompetent governmental and commercial systems (ICIHI, 1985; Von Braun, 1991; Von Braun et al., 1992; Bunting, 1992a).

ACKNOWLEDGEMENTS

We are most grateful to Professor A. H. Bunting, CMG, for his many insightful ideas and suggestions during the preparation of this paper. We express our appreciation to Professor R. Dudal, Dr J. Monyo and Dr G. Gryseels for their helpful comments on an earlier draft of the paper.

REFERENCES

ACC/SCN, 1992. *Second Report on the World Nutrition Situation—Vol.I. Global and Regional Results*. UN Administrative Committee on Coordination, Sub-Committee on Nutrition, Geneva.

Anderson, D. and Ahmed, K., 1993. Where we stand with renewable energy. *Finance and Development* **30**: 40–43.

Brammer, H., Antoine, J., Kassam, A. H. and Van Velthuizen, H. T., 1988. Land Resources appraisal of Bangladesh for Agricultural Development. Technical Reports 1–7, FAO/UNDP:BGD/81/035, Agricultural Development Advisor Project, Dhaka, Bangladesh.

Brown, L. R., 1987. *Sustaining World Agriculture. State of the World.* Worldwatch Institute, Washington, DC.

Bunting, A. H., 1992a. Feeding the world in the future. In: Spedding, C. R. W. (ed), *Fream's Principles of Food and Agriculture*, 7th edn, pp. 256–290. Blackwell Scientific, Oxford, UK.

Bunting, A. H., 1992b. *A Prospect of Africa.* Mimeograph Report, ISNAR, The Hague, The Netherlands.

Bunting, A. H., Kassam, A. H. and Abernethy, C., 1986. Development of irrigation in Africa: needs and justification. Consultants' report prepared for the FAO Consultation on Irrigation in Africa. 21–25 April 1986, Lomé, Togo.

Cesal, L. and Rossmiller, E., 1989. Development assistance and trade: the way it was, the way it is, and what the difference means. In: Clubb, D. and Ligon, P. C. (eds), *Food, Hunger and Agricultural Issues*, pp. 225–239. Proceedings of a colloquium on future US development assistance. Winrock International Institute for Agricultural Development, Arkansas, USA.

Dudal, R., 1987. Land resources for plant production. In: McLaren, D. J. and Skinner, B. J. (eds), *Resources and World Development*, pp. 659–670. John Wiley, Chichester, UK.

Dudal, R., 1992. Sustainability in irrigated land-use systems. In: Feyen, J., Wendera, E. and Badji, M. (eds), *Advances in Planning, Design and Management of Irrigation Systems as Related to Sustainable Land Use*, pp. 575–585. Proceedings of an International Conference organized by the Centre for Irrigation Engineering of the Katholieke Universiteit, Leuven, in cooperation with the European Committee for Water Resources, 14–17 September 1992, Leuven, Belgium.

Dudal, R., Higgins, G. M. and Kassam, A. H., 1982. Land resources for the world's food production. In: *Managing Soil Resources to Meet the Challenges of Mankind*, pp. 57–68. Proceedings of the 12th International Congress of Soil Science, 8–16 February 1982, New Delhi, India. Indian Society of Soil Science, New Delhi, India.

FAO, 1978–81. *Report on the Agroecological Zones Project.* World Soil Resources Report 48, Vol.I–IV. FAO, Rome.

FAO, 1982. *Potential Population Supporting Capacities of Lands in the Developing World.* FPA/INT/513. FAO, Rome.

FAO, 1984. *Land, Food and People.* FAO Economic and Social Development Series 30, FAO, Rome.

FAO, 1993. *Agroecological Assessment for National Planning: The Example of Kenya.* FAO Soils Bulletin 67, FAO, Rome.

Gouse, S. W., Gray, D., Tomlinson, G. C. and Morrison, D. L., 1992. *Potential World Development Through 2100: The Impacts of Energy Demand, Resources and the Environment.* Proceedings of the 15th World Energy Council Congress. Madrid, Spain, 20 September 1992. World Energy Council, London.

Hayward, J. A., 1987. Issues in research and extension. In: Davies, T. J. and Schirmer, I. A. (eds), *Sustainability Issues in Agricultural Development*, pp. 143–161. Proceedings of the Seventh Agricultural Sector Symposium, The World Bank, Washington, DC.

ICIHI, 1985. *Famine: A Man-made Disaster?* Independent Commission on International Humanitarian Issues, London.

Kassam, A. H., Van Velthuizen, H. T., Higgins, G. M., Christoforedes, A., Voortman, R. L. and Spiers, B., 1982. *Assessment of Land Resources for Rainfed Crop Production in Mozambique.* Field Documents 32–37. Land and Water Use Planning Project, FAO:UNDP/MOZ/75/001. Ministry of Agriculture, Maputo, Mozambique.

Kassam, A. H., Shah, M. M., Van Velthuizcn, H. T. and Fischer, G. W., 1990. Land resources inventory and productivity evaluation for national development planning. *Philosophical Transactions of the Royal Society, London* **329**: 391–401.

Kennedy, P., 1993. *Preparing for the Twenty-First Century*. Harper Collins, Glasgow.
Pardey, P. G. and Roseboom J., 1991. *National Agricultural Research from a Regional and Agroecological Perspective*. Working Paper No. 40, ISNAR, The Hague, The Netherlands.
Sarma, P. B. S., 1986. Water resources and their role in food production, In: *Global Aspects of Food Production*. Natural Resources and the Environment Series, Vol.20. IRRI, Los Banos, Philippines.
Shah, M. M., Fischer, G. W., Higgins, G. M., Kassam A. H. and Naiken, L., 1985. People, land and food production—potentials in the developing world. Collaborative Paper. IIASA, Laxenburg, Austria.
TAC, 1988. *Sustainable Agricultural Production: Implications for International Agricultural Research*. TAC Secretariat, FAO, Rome.
TAC, 1992. *Review of CGIAR Priorities and Strategies*. TAC Secretariat, FAO, Rome.
Von Braun, J., 1991. *A Policy Agenda for Famine Prevention in Africa*. IFPRI, Washington, DC.
Von Braun, J., Howarth, G., Kumar, S. and Pandya-Lorch, R., 1992. *Improving Food Security of the Poor: Concept, Policy and Programmes*. IFPRI, Washington, DC.

CHAPTER 7

The Future of the Land Lies in the Capability of its People and their Institutions

D. L. Dent,[a] D. B. Dalal Clayton[b] and R. B. Ridgway[c]

[a]University of East Anglia, Norwich, UK. [b]International Institute for Environment and Development, London, UK. [c]Natural Resources Institute, Chatham Maritime, Kent, UK

INTRODUCTION

Land use planning has been promoted as a rational way to manage land resources but often expectations have been disappointed. The most consistent feature of land use planning to date has been the failure to implement plans in anything like the shape envisaged by the planners. A review of land resource information and its use in developing countries (Dalal-Clayton and Dent, 1993) and case studies in Sri Lanka and Tanzania (Dent and Goonewardene, 1993; Kauzeni et al., 1993) highlight a variety of reasons for this.

It is a prerequisite of successful planning that there is a measure of agreement about the goals between policy-makers, planners, land users and other stakeholders. Their various, often conflicting interests must be established and respected if a fair bargain is to be struck between them. There must also be good baseline information about land resources and, building on this, the identification of a range of management options according to the agreed goals. The decision-makers need predictions of the economic, social and environmental consequences of implementing each option. Finally, there must be the institutional capability to make use of this information, choose the best option and put the plan into action. Rarely have all these criteria been satisfied.

LAND USE PLANNING METHOD

Figure 7.1 presents a procedure for land use planning that might be applied either for framing land use policy or, at the grass roots, for managing the land. The key question is, who shall be responsible for each step? Several case studies in Sri Lanka and Tanzania

The Future of the Land: Mobilising and Integrating Knowledge for Land Use Options
Edited by L. O. Fresco, L. Stroosnijder, J. Bouma and H. van Keulen.

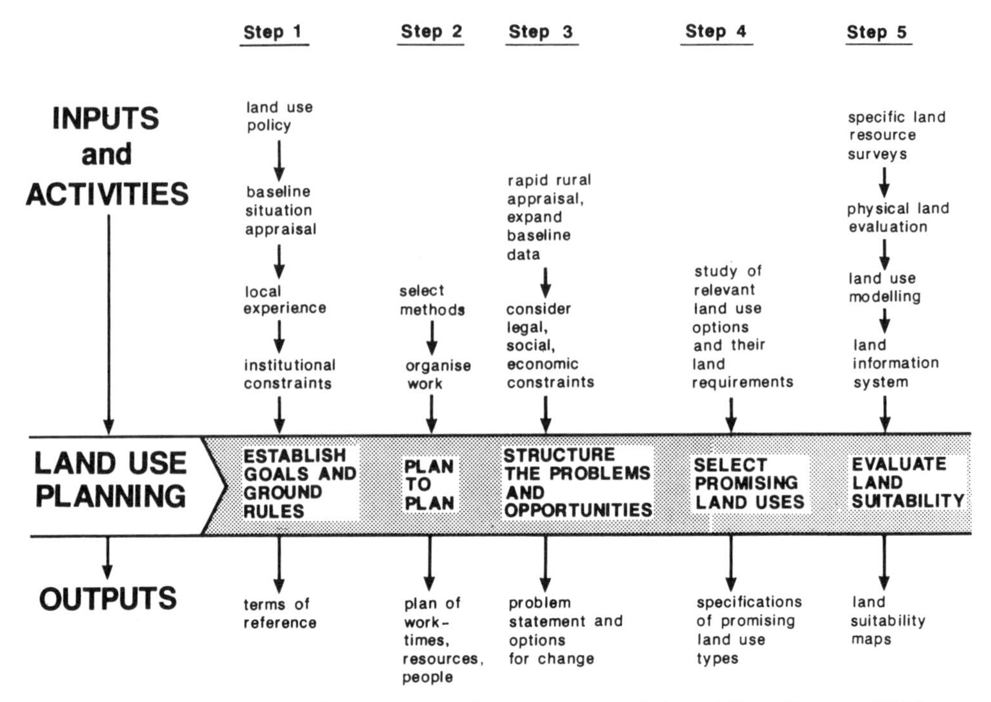

Figure 7.1 A synoptic view of the land use planning process. Adapted from Dent and Ridgway (1986)

have revealed cumulative failings all along the way. These failings arise from the kind of natural resources data available to planners and from the centralised nature of land use planning.

Usually, the process has been wrong-footed at the first step. Land use planning has been a top-down procedure and the goals, whether explicit or not, have been those of government or development agencies. The goals of the land users whom, in the last analysis, have to implement the plan, may be different. To the extent that these conflicts of interest are not resolved, plans stay just plans.

THE INFORMATION GAP

Most land users rely on traditional knowledge and their own experience, both won by trial and error. Their information may be apt and their judgements canny but neither can be transferred to other areas or cope with rapidly changing circumstances—ecological, technical or economic. In developing countries, the independent information needs of this huge group of managers and decision-makers have been neglected by land resources professionals.

For planners working on a grander scale, baseline information has been provided by professional surveys, often carried out by expatriate technical staff. Sometimes the

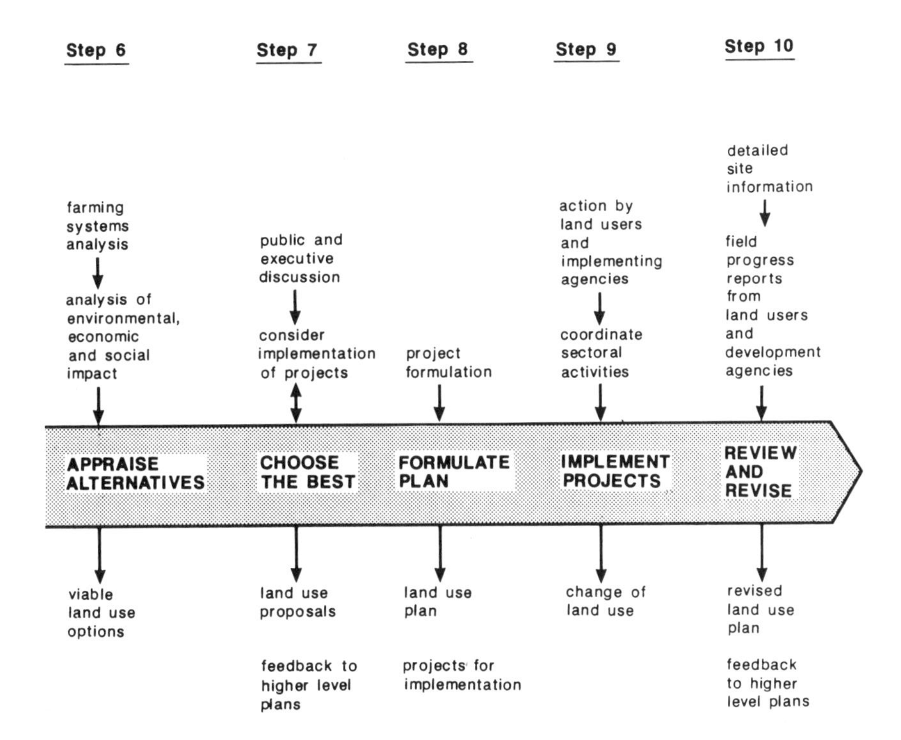

relationship between the two groups of professionals has been fruitful but, often, it is marked by mutual frustrations: the planners complain 'What we are wanting, we're not getting, and what we are getting, we're not wanting'; while the surveyors complain, 'We have been pouring information into the sand!'

Looking at the terms of reference of surveys, we often find no mention of who, in particular, will have to use the data and for what particular purpose. No account is taken of the capacity of planners and decision-makers to make use of the data. At the time, it was simply obvious that some information on land resources would be needed.

Under these circumstances, natural resources specialists have followed their own agenda. Taking soil survey as just one of many possible examples, surveyors have systematically collected landform and soil morphological data and representative sites have been further characterised by a wide range of *in situ* or laboratory measurements. World-wide, the items of data are remarkably standard. Where not of direct relevance to land use, they are assumed to serve as surrogates for relevant land qualities that cannot be measured directly or routinely, but evidence on the reliability and practical value of the assumed relationshps is conflicting (Mackenzie, 1993). In fact, very few of the data collected are used in land use planning or management.

COMPREHENSION AND UTILITY

Land resources data are under-utilised because other professionals, let alone policy-makers and land users, do not appreciate their utility and are not at ease with them.

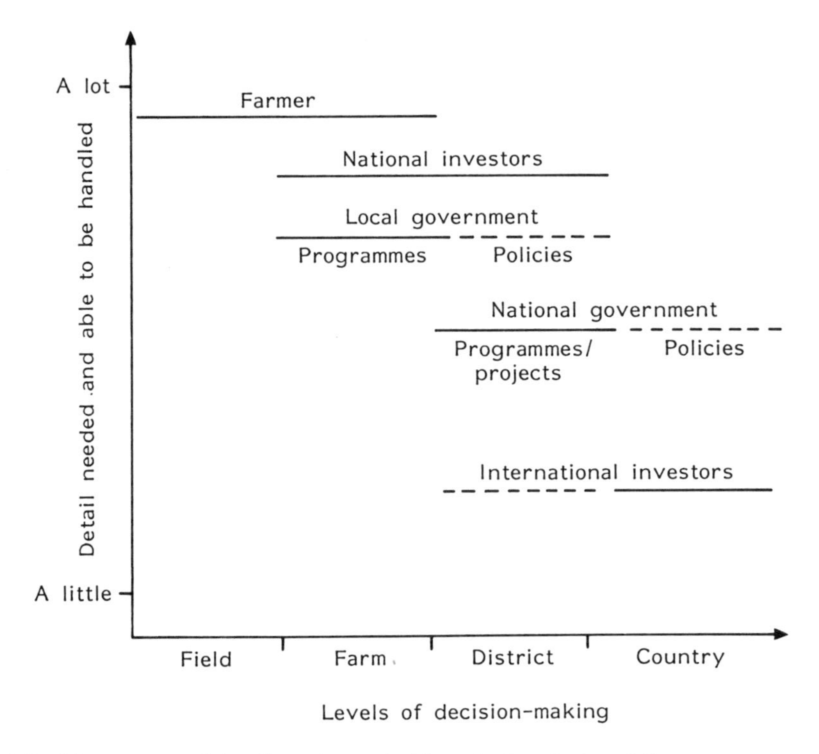

Figure 7.2 The amount of detail needed and able to be handled at different levels of planning

Jargon and the intimidating welter of detail are obvious reasons for this. More importantly, they may not match with the decision-makers' way of looking at the land; may not provide information at the appropriate level of detail or generalisation to support the decision in hand (Figure 7.2); or simply do not answer the questions they are asking. There are many decision-makers and a standard survey cannot meet all their different needs.

Integrating diverse information is difficult. Usually, decision-makers do this intuitively, placing greatest reliance on the information that is best understood.

That there is a problem of comprehension is well understood. In response, survey interpretations have been provided both to simplify the data and to integrate them. Examples include the Storie Index, Land Capability Classification and Land Suitability Evaluation. Each of these well-known interpretations was designed for a specific purpose which it fulfills very well. Now it is offered 'off the peg' by survey organisations or demanded by clients simply because it is less intimidating than a sheaf of soil maps.

In truth, the utility of standard, off-the-peg surveys and interpretations is very limited, especially in poor countries that are already farmed to the limits, or beyond the limits, of the present capacity of the land. It is not helpful to a subsistence farmer to tell him that his land is no better than S3 for his staple crop! Decisions already taken and acted upon severely restrict the room for manoeuvre. If they are undertaken mechanically,

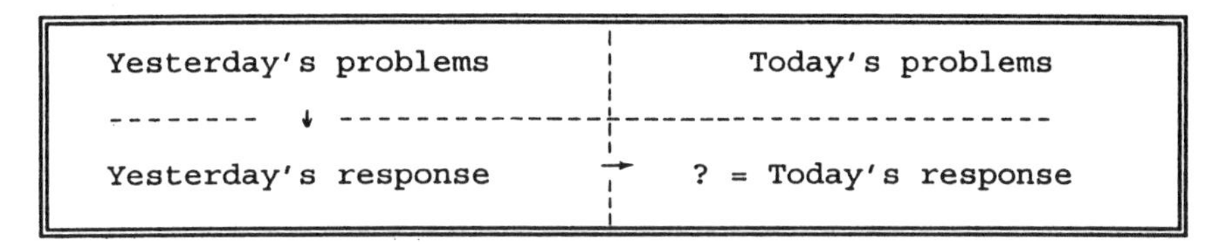

Figure 7.3 Problems and responses

natural resources surveys and standard interpretations are addressing yesterday's problems (Figure 7.3).

Decision-makers need dynamic interpretations of data on natural resources that keep pace with the changing nature of questions asked of them. Since land is not all the same, answers need to be area-specific, so that particular areas, people and projects can be identified and, usually, answers need to be quantitative. Ideally, policy-makers and managers need to use a bank of data iteratively in the course of evolution of projects or development programmes. A static 'one shot' survey cannot meet this need.

THE CAPABILITIES OF INSTITUTIONS

A review of many institutions and projects in poor countries reveals that there has been little dialogue between policy-makers or planners and, on the one hand, the supposed beneficiaries and, on the other hand, land resources specialists.

Planning has mostly been a centralised, top-down activity. Brinkman (see Chapter 1) points to the need for a platform for negotiation between all the parties involved, rather than attempting to impose the will of governments and development agencies on unwilling people and unyielding landscapes.

Governments are neither omnipotent nor omniscient. The loads they impose upon themselves in attempting to plan, administrate and implement land use soon exceed their administrative and logistic capabilities. They must learn when and where to hand over responsibility.

Participatory planning is now promoted as an alternative to the top-down model (FAO, 1993; Dalal Clayton and Dent, 1993). Participatory planning also faces problems of information and institutions. At local level, local institutions that can take and implement land use decisions have to be built and supported. In Sri Lanka, for example, village water management and cultivation groups have survived decades of paternal government and could take on a wider role. At present, however, we have no convenient models for extending the participation of land users and other stakeholders in higher levels of planning and policy-making.

Especially in developing countries, institutions responsible for land use are fragmented and compartmentalised. Land resources information, especially spatial information, has no established place in policy-making and this information is rarely carried through to the point of decision by land resources specialists. Even where useful data exist, they are often discarded on the way.

Without a strong institutional capability in land use planning, decisions will continue to be made on short-term political and financial grounds, and potentially useful natural resource information will be unused and, therefore, lost. Within natural resources survey institutions, there is limited capability for updating, upgrading or re-interpreting the database to meet new demands. Inadequate archival services and short institutional memory resulting from frequent changes of staff contribute to the loss of information.

We see two new roles for natural resources specialists: not in survey, nor even in interpretation, but in building up the capabilities of institutions to use their information.

At policy-making level, line ministries, sectoral agencies and, especially, inter-sectoral planning institutions need long-term relationships to build up the skills and experience needed to identify the key information and carry it through to the point of decision and, beyond, to learn from mistakes. This is work for decades, not for the traditional cut-and-run approach of development aid. Investment in surveys, for example, is wasted if there is no one who can make use of them.

Support for the grass roots planners. There will never be enough natural resources specialists to go around. Decision-makers at the local level will have to fend for themselves in respect of collection and interpretation of the detailed information they need and can use. Research is needed to find out what approaches to resource assessment and planning procedures these communities already use, and to determine what can be learnt from these and whether they can be transferred. Building on this, we must find out what other natural resources information is (1) critical, (2) can be used right now by the local decision-makers, and (3) can be collected and interpreted by the people who need it.

Do-it-yourself survey and calculation kits can then be developed and tested, and knowledge-based decision-support systems built to enable local decision-makers to gather and make use of the information they need.

ACKNOWLEDGEMENTS

Our research in Sri Lanka and Tanzania was funded by the UK Overseas Development Administration. We are grateful to Dr R. Brinkman for his pertinent comments on the poster from which this paper is derived.

REFERENCES

Dalal Clayton, D. B. and Dent, D. L., 1993. *Surveys, Plans and People: A Review of Land Resource Information and Its Use in Developing Countries*. Environmental Planning Issues No. 2, International Institute for Environment and Development, London.

Dent, D. L. and Goonewardene, L. K. P. A., 1993. *Resource Assessment and Land Use Planning in Sri Lanka: A Case Study*. Environmental Planning Issues No. 4, International Institute for Environment and Development, London.

Dent, D. L. and Ridgway, R. B., 1986. *Sri Lanka Land use Planning Handbook*. UNDP, Colombo, Sri Lanka.

FAO, 1993. *Guidelines for Land Use Planning*. FAO, Development Series 1, Rome.

Kauzeni, A. S., Kikula, I. S., Mohamed, S. A., Lyimo, J. G. and Dalal Clayton, D. B., 1993. *Land Use Planning and Resource Assessment in Tanzania: A Case Study*. Environmental Planning Issues No. 3, International Institute for Environment and Development, London.

Mackenzie, N., 1993. *Are Soil Surveys Gathering the Right Data?* CSIRO Division of Soils Seminar Series, CSIRO, Canberra, Australia.

CHAPTER 8

Selected Case Studies at Institutional Level

IMPACT OF INCREASED ATTENTION FOR NATURAL RESOURCES MANAGEMENT ON RESEARCH MANAGEMENT

P. Goldsworthy and L. Boerboom

International Service for National Agricultural Research (ISNAR), Den Haag, The Netherlands

In many developing countries agriculture is still the main user of land and water resources. Agricultural policies and the actions of poor farmers who have to give priority to survival have often combined to degrade natural resources and threaten future production. Concern has led to a shift in international and national research from a commodity, production orientation to a wider systems perspective concerned with sustainability. ISNAR's aim is to advise national agricultural research systems (NARS) on how to incorporate natural resources management and sustainability objectives into their research agendas. Other organisations concentrate on technical aspects of the new approach; ISNAR's special role is to advise NARS on the policy, institutional and research management changes that will be required.

Some of these changes include:

- *changes in research policy:* balancing production and environmental goals, identifying optimal natural resource uses, and costing of impacts of agricultural activities on the natural resource base;
- *changes in organisation:* cross-sectoral planning, a more unified research system, better links between institutions and participation of local resource managers, both individuals and communities;
- *new management tools:* for setting research priorities, for designing interdisciplinary research projects, for lengthening research planning horizons and for developing human resources; and finally
- *better information:* on the state of natural resources and for the management of research.

The Future of the Land: Mobilising and Integrating Knowledge for Land Use Options
Edited by L. O. Fresco, L. Stroosnijder, J. Bouma and H. van Keulen.

A NEW MODEL FOR LAND USE PLANNING IN RUSSIA

V. S. Stolbovoy
Dokuchaev Soil Science Institute, Moscow, Russia

OBJECTIVES

Land use planning in Russia is developing in two directions: (1) modernisation of the existing highly centralised (national level) planning system serving (collective) state agricultural enterprises, and (2) the formation of a new planning system which will serve private farmers (landowners) on the local (village) level. Land use planning problems encountered in Russia are caused by, for example, an extreme variety of natural conditions, the existing infrastructure, agricultural traditions, and social and national customs. Therefore, a universal model for land use planning does not exist. The critical attitudes of private farmers towards government efforts pose additional problems. However, the efforts of the administration are not sufficient. It may be that a revision of the role of soil surveying in land use planning provides a solution.

METHODS

The proposed principal innovations in soil survey consist of separated development of land evaluation methods and the introduction of methods of interpretation of soil-ecological data. We transformed for this purpose the FAO approach as well as criteria used by the US soil conservation survey. Taking into account the very limited technical possibilities of recent soil surveys in Russia, we developed both a manual version and an automated (GIS) one.

DISCUSSION

Land management in the former USSR was based on the general idea of centralised management as the only possibile way of changing the old pre-revolutionary system into a system based on a communist social philosophy. The Government made a plan, with figures representing the amount of production, and passed it on to republics, districts, administrative regions and finally to collective and state farmers. These figures had the status of a state law and had to be fully respected by everybody, regardless of whether or not the natural resources for the planned kind of activity were available. In order to facilitate the planning procedure, the Government developed the following basic objectives: to create and enlarge collective farms, to homogenise soil cover on the basis of amelioration, and to unify all technology. Realisation of production targets was reached without taking into account the economic interest of the workers. There was complete state control of all steps of the activity.

This planning system led to contradictions. Some of them were caused by the enormous size of the country and the great variety in natural conditions. There were no technical facilities to store, analyse and manage the huge amount of information. There were other contradictions between the general idea of a nation-wide, obligatory land use plan and the local needs, possibilities and wishes; be simplified views from the top (central

Table 8.1 Elements of previous and proposed approaches of land use planning

	Models	
Characteristics	Previous	Proposed
Basis	State monopoly on land ownership	Different forms of land ownership
Aim	Maximum production of a specific terrain unit	To provide sustainable development
State responsibility	Planning and management	Environmental control, market regulations by subsidies and taxation
Owners activities	Fulfilling plan	Realising his own land use options
Duty soil survey to state	To provide information (limited use in practice): • to make decisions from top level down to base • to serve administrative structure	To provide essential information for: • state environmental land use policy
Duty soil survey to owners	Practically non-existent	To provide information for: • owners' creative initiatives

planning offices) on local problems as the only direction of communication, and the absolute impossibility of communication from the base to the top; the absence of a link between actual land use practices and natural resources and their possibilities. These contradictions as well as some other discrepancies today form our inheritance and have to be solved in the future.

The necessity of modernisation of land use planning in Russia caused by land reform started a few years ago. Changes taking place in this country affect almost all aspects of land relations. The principal elements of innovations are shown in Table 8.1.

As can be seen from Table 8.1, the differences between the principal characteristics (land ownership, declared aims, responsibilities, etc.) of the previous and proposed models are very significant. Without any doubt, converting the previous system into the proposed one will take time and should be realised gradually rather than abruptly. One of the important issues will be modernisation of the state (government) soil survey. This modernisation might be a basis for a new model of land use planning.

It consists of the following blocks:

(1) *Inventory of soil-ecological resources on a local level.* We use ordinary soil maps compiled a 1:10 000–1:25 000 scale, supplemented with some topographic, land use and other data. This information is available for all agricultural lands of Russia, which the proposed model should use. Some of the lands have been mapped repeatedly, providing dynamic data.
(2) *Evaluation of land suitability for different kinds of activities* (agricultural and non-agricultural). We modernised the FAO model and created a system of special criteria.
(3) *Interpretation of soil-ecological conditions.* Interpretation begins with ecological zoning. The principal task of the procedure is to separate government and owners' competence. In accordance with recent laws, owners are independent in their activity. On the other hand, government is responsible for environmental protection. This

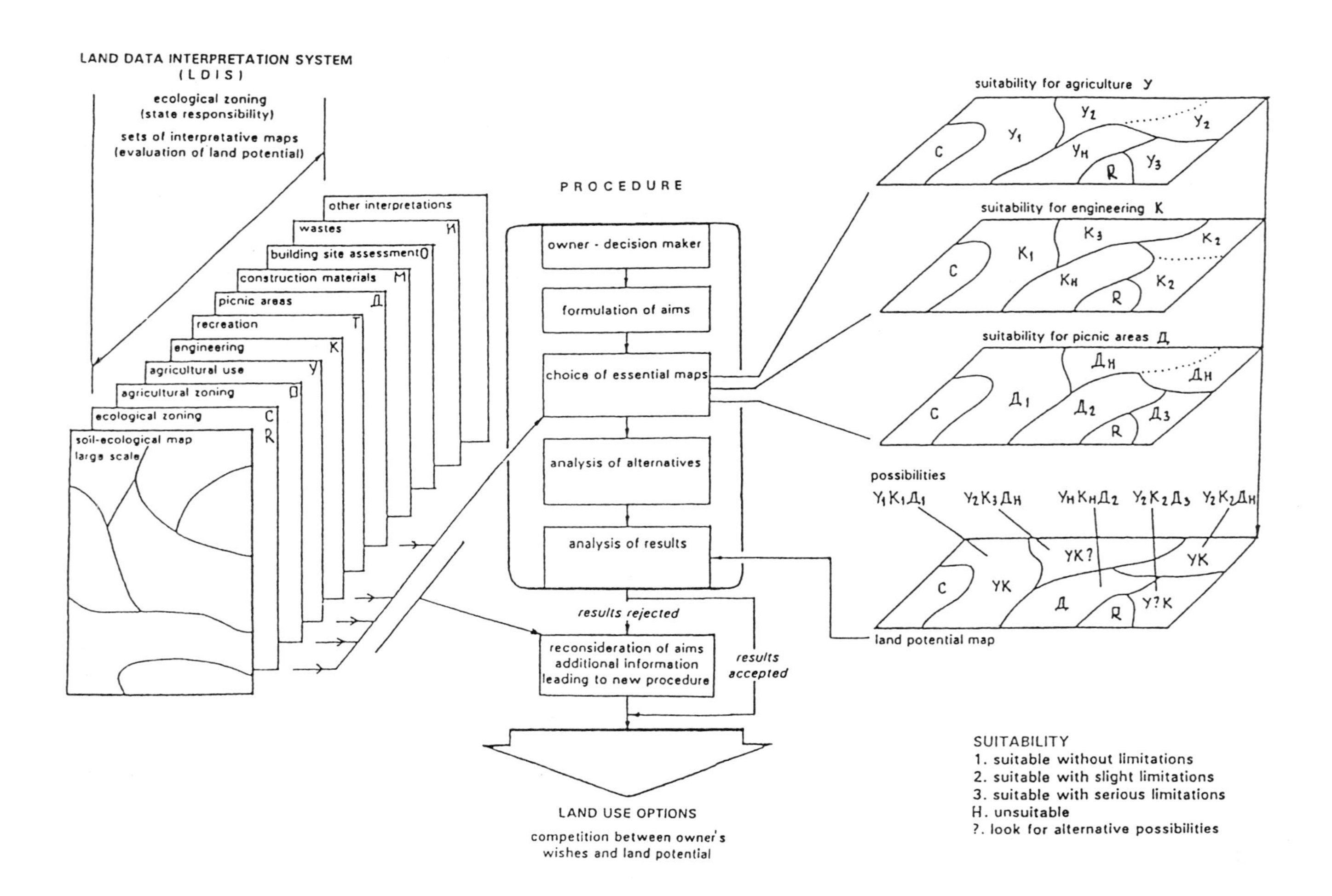

Figure 8.1 Land use planning on a local level based on a state soil survey supported system

means that government should limit owners. Ecological zoning divides all mapping units of a given area into three zones which: (1) must be conserved, (2) need to be rehabilitated, and (3) may be used. The balance of these zones defines landscape stability. Conservation means land use alike the natural ecosystem, for which there are (i.a.) two reasons: to conserve soils as a natural body (mainly potential erodible soils) and to conserve soil-ecological functions (hydrology, salt accumulation, etc.). Consequently, rehabilitation should be directed at reconstructing soils as bodies, functional holders or both. On the other hand, ecological zoning helps landowners to make reductions in tax payments as well as to get financial support from the government for rehabilitation. Ecological zoning will also assist in defining regional ecologically-based land use management policies.

We suggest creating seven sets of land suitability maps (a total of 38 maps) for those areas which may be used, based on the USDA data-interpretation system.

(4) *Planning of activities*. Private farmers should play the main role in this process. Our system exposes the soil-ecological potential of their lands. This information will help them make business decisions and create their own farming system. See Figure 8.1 for a diagram of the use of land use data-interpretation system to ascertain land potential.
(5) *Land management*. It is realised partly by government control on areas that are subject to ecological zoning. Government administration should protect the environment. Private farmers should use their creative initiative to formulate their own land use options.
(6) *Reinventorisation and revision of activity*. It may be necessary for private farmers to change their previous decisions.

CONCLUSIONS

A new model of land use planning for Russia is proposed. The goal of this system is to promote sustainable development based on harmonisation of state responsibility and private initiative. The elements of the system are a new approach to evaluation of land suitability, ecological zoning in combination with state responsibility for environmental conditions, realisation of the owner's right of free choice in business and stimulation of the owner's creative initiative to create land use options.

PART II

SUPRANATIONAL LEVEL LAND USE PLANNING

CHAPTER 9

'Ground for Choices': A Scenario Study on Perspectives for Rural Areas in the European Community

R. Rabbinge,[a] C. A. van Diepen,[b] J. Dijsselbloem,[c] G. H. J. de Koning,[d] H. C. van Latesteijn,[a] E. Woltjer[c] and J. van Zijl[c]

[a]Netherlands Scientific Council for Government Policy, Den Haag, The Netherlands. [b]The Winand Staring Centre for Integrated Land, Soil and Water Research, Wageningen, The Netherlands. [c]Social Democratic Party (PvdA), Den Haag, The Netherlands. [d]Research Institute for Agrobiology and Soil Fertility, Wageningen, The Netherlands

INTRODUCTION

European agriculture is going through a period of considerable change. The Common Agricultural Policy (CAP) proved to be very successful. For most agricultural commodities the Community has reached self-sufficiency. In some cases this has even led to substantial surpluses. The rise in productivity may continue at many places as attainable yields are still much higher than actual yields. This implies that the success of the CAP may turn out to be the major reason for a drastic change.

These developments may have a considerable effect on future land use in the EC. Not only surplus production induces changes, but other objectives have been put on the agenda. Social objectives, such as employment and income, economic objectives such as productivity and minimal costs, agricultural objectives such as efficient use of inputs, and environmental objectives such as minimisation of emissions all have consequences for land use. Next to these, forestry and nature conservation also claim land.

As a result, land use may change in the near future, both in size and in quality. How land use will change depends on the possibilities within the agricultural sector and the priority for various goals. Priority setting can be helped if some information is available on the possible options for future changes in land use and its consequences.

The Future of the Land: Mobilising and Integrating Knowledge for Land Use Options
Edited by L. O. Fresco, L. Stroosnijder, J. Bouma and H. van Keulen.

The Netherlands Scientific Council for Government Policy executed a study to explore options for land use. A land use allocation model was developed and a number of detailed studies on production potentials were conducted. In this chapter the methodology and some of the results are presented in the first section by Van Latesteijn and Rabbinge. In the second section Van Diepen and De Koning elaborate on one of the building blocks of the study: a qualitative and quantitative land evaluation is described. The chapter is concluded with a statement from the recipients, the policy-makers. Dijsselbloem, Van Zijl and Woltjer explain how they are using the results of the study and demonstrate its usefulness.

POSSIBLE CHANGE IN FUTURE LAND USE IN THE EUROPEAN COMMUNITY

Introduction

Agriculture in the European Community is becoming ever more productive. The combination of better production conditions, improved management and high-yielding varieties has led to a continuing period of growth. Even a greater rise in productivity may be expected in the future as a result of (bio)technological innovations. A positive result of this development is the achievement of food security, the primary objective of the CAP.

However, a dramatic rise in the costs of the agricultural policy conflicts with important trading partners over the subsidised dumping of EC surpluses on the world market; market distortion mainly to the detriment of developing countries; and increasing environmental problems resulting from current production methods. Without change these problems will become intractable.

It is therefore generally recognised that the CAP must be reformed; however, it is not clear in what form. The reforms recently agreed upon have been hailed as a breakthrough (CEC, 1991). This certainly holds for the pricing policy, i.e. a 29% fall in grain prices over three years is considerable and would bring European prices in line with those on the world market. However, the compensation scheme for set-aside land does not address the basic problem, since there was no fundamental debate on the *aims* of the policy, but it was limited to the instruments used. There was inadequate discussion on the extent to which these goals—and/or any adjustments deemed necessary—require a policy review.

Such a discussion would require from the member states of the European Community, and therefore also from the Dutch Government, strategic choices on the future of agricultural areas. In its study 'Ground for Choices', the Netherlands Scientific Council for Government Policy focused attention on the goals of agricultural policy for three reasons (WRR, 1992a):

(1) the widespread increase in agricultural productivity appears to continue, i.e. growing surpluses are being produced on the land already under cultivation;
(2) the anticipated growth in the budgetary burden on the Community if policy is not amended;

(3) the increasing social pressure for attention to aspects other than productivity, such as environmental protection, nature and landscape.

In this chapter we describe the approach adopted to investigate possible future changes in land use and present some of the results.

The study presents an analysis of possible variations in land use within the EC up to the year 2015. We developed the linear programming model GOAL (General Optimal Allocation of Land use) to examine where, depending on various policy options, land should be used for agriculture and forestry, and what methods should be employed to achieve certain combinations of policy goals as effectively as possible. The allocation of land use is thus guided by the relative value attached to different policy goals if priority is given to varying policy aims such as employment, the environment and economics, assuming a certain level of demand for agricultural products and use of the best technical means currently available. This gave rise to a sometimes radical reallocation of production and land use.

Since the various values attached to goals determine the outcome, this approach allows examination of possible scenarios corresponding to contrasting political philosophies about land-based agriculture and forestry in the EC. A philosophy can be defined in this context as a cohesive set of preferences with regard to a number of goals. The core of this study comprises four such scenarios. Besides agricultural production, they also encompass aims relating to socio-economics, the environment and nature conservation and development.

The four scenarios

Four contrasting philosophies have been devised on the basis of the main movements in the current debate on agriculture. These are extreme philosophies, in which the ideas put forward in the debate are taken to their logical conclusions. They determine the order of policy goals which form the basis of scenarios.

Scenario free market and free trade (FF)

Under this scenario agriculture is treated as any other economic activity. Production is as low-cost as possible. A free international market for agricultural products has been assumed, with a minimum of restrictions in the interests of social provisions and the environment. The philosophy represented by this scenario is similar to the American approach to the current negotiations on the General Agreement on Tariffs and Trade (GATT).

Scenario regional development (RD)

This scenario accords priority to regional development of employment within the EC, which creates income in the agricultural sector. The predominant philosophy can be regarded as a continuation and extension of current EC policy.

Scenario nature and landscape (NL)

Under this scenario the greatest possible effort is made to conserve natural habitats, creating zones separating them from agricultural areas. Besides protected nature reserves, areas would also be set aside for human activity. Nature conservation groups are exponents of this philosophy.

Scenario environmental protection (EP)

The primary policy aim under this scenario is to prevent alien substances from entering the environment. In contrast to scenario NL, the main aim is not to preserve or stimulate certain plant and animal species, but to protect soil, water and air. Natural and agricultural areas are therefore not physically separated but integrated. Farming may take place anywhere, but subject to strict environmental restrictions. This philosophy is in line with the concept of integrated agriculture as developed during the last decade, partly at the instigation of the Council (Van Der Weiden et al., 1984).

Land requirement assessment

The calculations with the GOAL model do not comprise all the problems dealt with in this study. Goals relating to nature and landscape cannot be expressed in figures that the model can interpret. Therefore, maps have been drawn representing the best division of land from the point of view of landscape and nature conservation. The results of the model were assessed on the basis of these maps, so that they may have to be modified as new space requirements arise.

Role of the scenarios

In the report 'Ground for Choices' the GOAL model and the needed input are described in detail, hence here we only give an indication of how the model works and what results are obtained.

The model does not produce a forecast. The scenarios explore options of technical possibilities based on a series of well-founded assumptions and presuppositions; however, such factors as price changes, assumptions about the behaviour of actors and institutional obstacles are excluded. Hence, this is not a study of the effects of possible amendments to the CAP, although its results indicate the technical limitations to such changes. In many other policy areas such a definition of technical limitations would be impossible (for example, when should a country be considered 'full', or what level of prosperity is 'enough'?). This is possible for land-based agriculture in the EC, though, because it can be based on well-known quantitative data (demand for agricultural products, technologies, possible use of land, etc.).

Policy-making can benefit from this type of information, because the options can be used to determine to what extent current policy can cope with the major developments generated in the scenarios (particularly the continuing rise in productivity and the associated decrease in employment in land-based agriculture). An estimate can, therefore, be made of the effort required to achieve goals, depending on whether we will have to 'go against the tide' or simply go with it. Hence, the results can serve as guidelines for future policies. If they all point in the same direction, there is clearly conflict between the technical possibilities and policy that aims at something else. Variations in the results can point to unsuspected potential in certain areas. They can also show extra possibilities by indicating when certain developments can be substituted for others.

One possible source of conflict might be the fact that in all four scenarios agricultural land use is much lower than the 127 million hectares currently in use in the EC. Would the great effort needed to maintain the current area of agricultural land in the long term be worth it? Should not other goals be given preference? Such questions arise from simply defining technical possibilities.

The scenarios are designed to promote debate on policy options at various levels. First, they demonstrate the possibilities for achieving the goals considered important in the various philosophies. These are results at European Community level. They also show the areas most suitable for agriculture in the EC, the type of agriculture most effectively pursued in each area (arable farming, livestock, permanent crops or forestry) and the methods that should be used (geared towards highest production efficiency, environmental protection or maximum use of land). These results have an effect at regional level. If the results at EC and regional level have consequences for certain countries, they will affect policy at national level as well.

Development of the GOAL model

The GOAL model is a linear programming model that can optimise land use to meet a policy goal, given a limitative set of types of land use and an exogenously defined demand for agricultural and forestry products. A number of policy goals are coupled to types of land use as objective functions, e.g. maximisation of efficiency of inputs for agriculture, minimisation of regional unemployment in land-based agriculture, and minimisation of the use of pesticides. Political philosophies can be formulated by assigning different preferences to the objectives by restricting the objective functions to a certain domain; for example, the total labour force cannot be less than a minimum level. In this way scenarios can be constructed that show the effects of policy priorities; for example, to maintain the labour force, types of land use will have to be selected with a relatively high input of labour.

The types of land use that the model can select are defined in quantitative terms. Because we want to explore possible long-term options, current agricultural practice in Italy or East Anglia should not be used as a reference, because it reflects current conditions, not those of the future. Therefore, we must define types of land use that might be effectuated in all regions of the EC in the future. For that purpose the concept of best technical means is used, i.e. agriculture takes place according to methods already operational in plant testing stations, experimental farms and many advanced farms. This does not imply predescribed agricultural practice, but gives input–output ratios representing the highest possible efficiency under the prevailing biophysical conditions. Basically, three types of production techniques are distinguished:

- yield-oriented agriculture, aiming at maximum efficiency of inputs per unit product,
- environment-oriented agriculture, aiming at lowest emissions and immissions per unit area, and
- land use-oriented agriculture, aiming at maximum land use.

These forerunners are used as a reference for future developments, thus ensuring consistent calculations across all member states of the EC. Three levels of analysis were necessary to construct the GOAL model.

Crop level

Plant properties, soil properties and climate properties determine the potential crop yield at a given location (Figure 9.1). First, the suitability of the soil for a certain crop is assessed to exclude all units where it cannot be grown (e.g. wheat on steep slopes and maize on clay soils). This can be denoted as qualitative land evaluation. Next, by means of a simulation model, potential yields are calculated for the suitable areas. This can be denoted as quantitative land evaluation (Van Lanen, 1991).

The qualitative land evaluation of the EC is based on the use of a geographical information system (GIS) (Van Diepen et al., 1990), and is executed at the level of land evaluation units (LEUs), representing combinations of soil and climate conditions considered to be homogeneous (22 000 units to cover the EC). By looking at factors like steepness, salinity and stoniness of the soil, the suitability for mechanised farming is assessed.

The quantitative land evaluation is based on the use of the WOFOST crop growth simulation model (Van Keulen and Wolf, 1986), applied to calculate the potential yields of winter wheat, maize, sugarbeet, potato and grass. Required inputs are technical information on regional soil (such as water-holding capacity) and climate properties and relevant crop properties (such as phenological development, light interception, assimilation, respiration, partitioning of dry-matter increase over plant organs and transpiration).

Two degrees of water availability are distinguished: rainfed and irrigated. In the rainfed situation, yield potential can be limited by the availability of water at any point during

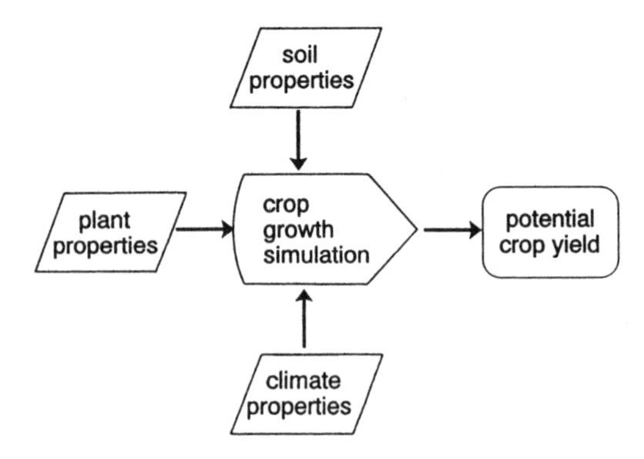

Figure 9.1 The inputs and outputs of the analysis at individual crop level

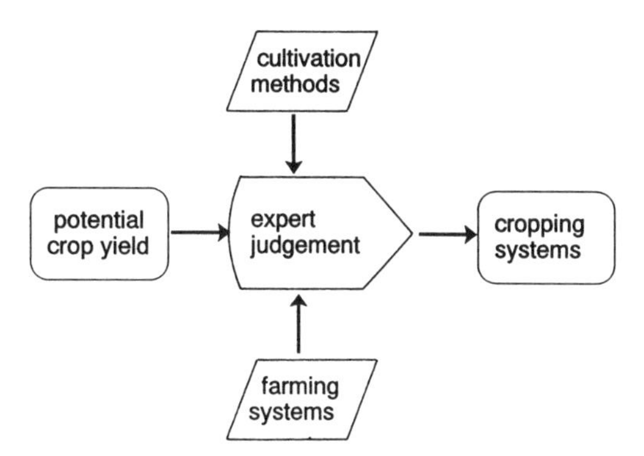

Figure 9.2 The inputs and outputs of the analysis at the level of cropping systems

the growing season. The attainable yields in that situation are referred to as water-limited yields. In the irrigated situation, crop yields are fully determined by climate and properties of the crop. The model results give an indication of the maximum attainable yield at a given location, referred to as potential yield.

The water-limited and potential yields are used as input at the next level of analysis.

Cropping system level

To examine land use possibilities in the future, information on individual crops is not sufficient. All crops are grown in a cropping system that defines all inputs and outputs. Moreover, in most cases monocropping is not a sustainable system and only a limited number of crop combinations can be used in practical cropping systems. Therefore, potential yields of indicator crops are translated into cropping systems characterised by a certain rotation scheme, certain management decisions and a certain use of inputs (Figure 9.2). It is striking that at this level the only viable method is expert judgement. From experience, both in practice and in experiments, the expert can deduce input and output coefficients of cropping systems. Yield levels are different from the potential level, and maximum efficiency depends on soil and location. These systems are not widely practiced yet, but are available at experimental farms and at some advanced farms throughout the EC. This element in the analysis is crucial, yet open to debate due to the subjective choices that are involved (De Koning et al., 1992).

Land use level

At the level of land use possibilities, all information is combined. Requirements for various goals related to land use together with alternative cropping systems and a demand for agricultural produce are fed into the GOAL model to generate scenarios of different options for land use at the level of NUTS-1 regions within the EC (Figure 9.3).

An IMGP (interactive multiple goal programming) procedure is used to optimise a set of objective functions incorporated in the model. In this procedure restrictions are

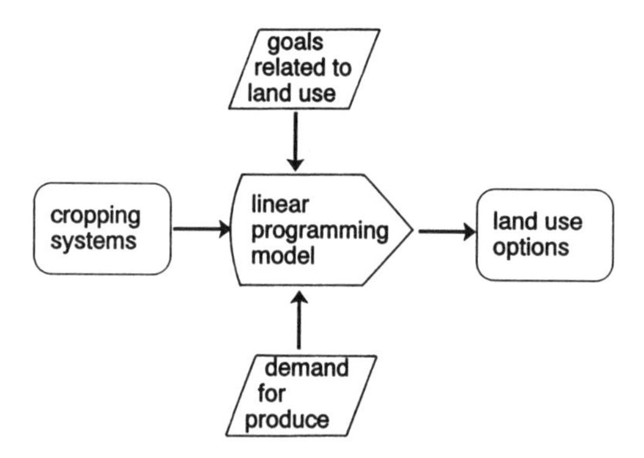

Figure 9.3 The inputs and outputs of the analysis at land use level

put to the objective functions to express preferences in policy goals. Hence, the four different scenarios (FF, RD, NL and EP) are characterised by different restrictions to the objective functions and by varying the demand. A few examples can illustrate this.

In FF the costs of agricultural production are minimised, without other restrictions on the objectives. Moreover, free trade implies that import and export is allowed, so the demand for agricultural produce from within the EC is modified according to expectations regarding new market balances. The model will now select the most cost-efficient types of land use and allocate those to the most productive regions.

In EP again the costs of agricultural production are minimised, but here strict limitations are set to the objective functions representing the use of fertilisers and pesticides, while the demand for agricultural produce represents self-sufficiency. The model will now select types of land use that agree with the imposed restrictions.

Results at the level of the European Community

Contrasts among the scenarios

The values of the individual goals differ dramatically among the four scenarios and from one area of policy to another. For land use the highest and lowest values differ threefold. The difference is twofold for land-based agriculture, employment and use of nitrogen (total and per hectare). Highest values for use of crop protection agents per hectare are four times the lowest, while the totals differ by a factor of seven.

The first conclusion that can be drawn from these significant differences is that there is scope for a clear policy to be pursued

Land use

In all four scenarios agricultural land use is considerably lower than at present (Figure 9.4). The highest land productivity is achieved in scenario NL, where the area of agricultural land is smallest. The discrepancy between the area of land currently in use

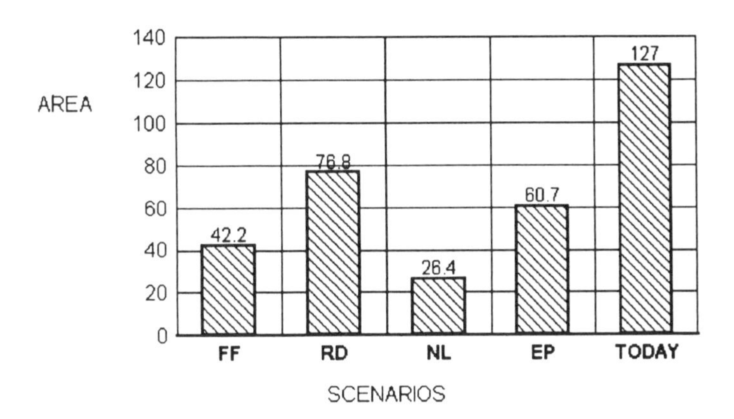

Scenario FF - Free market and free trade
Scenario RD - Regional development
Scenario NL - Nature and landscape
Scenario EP - Environmental protection

Figure 9.4 Land use in the different scenarios compared with current land use in the EC (in million hectares). Source: WRR

and the area technically necessary for food production shows that the present set-aside schemes can only be the very beginning of structural changes.

The second conclusion is that there is little scope for a policy aiming at maintaining all current agricultural land in use.

Employment

In all scenarios agricultural employment is much lower than the current level (Figure 9.5). Even in scenario RD, where the objective is to keep as many people as possible employed in land-based agriculture without subsidies, employment declines, i.e. from 6 million to 2.2 million manpower units (MPUs, 1988/89). These results indicate that preserving the current level of employment means maintaining hidden unemployment (in some regions up to 50%) at high costs. Moreover, the current loss of jobs in the agricultural sector of 2–3% per year will result in a decline of about 40% in 15 years time, despite all the measures taken.

The third conclusion is that in all cases considerable effort is required to accommodate the wastage of labour in agriculture.

Environment

The impact of agriculture on the environment is affected mainly by the use of crop protection agents and artificial (nitrogen) fertiliser. It is technically possible to significantly reduce the use of both without adversely affecting production (Figures 9.6 and 9.7). In particular, crop protection offers considerable scope.

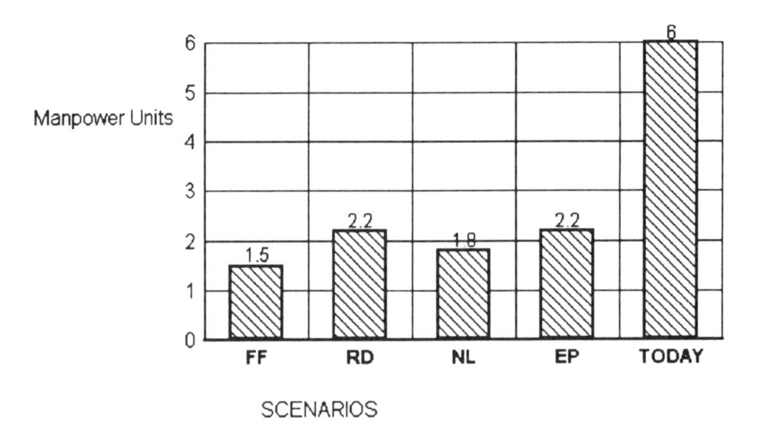

Scenario FF - Free market and free trade
Scenario RD - Regional development
Scenario NL - Nature and landscape
Scenario EP - Environmental protection

Figure 9.5 Employment in the different scenarios compared with current employment in the EC (in millions of manpower units). Source. WRR

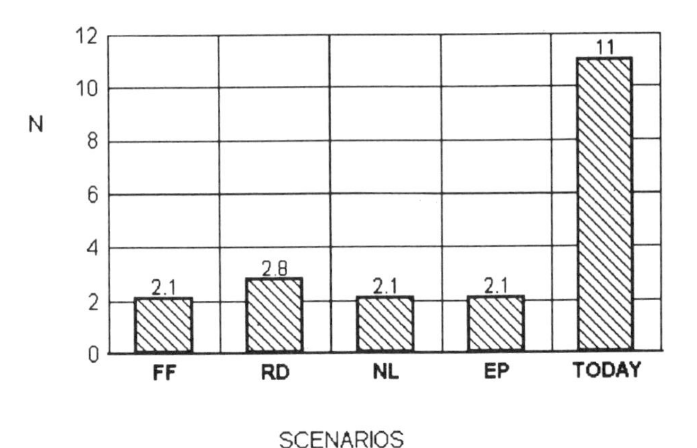

Scenario FF - Free market and free trade
Scenario RD - Regional development
Scenario NL - Nature and landscape
Scenario EP - Environmental protection

Figure 9.6 Surplus of nitrogen fertiliser in the different scenarios compared with current surpluses in the EC (in million tons). Source: WRR

A reduction in the use of fertilisers and pesticides is considered in current European policy as a service that farmers render to society. It is assumed that as a result they will suffer a loss of income and must therefore receive compensation. However, the results of the scenarios show that the surplus of nitrogen and the use of crop protection

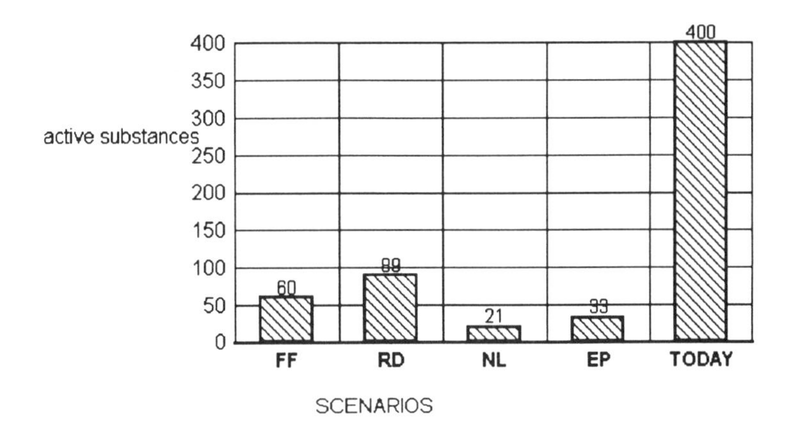

Scenario FF - Free market and free trade
Scenario RD - Regional development
Scenario NL - Nature and landscape
Scenario EP - Environmental protection

Figure 9.7 Crop protection. Source: WRR

agents can be sharply reduced without loss of production. Generally speaking, therefore, there is no need for compensation. None the less, considerable regional differences exist with respect to the environment. In the north-western corner of Europe in particular, where the use of pesticides and nutrients is highest (overuse, from the standpoint of rational and efficient management), application can be reduced without necessarily leading to a lower level of production. These results show that taking general policy measures with regard to a highly differentiated, regional activity such as agriculture is precarious.

The fourth conclusion is that policy measures can successfully promote more environmentally friendly production methods by limiting the use of nitrogen fertiliser and above all by reducing the large-scale use of crop protection agents.

Results at the level of individual regions

In addition to information on objectives at EC level, the scenarios also provide information on the partitioning over its individual regions. Each scenario shows a different regional land use pattern.

In scenario FF agriculture is confined mainly to the north-west of the EC; in scenario RD agricultural activities are distributed fairly evenly throughout the EC; in scenario NL many agricultural activities shift to the southern regions; in scenario EP agricultural activities are fairly evenly spread over the EC, with the exception of the Benelux and Ireland. These differences in spatial distribution of agricultural activities are connected to differences in policy goals. For instance, in scenarios FF and NL the distribution of employment over the regions is extremely uneven.

It is interesting to compare these results with the existing distinction between strong and weak regions in the EC (weak regions are those with a low score in terms of production, productivity and employment rate). From the weak regions in scenario FF, only Ireland retains a substantial share of employment in arable farming (in this scenario the creation of labour is relatively expensive in the southern regions). In scenario NL, Spain and Italy retain 40% and 34% of the current employment, respectively, and Portugal only 14%. (In scenario NL it is assumed that agriculture takes place on the smallest possible area of land and therefore gives the highest productivity. In this scenario the creation of jobs is relatively expensive in Greece and Ireland.)

The significant differences among the scenarios show that regions have different potentials for productivity increases. Weak regions in scenario FF are strong in scenario NL. In the latter scenario, which seeks to minimise the area of agricultural land in favour of large nature areas, land-based agricultural activities virtually disappear in a number of regions with a strong position at present. In this scenario, production on a limited area of land is given preference over production at minimum costs. This shows the relative value of the term 'weak' and the importance of policy objectives for the future of rural areas in the EC. Development of highly productive, irrigated agriculture in southern Europe may cause land use and agricultural employment problems in the northern member states.

Scenarios RD and EP give a relatively uniform distribution of land use over the EC. In scenario RD this is a result of the condition that maximum employment must be retained in all regions, which results in 29% of the current level of employment in all regions. Since the same percentage of employment is maintained in all regions, those with a high level of employment at present (such as the Mediterranean regions) enjoy a relative advantage. In scenario EP, 50% of the present level of employment is retained in Spain, 14% in southern Italy, 11% in Greece, and 10% in Portugal. Restrictions other than costs in these two scenarios result in a shift of agricultural activities to southern Europe (provided the necessary irrigation development takes place).

For the strong regions, mostly situated in the north-western part of the EC, the Netherlands is representative. In scenario FF, only 5% of employment in land-based agriculture is retained in the east of the Netherlands (the minimum allowed in any scenario), 18% in arable and livestock farming in the south, 26% in the west, and 36% in the north. In scenario RD, 29% of employment is retained in all regions; a condition imposed in this scenario. In scenario NL, land-based agriculture disappears from the Netherlands almost completely; the remaining 5% employment is provided by forestry and some livestock farming in the south. In scenario EP, the same picture emerges: 5% employment remains in arable farming in the north, east and south and in forestry in the west. Similar effects occur in Denmark, Germany, Belgium and Luxembourg. These results show that 'strong' is also a relative term.

Regional shifts also occur when the scope is examined for using agricultural land which can no longer be exploited profitably, for creating a network of protected areas in the EC. To that end a separate study was executed aiming at devising an 'ecological network' for the EC (Bischoff and Jongman, 1993), based on ecological principles and the current state of protection of different areas throughout the EC. The results show that roughly 36% of the total area must be reserved for nature protection to safeguard a healthy natural environment. Compared to the current 2%, this would require a major expansion in nature conservation areas. However, these results are only tentative, therefore we have chosen the acronym TEMS (Tentative Ecological Main Structure) to denote the 'necessary area' for nature.

In all four scenarios, sufficient land is in principle available in most regions to allow a significant area to be used for this purpose in addition to arable farming and forestry. Scenarios FF and NL are particularly attractive for nature development. It is, however, surprising that the 'surplus areas' occur mainly in the central part of the EC rather than in the Mediterranean areas, where at present low productivity, an ageing population and emigration result in substantial land abandonment. The scenarios therefore, indicate the scope for a different type of development in the Mediterranean area.

With respect to the costs of agriculture, a difference of 20 billion ECU exists between scenarios FF and NL, in both of which agricultural products may be imported from outside the EC. This difference can be seen as the price to be paid for converting large areas into protected nature areas (minus acquisition and development costs; it should be borne in mind that the additional costs in NL are moderated by the benefits arising from increased employment and reduced use of crop protection agents; production on a smaller area will also affect costs). The difference in costs between RD and EP is difficult to attribute to a single factor. However, the uniform distribution of employment required in RD offsets the lower use of nitrogen in EP. Even distribution of employment or a relatively low level of environmental pollution can be achieved at comparable cost.

Scope for other policies

The driving force behind change in land use and land productivity is technological progress. The analysis shows that the direction and pace of change is influenced by policy measures. Improvements in production conditions, price guarantees, research, information campaigns and education promote technological development. Adjustments can be made by modifying production conditions and product requirements as will be outlined below.

Use of rural areas

At EC level an integrated policy for physical planning does not (yet) exist; physical planning policy in rural areas is mainly indirect, incorporated in agricultural policy, regional policy or environmental policy. The scenarios show that, in the absence of an integrated policy, regional conflicts will increase rather than decrease. Growing incompatibility among European, national and regional policy seems unavoidable. A general European policy, indicating what areas should be used, is therefore required. Such indications could serve as a frame of reference for decisions on grant requests for European funds to stimulate structural improvements in production conditions (irrigation, rural development projects or other infrastructural works).

There seems to be scope also for a nature development policy at EC level (but has not (yet) been utilised). European landscapes and nature parks are few in number at present. Concerted action by European and national authorities and nature conservation groups may get things moving.

Setting aside agricultural land for alternative use

Under the present set-aside scheme there is not yet much scope for setting aside productive agricultural land as it must be kept for agricultural purposes, and the scheme assumes that productivity increases will be halted or even reversed. The results of this study show the contrary. If productivity steadily increases, a set-aside scheme becomes extremely expensive. It seems improbable that this will receive much political support, especially since land and income supports and other measures will also make demands on European funds.

Our calculations indicate the possibilities for reduced production capacity through alternative use of agricultural land such as nature development and recreation. There is also scope for agrification, where preference must be given to activities requiring extensive areas, such as energy recovery. Scope exists at European level, but it is not yet very attractive economically. However, a study by the Netherlands Energy and Environment Company (NOVEM) confirms the results of earlier studies that energy recovery on arable land has perspectives in the long term, provided the energy is refined (electricity, gasification, etc.) (NOVEM, 1992).

Regional development and employment

As already stated, in all scenarios employment in land-based agriculture is much lower than at present. European policy attempts to counteract the loss of jobs by improving the structure of agriculture. An evaluation of the impact of the structural funds intended for this purpose has shown that even now their effects are absent or even counter-productive (Van Der Stelt-Scheele, 1989). A policy that takes account of changes resulting from technical progress could make better use of the funds and alleviate the adverse effects.

The same applies to some degree to income support. If, for social reasons, supplementing farmers' incomes is considered, various ways exist. If support is linked to individuals, it amounts to a Community assistance scheme. If it is linked to land, it cannot be confined to agricultural land only, since that hampers land mobility. By granting a support for land put to alternative use, a basic financing system will be created for other purposes, such as nature conservation. Such ideas require further consideration. The scenarios show that current plans, involving the use of structure funds, amount to 'carrying coals to Newcastle'.

Conclusions

Research agenda

The current study required a considerable research effort. In developing the methodology and formulating the GOAL model, problems were encountered that are of sufficient interest to be referred to again here to facilitate similar studies in the future.

The analysis focused on the 12 EC member states, and can be extended in two directions. First, the study included only the territory of the EC before the unification of Germany. If countries with a large agricultural potential, such as most of central and eastern Europe, join the Community, the need for a review of the objectives of European agricultural policy becomes even more pressing. The GOAL model can be used to examine the consequences of such developments. Second, follow-up studies at regional level may provide information on the prospects for specific regions within the conditions set by the scenarios. Greater attention can then be devoted to other economic sectors.

One of the key assumptions in the model is that agriculture throughout the EC takes place with the best available techniques and without wastage. The best regional specification, permitted by current knowledge has been applied. A more detailed description of production techniques differentiated to specific regional conditions may be worthwhile.

The study does not deal with the financing of policy on rural areas. Only the total costs of agriculture are considered, which already show substantial differences between scenarios. The partitioning of costs between producers and authorities was not examined, nor were the consequences for European taxpayers. This information is essential if policy alternatives are to be developed further.

The financing structure of nature conservation policy has not been considered either. An attempt to distinguish between different forms of nature management with the aim to safeguard the various ecological values at minimum cost has not led to directly applicable results (Creemer, 1990). The positive response of nature conservation groups to this first attempt warrants further effort in this direction.

To make the study more specific, a tentative network of protected areas in the EC has been developed. Although this approach proved very useful in interpreting the results of the scenarios, it requires further development if it is to be used as a basis for a future European nature conservation policy. That would have to be an all-EC effort, since the necessary criteria must be agreed. In addition, regional input is needed to indicate areas suitable for inclusion in a network of protected areas.

Policy agenda

The results of this study suggest a clear policy agenda. They indicate that radical changes in the rural areas in the EC are possible. EC policy in this field is developing rapidly. National governments can use the scenarios as a guide in their contribution to this policy. Some general conclusions regarding future policies can be drawn from the scenarios.

The objectives should form the starting point in all proposals, surveys and analyses on reorganisation of European agricultural policy. The GOAL model could be used for this purpose. Policy goals must dictate the choice of instruments. Discussion on these goals must be conducted explicitly, not through policy instruments. The selected goals must serve as the background for policy formulation. Although other considerations will undoubtedly play an important role in the negotiating process, a situation should be avoided where the combination of goals and instruments leads to conflicting instruments, as is now often the case.

All the land use options in the 12 EC member states studied, show considerable surpluses of agricultural land, though their size and regional distribution vary among scenarios. This means that a policy aimed at maintaining the use of land for agricultural purposes in the long term (either directly through extensification, for example, or indirectly through set-aside schemes) will meet increasing resistance. The costs of such a policy may rise sharply and the eventual results will sometimes be incompatible with other goals (e.g. nature conservation and environmental goals).

All the options studied show that far fewer jobs are required in agriculture than at present. Even today a high level of hidden unemployment exists in many regions of the EC, and this level will rise sharply if the present number of jobs is maintained. Measures can be devised to mitigate the adverse consequences of this loss of jobs, but the artificial maintenance of maximum employment in agriculture is unaffordable and impracticable.

The environmental impact of agriculture in many areas of the EC is very serious, especially in the Netherlands. This study indicates great technical potential for tackling this problem, and policy could be formulated to realise this potential, as suggested in earlier reports: levies on pesticides; promoting research and information campaigns on integrated cultivation systems; improving production conditions in areas intended for agriculture; training; certificates for workers in the crop protection sector; deposit systems for plant nutrients, etc. (WRR, 1992b). None of these proposals are new. However, they should be introduced at European level, and the fact that this will benefit both the environment and production should be an incentive to do so.

Possibilities for an active European nature conservation policy certainly exist in view of land use and there seems to be little conflict with agriculture. At European level the Netherlands could encourage the further development of a network of protected areas. A precondition is that a financing structure must be established for European nature conservation policy. A combination of government funds and private financing ('bonds for nature') is an obvious choice.

QUALITATIVE AND QUANTITATIVE LAND EVALUATION IN THE EUROPEAN COMMUNITY

Introduction

In 1989 the Netherlands Scientific Council for Government Policy (WRR) requested the DLO-Winand Staring Centre (SC-DLO) to make an assessment of the crop production potential of the regions of the European Community (EC) for various kinds of land use. The aim was to estimate the scope for growth in agricultural production in the future, by quantifying the biological production ceilings that could be reached under continuing technological development, and by indicating the distribution of this yield potential over the regions. As yield potential depends also on yield variability, this implied that effects of heterogeneity of the crop growing environment in space and in time had to be taken into account.

The study should provide unbiased information for quantifying regional production volumes per land allocation scenario and should indicate regional differences in yield potential under well-defined production situations. The Netherlands Scientific Council for Government Policy stipulated that the method should be scientifically sound and objective, that its results were reproducible and consistent, and that the study would explore the biophysical possibilities for regional crop yield potentials.

The working hypothesis was that throughout the EC the same high technological level would prevail, so that regional differences in crop production potential could be attributed solely to differences in agro-ecological conditions, disregarding current comparative advantages of some regions with a favourable socio-economic production structure.

The crop production potential was to be characterised by a theoretically attainable yield level for a number of indicator crops, including the annual field crops winter wheat, grain maize and silage maize, oilseed rape, potato, sugar beet, and also grass, forests and fruit trees, in each of the EC regions. For a given location this yield level depends on crop properties, weather and soil conditions, and its assessment is achieved by applying a land evaluation procedure. The choice of a procedure depends on the information requirements of the user of the evaluation results. The methodological options available range from qualitative based heavily on judgement to quantitative based on crop growth simulation (Van Diepen et al., 1989). This section starts with the breakdown of the EC land area into small land units, serving as geographical unit areas for evaluation, before focusing on the choice of land evaluation procedures for the assessment of regional crop yield potential across the European Community. Alternative methods will be presented and discussed with emphasis on the relation between analytical complexity and discriminative power of the various methods, in view of their relevance for supporting policy formulation.

Methods and data

Approach

The assessment has a geographical and an analytical aspect and involved the agro-ecological zoning and the evaluation of the production potential per zone. This required the use of a GIS and land evaluation models.

It is difficult to study spatial and temporal variability simultaneously, therefore, they were dealt with separately in the present study. Spatial variability was characterised using static data, e.g. climate zones are distinguished using long-term mean values, and temporal variability was studied within a spatially homogeneous zone using dynamic data, e.g. time series of monthly weather were used to generate time series of annual yields per land unit. Consequently, within this study there are no dynamic processes crossing spatial boundaries.

Zoning and the role of the GIS

Spatial variability was taken into account by the identification of geographically homogeneous zones, serving as discrete unit areas for land evaluation—the so-called land evaluation units (LEUs). A LEU is a land unit defined as a unique combination of soil unit, agro-climate unit and administrative unit. The soil and climate unit together form an agro-ecologic unit, and the administrative unit is needed to allow comparison with official regional production statistics. The LEUs were identified with a GIS by overlaying the EC soil map (CEC, 1985), an agro-climatic map (Thran and Broekhuizen, 1965) and a map of EC-administrative regions at NUTS-1 level (Nomenclature des Unites Territoriales Statistiques). The LEU map counted some 4200 different units, distributed over some 22 000 map polygons.

The production potential for each land evaluation unit is determined in an analytical procedure using attribute data of the LEU. In the geographical sense, this evaluation is a point analysis. The overall regional potential is obtained by aggregation of yield data from LEU-level over agro-climatic regions or over NUTS-1 regions. The use of a computerised GIS is a quantitative method and it is the only way to handle accurately large quantities of geographic data. Its qualitative alternative of estimating areas by eyeballing has never been considered seriously.

Attribute data per zone

A climatic zone was characterised with monthly weather data of a weather station that was considered as representative because it was selected from the list of stations of which the data had been used for the compilation of the agro-climatic regions. Long-term average data were available for all 109 weather stations, and, in addition, for 81 of these stations, 26 years of monthly temperature and rainfall data. The information on the 546 units of the EC soil map consists of the name of the dominant soil unit, its texture and slope class. The soil name expresses the soil's genesis and morphology. In addition a soil phase (e.g. gravelly) may be indicated on the soil map. Other soil data needed in the land evaluation procedure had to be derived through subjective interpretation of soil unit definitions. This includes the estimation of the groundwater influence, effective rooting depth, and water-holding capacity.

Choice of evaluation procedures: quantitative or qualitative

Option of quantitative crop-growth simulation model

It was assumed that the best way to comply with the wishes of the Netherlands Scientific Council for Government Policy was through application of dynamic crop growth simulation models. This assumption was founded on the idea that a simulation model is the only tool to determine quantitatively the effect of variability in weather and soil conditions during a growing season and over a series of years.

The WOFOST model, a universal model for annual field crops and grass, was used to calculate the potential and water-limited yield levels for all relevant land units and an estimate was made of the crop nutrient requirements needed to obtain these theoretical yield levels.

The WOFOST crop growth model integrates the effects on crop growth of all relevant environmental factors such as radiation, temperature, rainfall, evaporation and soil moisture conditions during the complete growth cycle of the crop from a knowledge of the growth process. If the processes are correctly described, the model can be applied to environments other than that for which it was originally developed.

Option of qualitative methods

A completely different approach is the application of purely qualitative evaluation methods, i.e. methods relying on judgement, with or without computer aid. Such procedures are proposed in the 'Framework for Land Evaluation' (FAO, 1976).

The only existing land evaluation study at EC level was a qualitative one carried out by Lee (1986) on 10 EC countries and focused on the identification of the most suited areas for crop production. The evaluation followed the usual procedure to classify the suitability of a zone by combining separate climate and soil suitability ratings into one overall suitability rating. The soil limitations considered were drought, wetness/tilth, topography and rockiness. In the evaluation by Lee, topography and rocks marked the separation between suitable and unsuited soils, and the three factors, drought, wetness/tilth and topography, lead to a differentiation in three grades of soil suitability.

After combining with the climate grading, the final results are obtained. The yield projections for various crops are made on the basis of the observed current yield level in the best areas and extrapolated as a time trend until the year 2000.

Methods applied for land evaluation of the European Community

A mixed qualitative/quantitative approach for annual crops

We followed a mixed qualitative/quantitative approach for annual crops (Van Lanen, 1991) consisting of two steps. The first qualitative step consists of sieving all LEUs for their suitability for a given crop. All obviously unsuited LEUs are excluded from the second quantitative step.

The exclusion criteria are quite trivial: steepness, rockiness, stoniness, gravelly soils, salinity, heavy clay texture and poor drainage conditions. However, drought was not considered, as its effects would be evaluated quantitatively. The exclusion criteria are land use specific, and increasingly severe criteria were applied to grass, cereals and root crops.

To the remaining suitable soils the crop simulation model is applied to quantify the effects of some limitations on yield. The effects of temperature, sunshine and latitude are accounted for in the model by a description of the growth processes: light interception, CO_2 assimilation, assimilate partitioning and respiration. The effects of drought and of wetness are quantified by tracking the soil moisture condition and by accounting for its effect on evapotranspiration and hence on growth rate, the effects of tilth by delaying the date of sowing, and the effects of topography by assuming less complete water infiltration into soils on sloping terrain.

This resulted, for each suitable LEU, in a series of simulated yields over 26 years, of which the mean served as indicator for the yield potential.

A qualitative approach for forests

For the evaluation of the production potential of forest the quantitative method could not be applied because tree growth models are not available. Therefore, a computer-aided qualitative procedure was applied. The advantage of a computer-aided procedure is, apart from working speed, that the evaluation procedure must be formalised, implying that the evaluation procedure uses specified input data and specified interpretation rules and factor ratings. Traditional qualitative land evaluations are, in practice, often loosely applied, and may even use information not included in the formal land database.

Results

The results of the land evaluation study were delivered as maps and tables of area extent and yield levels per region. The results of the qualitative assessment are given as the extent of areas excluded for mechanised cultivation of cereals and root crops respectively. There is a striking agreement between this assessment and the results given by Lee (1987) concerning suitability classes for cultivation (S3 = moderately/poorly suited and U = unsuited). Table 9.1 gives the area extent aggregated by country for both qualitative assessments.

Tables 9.2 and 9.3 allow a comparison of the estimates of the yield potential by region according to the qualitative method by Lee and the crop simulation model, for the crops winter wheat, grain maize, sugar beet and potato. The combined rating of soil and climate conditions by Lee has resulted in concentrating the projected production on the most suited land in a few regions and consequently in discarding many other regions as being less favourable, without yield estimate. Some regions seem to have no potential at all for any crop. It appeared that the most suited regions for temperate crops were in northern France, Belgium, the Netherlands and England, and for more heat-requiring crops the Po Valley emerged as the best region. The projected yield level under high management in the year 2000 does not vary over the regions labelled as most suitable. On the other hand, the crop simulation model gives results for a wide range of soil and

Table 9.1 Extent of land (% of total area) unsuited (U) and poorly suited (S3) for mechanised arable farming according to Lee and excluded land for cereal growing (cer) and for root crop cultivation (root) in WRR study

			Land excluded for	
	U (%)	S3 (%)	cer (%)	root (%)
West Germany	4	53	48	57
France	22	35	43	61
Italy	45	32	73	88
Netherlands	9	33	18	38
Belgium	3	35	36	42
Luxembourg	8	77	69	85
United Kingdom	54	22	66	72
Ireland	59	13	53	53
Denmark	3	0	10	11
Greece	75	14	87	92
Spain			58	88
Portugal			67	84

Source: De Koning and Van Diepen (1992) and Lee (1987).

Table 9.2 Projected yield (ton/ha harvested weight) at high management level in the most suited areas (on well-suited soils under favourable climate) in the year 2000 according to Lee for some crops in some selected regions in the EC

	Winter wheat	Grain maize	Sugarbeet	Potato
Niedersachsen	8.5–10.9	nil	nil	nil
Bayern	nil	nil	nil	nil
Bassin Parisien	8.5–10.9	8.5–11.0	73–92	55–73
Sud-Ouest France	nil	8.5–11.0	nil	nil
Emilia Romagna	nil	8.5–11.0	73–92	nil
It south of Roma	nil	nil	nil	nil
Vlaams gewest	8.5–10.9	nil	73–92	55–73
East Anglia	8.5–10.9	nil	nil	55–73
Scotland	8.5–10.9	nil	nil	55–73
Ireland	8.5–10.9	nil	nil	55–73
Denmark	8.5–10.9	nil	nil	nil
Greece (North)	nil	8.5–11.0	73–92	nil

Source: Lee (1987).

Table 9.3 Simulated water-limited yield (ton/ha dry matter)[a] averaged over all suited soils for some crops in some selected regions in the EC

	Winter wheat	Grain maize	Sugarbeet	Potato
Niedersachsen	6.62	nil	14.1	11.9
Bayern	7.48	11.15	17.5	12.3
Bassin Parisien	6.80	8.18	14.5	10.3
Sud-Ouest France	7.70	8.34	13.4	9.8
Emilia Romagna	6.14	6.81	11.3	8.9
Abbruzi	5.29	2.93	7.1	5.9
Vlaams gewest	6.99	nil	14.8	12.3
East Anglia	7.37	nil	12.8	10.2
Scotland	7.83	nil	11.1	12.8
Ireland	8.45	nil	13.9	11.8
Denmark	5.81	nil	12.3	9.6
Greece (North)	5.13	3.55	8.9	8.1

[a]Normal moisture contents: wheat, 16%; grain maize, 14%; sugar beet, 80%; potatoes, 78%.
Source: De Koning and Van Diepen (1992).

climate units and indicates small differences in regional yield potential in terms of average water-limited yield. All these differences can be related to differences in weather and soil conditions during the growth cycle.

Discussion

Qualitative methods

Qualitative land evaluation allows us to make a rather consistent selection of suited soils on the basis of simple soil criteria. But the claim by some proponents of this approach that qualitative procedures lead to a more balanced view of the complexity of the real world than computer models does not hold. Qualitative methods go awry when a multitude of factors play a role in varying intensity over large areas. Instead of clarifying, they obscure the complexities. The maximum yield levels were a mixture of observed yields and extrapolated trends in yields, without a relationship to varying environmental factors. The results did not meet the information requirements of the Netherlands Scientific Council for Government Policy.

The case of drought limitation may serve as a simple illustration of the problem of intuitive multifactor analysis. As a general rule, sandy soils have some drought limitation. In qualitative land evaluations they are therefore rated as moderately suitable for cropping. However, the severity of the problem depends also on the climatic water balance and groundwater influence. Under humid climates soil sandiness may be no problem at all; under dry climates it may be a severe problem. The estimation of the yield effect of the drought limation over a range of climatic zones and groundwater conditions is impossible to solve with qualitative methods. It becomes even more difficult when the effects of drought must be combined with those of limited opportunities for soil tillage.

Yet, many authors take it for granted that only qualitative methods are applied for evaluation of the productive capacity of land resources at the scale of large countries or continents. Such evaluations use strongly generalised data. In this respect they are opposite to the paradigm of De Wit and Van Keulen (1987), who propose to use undistorted input data, and 'to calculate first, average later'.

Quantitative methods: strengths and weaknesses of simulation models

The crop model produces yield estimates and simulated yield variability can be related to variability in observed environmental factors. However, the model is conceived as a closed, controlled system, while in reality a crop may be exposed to factors not included in the model. Even if the model results are plausible, the question remains how good is the model in predicting the regional yield potentials.

The purpose of the modelling is to obtain insight into a well-defined part of reality, to quantify effects on yield and to keep control of analytical complexity. To achieve this the model is deliberately designed as a schematisation in terms of uniform field conditions and by distinguishing the hierarchy in potential, water-limited and nutrient-limited production situations. The model does not include influences of weed competition, pests and diseases.

Yet, the model, and physical quantitative methods in general, cannot completely fulfil their purposes because of a number of limitations inherent to the model and to the data needed by the model. To mention a few:

- the modelled processes may be more complicated than described in the model, e.g. interactions between processes and feedback mechanisms;
- schematisation in terms of soil layers (one layer soil) and time step (one day) is too coarse for some processes;
- parameterisation of the model is not perfect, because not well known, e.g. maintenance respiration;

- the model contains too many variables for statistical validation;
- there is no correct description of the environment, e.g. the air temperature and crop temperature may vary quite independently;
- many site-specific model input data are not available and have to be guessed;
- the mechanisms of recovery of the crop after serious stress are not known;
- there may be errors in the weather, soil or crop data.

There is continuous debate, and need for research, on how to tackle these problems of the inadequacy of models: either by simplifying or refining the model or by introducing stochastic elements in the model or by using more generalised or more detailed data.

Expert judgement

It has become fashionable to speak about expert systems in the context of land evaluation and it is good to make a distinction between various aspects of expert judgement in land evaluation. On the one hand, expert judgement may be called in to add quality to quantitative procedures; on the other hand, expert judgement may replace quantitative procedures. In combination with the use of simulation models, expert judgement is needed to arrive at a meaningful zonation, to select zones for further quantitative analysis, to fill gaps in the required input data (e.g. by formulating so-called pedotransfer functions), and to judge the validity and consistency of input data. After model application it is necessary to interpret the model output and to combine it with agronomic and environmental information not taken into account in the model. This is the general working procedure in integrative land evaluation research. We try to answer a question by combining available information from many sources and fill the information gaps with intelligent guesses.

The application of purely qualitative methods may be justified if they are applied to solve routine problems. When the questions address new problems we may well try to speculate on the possible solutions, but we should avoid treating first approximations, hampered by lack of knowledge and data, as experts' judgements. The danger in the application of qualitative methods is that it tends to confirm the status quo of knowledge and to follow the consensus of public opinion, thereby blocking the development of ideas that the world may be different from that.

IMPLICATIONS FOR RURAL POLICY IN THE EUROPEAN COMMUNITY

The European Common Agricultural Policy has been readjusted many times over the last decade. Following attainment of the policy goals, set in 1958, agriculture in the EC arrived at a situation of structural production surpluses, accompanied by high budgetary costs and unacceptable claims on environment and nature. This situation still continues and urgently calls for a discussion on the future of Europe's rural areas. All policy adjustments so far have lacked the necessary reflection on the new function of the rural space.

This also holds for the latest, most far-reaching policy reform—the MacSharry Reform. In the discussion paper accompanying these measures, the Agricultural Commissioner emphasised the importance of efficient agriculture integrated with the other functions of the rural areas, such as forestry, tourism, nature, landscape and recreation. Unfortunately, the time pressure on the Council of Ministers of Agriculture restricted the discussions to policy instruments only. A missed opportunity, because despite justified criticism of the ideas of Commissioner MacSharry, his ideas deserved substantial attention.

But not everyone lets opportunities go by. In 1988 this challenge was taken on by the Netherlands Scientific Council for Government Policy in its study on the future of the rural areas in the European Community in the concrete form of a model-scenario study. Four scenarios were formulated, based on four coherent visions of rural and agricultural policy (for land-based agriculture and forestry in the EC) to approximately the year 2015.

The results of the study indicate how land in rural areas can best be used following different policy objectives, thus explicitising the consequences of political choices. Even more importantly, the results point out to politicians and policy-makers the need and the possible scope for choices.

These political choices will not be objectively, rationally the best choices. Policy is, and shall always be, a compromise between different interests: the interests of member states, agricultural lobbies, industrial lobbies, nature and environment lobbies. Moreover, we must be careful not to overestimate the intervention possibilities in the reality of today's society, albeit difficult to accept for social democrats.

The main conclusion drawn by the Council in its report is that in all four scenarios a surplus of cultivated land arises. Therefore: 'a general European policy, which indicates what areas should be used for what purpose, is required.' This is certainly no sinecure: in the Netherlands, governments have had to resign over land policy. Nevertheless, the Council's recommendation stands. If, because of 'Ground for Choices', the surplus of cultivated land will remain prominently on the policy agenda, the study will have rendered its service.

A politically important question is at which level of administration this new land policy should take form. Is the bureaucracy in Brussels able to formulate such a policy and, following a decision by the Council of Ministers, to implement this policy from a centralised level? A reasonable doubt seems justified.

In different regions of the Community different scenarios may have to be implemented. For example, for the north of the Netherlands the Regional Development scenario (RD) is very plausible and for the south the Environmental Protection scenario seems obvious. The Council hints at this possibility in the epilogue of 'Ground for Choices', when it suggests diverging developments of agriculture in the Community: 'on a relatively small area highly productive agriculture with the best technical means, satisfying the major demands for food. On the remaining area an extensive form of agriculture can take place with emphasis on nature and landscape.'

Therefore, broad agreement at the Community level on the choice of strategy in combination with more detailed choices at regional level would be more realistic. This would also be in line with the principle of subsidiarity: the Community should concentrate on issues that the member states themselves cannot deal with.

Taking on new challenges, first of all a well-balanced sustainable development of the rural areas in the different regions of the Community is no longer possible by developing one big package of European guidelines. This observation will be even more valid when the Community expands from 12 to 16 and later possibly 24 member states.

Within a rough and sometimes prescriptive framework, regions—possibly coinciding with states—can opt for a development fit for their specific situation. At Community level the framework comprises market and price policies as well as supplementary income policy.

A choice for a more market-oriented agriculture must be accompanied by an active restructuring of agriculture. The reformed CAP and EC structure funds have so far only tried to slow down such a process.

We make the political choice of maintaining agricultural production in all regions of the EC. This implies that regions will have to comply with production-control targets in a structural way. For this purpose, a scenario study of regional level can be an apt instrument. Regions must be enabled to take cultivated land out of production for purposes such as nature development, forestry, recreation, etc. By doing this the present inefficient method of individual set-aside can be abandoned.

There are more advantages to be gained by a strong regional policy component. The main advantage will be a higher efficiency and social acceptance of rural policy, on a level closer to the issues and the people involved.

Also, a regionally oriented policy may prevent the fate of most European agreements: more *ad hoc* solutions offering something to everyone.

This plea for regionalising the CAP within a Community framework does not answer the question of the future of rural areas and their populations. We should offer agriculture a future again: a future as supplier of food, raw materials for industry, of nature and landscapes. Such a development requires an economically viable agriculture that earns an income from fulfilling all these social and economic needs. Such an income will be composed of two elements: an economically fair price for the raw materials and an income supplement linked to the production of public goods. During the process of restructuring, direct income support should temporarily be given in addition to the two main components.

This is not a plea for re-nationalising agricultural policy but a plea for a policy based on smaller socio-geographical and administrative units; units that may cross state borders. The Community and its regions will have to invest in the future of their rural areas. Opportunities will have to be created at regional strong points. The strength of the European Community is its diversity, and that strength should be exploited much more.

Finally, it seems appropriate to quote from a letter that Sicco Mansholt, the founder of the European agricultural policy, recently wrote to the scientists responsible for 'Ground for Choices':

> This study is of major importance for the future development of the agricultural policy. Its strength lies in its presentation: Whatever scenario one chooses, the consequences are grave. Politicians can no longer postpone making choices, as that will lead to chaos. When I recall the sixties, when I had to make political choices, how I wished I would have had at that time 'Ground for Choices'.

REFERENCES

Bischoff, N. T. and Jongman, R. H. G., 1993. *Development of Rural Areas in Europe: The Claim for Nature*. WRR Preliminary and Background Studies no. V79, Sdu Uitgeverij, The Hague, The Netherlands.

CEC (Commission of the European Communities), 1985. *Soil map of the European Communities, 1:1,000,000*. Commission of the European Communities, Luxemburg.

CEC (Commission of the European Communities), 1991. The development and future of the Common Agricultural Policy. Proposals of the Commission; COM(91) 258 final, Brussels, 11 July. CEC, Brussels, Belgium.

Creemer, M., 1990. *Natuurbeheer in Europa, een inventarisatie van doelstellingen, methoden en kosten van inrichting en beheer in beschermde gebieden in de landen van de EG* (Nature conservation in Europe. An inventory of aims, methods and costs of arrangement and management of protected areas in the member states of the European Community). The Hague, report on a period of practical training. Netherlands Scientific Council for Government Policy, The Hague, The Netherlands (in Dutch).

De Koning, G. H. J. and Van Diepen, C. A., 1992. *Crop production potential of rural areas within the European communities. IV: Potential, water-limited and actual crop production*. Working Document W68, Netherlands Scientific Council for Government Policy, The Hague, The Netherlands.

De Koning, G. H. J., Janssen, H. and Van Keulen, H., 1992. *Input and output coefficients of various cropping and livestock systems in the European Communities*. Working Document W62, Netherlands Scientific Council for Government Policy, The Hague, The Netherlands.

De Wit, C. T. and Van Keulen, H., 1987. Modelling production of field crops and its requirements. *Geoderma* **40**: 253–265.

FAO, 1976. *A Framework for Land Evaluation*. FAO Soils Bulletin 32, FAO, Rome.

Lee, J., 1986. *The impact of technology on the alternative uses for land*. FAST Occasional Paper FOP 85, CEC-DG XII Science, Research, Development, Brussels, Belgium.

Lee, J., 1987. European land use and resources. An analysis of future EEC demands. *Land Use Policy* **4**: 179–199.

NOVEM (Netherlands Energy and Environment Company), 1992. *De haalbaarheid van de produktie van biomassa voor de Nederlandse energiehuishouding*. Eindrapport, NOVEM, Utrecht, The Netherlands.

Thran, P. and Broekhuizen, S., 1965. *Agro-climatic Atlas of Europe. Vol. 1: Agro-ecological Atlas of Cereal Growing in Europe*. Elsevier, Amsterdam.

Van Der Stelt-Scheele, D. D., 1989. *Regionaal beleid voor de landelijke gebieden van de Europese Gemeenschap; inventarisatie en evaluatie* (Regional policy for the rural areas of the European Community: inventory and evaluation). Working Documents no. W46, Netherlands Scientific Council for Government Policy, The Hague, The Netherlands (in Dutch).

Van Der Weiden, W. J., Van Der Wal, H., De Graaf, H. J., Van Brussel, N. A. and Ter Keurs, W. J., 1984. *Bouwstenen voor een geïntegreerde landbouw* (Building blocks for an integrated agriculture). WRR Preliminary and background studies no. V44, Staatsuitgeverij, The Hague, The Netherlands (in Dutch).

Van Diepen, C. A., Wolf, J., Van Keulen H. and Rappoldt, C., 1989. WOFOST: a simulation model of crop production. *Soil Use and Management* **5**: 16–24.

Van Diepen, C. A., De Koning, G. H. J., Reinds, G. J., Bulens, J. D., Van Lanen, H. A. J., 1990. Regional analysis of physical potential of crop production in the European Community. In: Goudriaan, J., Van Keulen, H. and Van Laar, H. H. (eds), *The Greenhouse Effect and Primary Productivity in European Agro-ecosystems*. Pudoc, Wageningen.

Van Keulen, H. and Wolf J., 1986. *Modelling of Agricultural Production: Weather, Soils and Crops*. Simulation Monographs, Pudoc, Wageningen.

Van Lanen, H. A. J. 1991. *Qualitative and quantitative physical land evaluation: an operational approach*. Doctoral thesis, Wageningen Agricultural University, Wageningen.

WRR (Netherlands Scientific Council for Government Policy), 1991. *Technologie en overheid. Enkele sectoren nader beschouwd* (Technology and government. A close inspection of some sectors). Report to the Government no. 39. Sdu Uitgeverij, The Hague, The Netherlands (in Dutch).

WRR (Netherlands Scientific Council for Government Policy), 1992a. *Ground for choices. Four perspectives for the rural areas in the European Community*. Report to the Government no. 42, Sdu Uitgeverij, The Hague, The Netherlands.

WRR (Netherlands Scientific Council for Government Policy), 1992b. *Environmental policy: strategy, instruments and enforcement*. Report to the Government no. 41, Sdu Uitgeverij, The Hague, The Netherlands.

CHAPTER 10

Experimental Silvopastoral Systems: The First Five Years

W. R. Eason,[a] J. Simpson,[b] R. D. Sheldrick,[c] R. J. Haggar,[a] E. K. Gill,[a] J. A. Laws,[c] D. Jones,[a] E. Asteraki,[c] P. Bowling,[a] J. Houssait-Young,[c] J. E. Roberts,[a] D. G. Rogers[d] and N. Danby[e]

[a]Institute of Grassland and Environmental Research, Aberystwyth, UK. [b]Forestry Authority Research Division, Midlothian, UK. [c]IGER, North Wyke, UK. [d]FARD, Kennford, Exeter, UK. [e]FARD, Talybont-on-Usk, Brecon, UK

INTRODUCTION

Agroforestry, the integration of agricultural and forestry practices on the same land area, offers one option for alternative use of current and predicted agricultural land surplus. Agroforestry can be more fully defined as land use systems where woody perennials are grown in association with herbaceous plants (crops, pastures) and/or livestock in a spatial and/or temporal arrangement, and in which there are both ecological and economic interactions between the tree and non-tree components of the system (Young, 1989). Computer models (Tabbush et al.,1985; Doyle et al., 1986) suggest that silvopastoral land use systems, in which animal and forest production are combined on the same land area, may be a viable alternative land use option for temperate European Community countries. Agroforestry systems would provide for a transitionary period from full agricultural production to full forest production. The duration of this period will depend on factors such as the spacing between trees, tree species used and management practice. This formed the basis of an investigation when in 1987 a network of silvopastoral experiments, covering a range of soil and climatic conditions, were established *de novo* in the United Kingdom (Sibbald and Sinclair, 1990). The results from this study will be used in the construction of biophysical and bioeconomic models. This chapter reports on the first five years post-establishment of two of these sites representing upland and lowland conditions.

The Future of the Land: Mobilising and Integrating Knowledge for Land Use Options
Edited by L. O. Fresco, L. Stroosnijder, J. Bouma and H. van Keulen.

Table 10.1 Altitude, annual rainfall and mean soil temperature at lowland (North Wyke) and upland (Bronydd Mawr) silvopastoral experimental sites (1992)

	North Wyke	Bronydd Mawr
Altitude	180 m	330 m
Annual rainfall	1008 mm	1471 mm
Mean soil temperature	10.9 °C	8.4 °C

MATERIALS AND METHODS

Sites

The two sites differ primarily with respect to altitude and climate (Table 10.1). Both sites are representative of current sheep-grazed pasture land in the UK.

Experimental design and management

Acer pseudoplatanus was planted at both sites, with additionally *Larix eurolepis* at the upland site, and *Fraxinus excelsior* at the lowland site. *Acer*, the common species, was selected because of its potential to produce high quality timber over a range of site conditions. This was also planted at other sites within the National Network Experiment (Sibbald and Sinclair, 1990). To allow a comparison between sites, therefore, most of the data presented here focus on the results from *Acer* treatments. Treatments include two agroforestry spacings (100 and 400 stems/ha with trees individually protected) planted in a regular grid pattern in sheep-grazed pasture, a forest control (2500 stems/ha with no individual protection), and an agricultural control with no trees, each replicated three times. Trees in agroforestry plots are protected from sheep damage by plastic tree shelters (Tubex Ltd, UK) (minimum height 1.2 m; diameter 8.4–11.4 cm) (Nixon et al., 1992). At the upland site these were replaced with larger mesh shelters in 1991 (Netlon Ltd, UK) (height 1.5 m; diameter 35 cm). A weed free area extending 0.5 m from the tree was maintained with herbicide application for the first three years.

Ryegrass-based pasture is grazed by a core group of sheep with a target sward height of 4–6 cm maintained by the addition or removal of sheep outside of the core group. All sheep are weighed on and off plots and core groups at monthly intervals. Grazed plots receive 160 kg N/ha per year as ammonium nitrate and 50 kg each of P and K. Forestry controls receive no fertiliser.

Assessments

A whole range of routine and isolated assessments are made, not all of which can be described here. The principal assessments include:

(1) Trees: dimensions; levels of damage (snapping of leaders and side branches). Tree root distribution has been determined using a modified logarithmic spiral trench (Huget, 1973; Laws et al., 1992).

(2) Animals: lamb production and condition score is recorded throughout the grazing season.
(3) Pasture: production is back calculated from animal output.
(4) Soil: changes in the pattern of nutrient and soil moisture availability. The compaction of soil as a result of animal traffic has been recorded inside and outside the weed-free zone adjacent to the tree.
(5) Climate: standard site meteorological assessments as well as more detailed microclimate assessments examining the influence of the tree canopy on light and temperature.
(6) Wildlife: avifauna and carabid beetle populations.

RESULTS AND DISCUSSION

Clearly it is not possible to present full detailed results on all assessments. Instead the most significant findings only can be presented here. These also largely focus on *Acer*, the common species.

Tree assessments

Establishment and survival has been good (generally over 90%). Tree growth at the wider spacing (100 stems/ha) was consistently reduced compared to closer spaced treatments although the difference was only marginal at the upland site (Table 10.2). Taking data from other Network sites, this tree density effect on growth is highly significant (Sibbald, 1991).

At the wider spacing, the sheep : tree ratio is increased and the increased attention that each tree receives in these plots may impact on tree growth. This is evidenced by increased soil compaction around trees at the wider spacing (Table 10.3). This may affect root development especially in the upper soil layers.

There were significantly fewer tree roots in grazed agroforestry than in forest control plots, where they were recorded at the lowland site (Table 10.4). Although reduction in the upper layers may be partly attributable to soil compaction effects, it is likely that a reduced rooting density is in part, at least, a response to the high levels of soil nutrients in the grazed plots compared to the unfertilised forest control treatment. Whilst high nutrient levels may result in greater shoot growth (Table 10.2), the root : shoot ratio may be reduced (Fitter and Hay, 1989).

Table 10.2 Mean tree height (cm) of *Acer* in agroforestry (100 and 400 stems/ha) and forest control plots (2500 stems/ha); data are from 1992

Treatment	North Wyke	Bronydd Mawr
Acer, 100	108.7 (7.6)	197.3(32.6)
Acer, 400	154.7 (3.3)	210.0 (4.2)
Acer, 2500	130.1(22.1)	82.7(13.4)

Standard errors in parentheses.

Table 10.3 Soil compaction (kilopascals) in the weed-free area adjacent to tree and in undisturbed pasture at the two agroforestry spacings of *Acer* (100 and 400 stems/ha) at Bronydd Mawr. Different letters preceding means, indicate differences between weed-free and pasture measurements; letters following means, indicate differences between spacing treatments within weed-free or pasture measurements; ($P<0.05$ by analysis of variance)

Treatment	Weed-free	Undisturbed pasture
Acer, 100	a58.0a	b36.7a
Acer, 400	a40.0b	b35.7a

Table 10.4 The effect of spacing on tree root number per m^2 in three soil layers. Data for *Acer* and *Fraxinus* at North Wyke lowland site. Means preceded by a different letter within a single soil depth are significantly different ($P<0.001$)

Depth	2500 Stems/ha	100 Stems/ha
0–10 cm	a474	b134
10–20 cm	a166	b44
20–30 cm	a39	b12

Although the plastic shelters have encouraged early rapid tree growth and prevented direct sheep damage, there has been a cost in terms of tree growth form and subsequent wind damage on the upland site. *Larix* have large canopies with a significant 'sail area' supported by long, thin stems, which has led to significant stem lean. Clearly the more exposed conditions on the upland site have aggravated the situation. In addition, emerging leaders of *Larix* and *Acer* rub against the plastic shelter so weakening the stem at a single point, later resulting in windsnap. The extent of the damage would appear to be related to the degree of exposure (Table 10.5). Trees at the lowland site however, notably *Fraxinus*, have developed normal growth forms and have not suffered significant damage.

Animal assessments

To date, there is no significant effect of tree planting on animal production (Table 10.6). As the tree canopy develops and impacts significantly on the microclimate to affect

Table 10.5 Tree damage (%) to leaders and side branches of *Acer* emerging from plastic shelters in replicates 1–3 at Bronydd Mawr in an upland site at the two agroforestry spacings (100 and 400 stems/ha). Exposure expressed as mean daily rate of attrition (cm^2) from tatter flags (means with standard errors)

Treatments	Replicate 1	Replicate 2	Replicate 3
Acer, 100 (leaders)	7.0	41.6	7.7
Acer, 100 (side branches)	9.9	50.5	0
Acer, 400 (leaders)	20.5	25.4	20.0
Acer, 400 (side branches)	31.8	29.0	25.8
Exposure	7.1(1.3)	10.3(0.5)	4.1(2.2)

Table 10.6 Mean liveweight of stock carried (tonne days/ha) for North Wyke site (data not available for Bronydd Mawr) for 1992. No significant differences between means

Treatment	Replicate 1	Replicate 2	Replicate 3	Mean
Agricultural control	281	275	290	290
Acer, 100	263	288	309	287
Acer, 400	273	318	356	316

Table 10.7 Counts of bird species visiting Bronydd Mawr agroforestry, over the period 1989–1991

	Winter		Spring	
Treatments	1989/90	1990/91	1990	1991
Agricultural control	9.0	10.0	6.0	10.0
Forestry control	8.0	8.7	14.3	15.7
Acer, 100	13.7	15.7	15.0	17.7
Acer, 400	11.3	13.7	9.3	14.7
S.E.D.	1.6	1.86	1.81	1.99

pasture production, it is anticipated there will be a concomitant fall in animal output. We are currently monitoring the influence of the trees on soil temperature and the photosynthetically active radiation received by the pasture. Initial results indicate that there are already measurable effects but that these are not limiting pasture or animal production.

Wildlife studies

Avifauna and carabid beetle populations have been examined at the upland site. Although it is too early to draw firm conclusions, both avifauna and beetle populations have been affected by tree planting. Some preliminary data for avifauna populations at the upland site are given (Table 10.7). As expected, at an upland site, birds of moorland and open country are the dominant species. However visits by woodland and scrub birds, notably *Sitta europea* and *Ficedula hypoleuca* are increasing, especially on plots adjoining wooded banks, whose flora provide corridors for accessibility onto the trial.

CONCLUSION

Valuable information on tree establishment in silvopastoral systems has been gained. For example, tree–animal interactions are apparent with increased sheep–tree contact in wider-spaced plots, as evidenced by effects on soil compaction and tree growth. It is also clear that much needs to be learnt about the growth of trees on previously agricultural land. It has been shown that root development is likely to be affected under such conditions. For wide-spaced tree planting systems like agroforestry, this could have implications for tree wind stability, particularly on more exposed upland sites. It is also possible that high soil fertility levels may be affecting tree growth form, most notably in *Larix*, although this needs further clarification. After four years, although there has been no impact on agricultural output, there is evidence that the presence of trees is now affecting microclimate with possible ecological benefits, at least in terms of bird and insect populations.

Agroforestry systems clearly provide one option for alternative land use within the European Community. A significant EC-funded agroforestry research programme involving a number of member states is now underway. The data from these and other UK sites will contribute to the construction of biophysical and bioeconomic models of this land use system.

ACKNOWLEDGEMENTS

These sites are jointly managed by the Institute of Grassland and Environmental Research and the Forestry Authority Research Division. This work is sponsored by the UK Ministry of Agriculture, Fisheries and Food, and the European Community.

REFERENCES

Doyle, C. J., Evans, J. and Rossiter, J., 1986. Agroforestry: An economic appraisal of the benefits of intercropping trees with grassland in lowland Britain. *Agricultural Systems* **21**: 1–32.

Fitter, A. H. and Hay, R. K. M, 1989. *Environmental Physiology of Plants*. Academic Press, London.

Huget, J. G., 1973. Nouvelle methode d'etude de l'enrancinement des vegetaux perennes a partir d'une tranchee spirale. *Annals of Agronomy* **24**: 707–731.

Laws, J. A., Houssait-Young, J. and Sheldrick, R. D., 1992. Assessment of the rooting patterns of young sycamore and ash using a logarithmic spiral trenching technique. *Agroforestry Forum* **3**(3): 22–25.

Nixon, C. J., Rogers, D. G. and Nelson, D. G., 1992. The protection of trees in silvopastoral agroforestry systems. Forestry Commission Research Information Note 219, Forestry Commission, Edinburgh.

Sibbald, A., 1991. Silvopastoral National Network Experiment—Annual Report 1991. *Agroforestry Forum* **3**(2): 2–5.

Sibbald, A. and Sinclair, F. 1990. A review of agroforestry research in progress in the UK. *Agroforestry Abstracts* **3**: 149–164.

Tabbush, P. M., White, L. M. S., Maxwell, T. J. and Sibbald, A. R., 1985. Tree planting on upland farms—Study Team Report UK. Forestry Commission and Hill Farming Research Organisation. Hill Farming Research Organisation, Edinburgh.

Young A., 1989. *Agroforestry for Soil Conservation*. CAB International, Wallingford, UK.

CHAPTER 11

Land Evaluation for Land Use Planning—A Southern African Development Community Experience

C. Patrick

Department of Agricultural Research, Gaborone, Botswana

INTRODUCTION

The Southern African Development Community (SADC) is an economic grouping of 10 independent states covering a vast part of the Southern African Subcontinent, ranging from the Equator in the north to the Tropic of Capricorn, in the south. The community (formally the Southern African Development Coordinating Conference) was officially launched by the signing of the Lusaka Declaration in April 1980. At the time, nine countries were involved, i.e. Angola, Botswana, Lesotho, Malawi, Mozambique, Swaziland, Tanzania, Zambia and Zimbabwe. The tenth member, Namibia, was an observer until awarded full membership at independence in 1990. All member states were given sectors of economic importance to coordinate: energy conservation and development, Angola; agricultural research and animal disease control, Botswana; soil and water conservation and land utilisation together with tourism, Lesotho; inland fisheries, wildlife and forestry, Malawi; transport and communications, Mozambique; manpower development, Swaziland; industrial development, Tanzania; mining, Zambia; and food security, Zimbabwe. Namibia was given marine fisheries when it became a member.

The theme of the future of the land—mobilising and integrating knowledge for land use options—has a direct bearing on the work of the Food, Agriculture and Natural Resources Sub-Sector. The policy and development strategy is aimed at improving the living conditions of the population and enhancing national and regional food security. These depend on the wise use and long-term conservation of the region's natural resource base. It is, therefore, essential that SADC's policy for the conservation and sustainable

The Future of the Land: Mobilising and Integrating Knowledge for Land Use Options
Edited by L. O. Fresco, L. Stroosnijder, J. Bouma and H. van Keulen.

Table 11.1 Land use in the SADCC Region (1980)

Country	1990 Population (millions)	Land area (1000 hectares)	Arable land[a] (%)	Perm. crop (%)	Pasture (%)	Forest and woodlands (%)	Other land[b] (%)
Angola	9.7	124 670	2.4	0.4	23.3	43.3	30.5
Botswana	1.3	58 537	2.3	—	75.2	1.6	20.9
Lesotho	1.7	3 055	9.6	—	65.9	—	24.5
Malawi	7.9	9 408	24.4	0.2	19.6	47.5	8.3
Mozambique	15.5	78 409	3.6	0.3	56.1	25.8	14.2
Swaziland	0.7	1 720	11.6	0.2	72.7	5.8	9.7
Tanzania	26.0	88 604	4.6	1.2	39.5	47.6	7.1
Zambia	7.6	74 072	6.9	0.1	47.2	27.6	18.2
Zimbabwe	10.1	38 667	6.4	0.2	12.6	61.5	19.3
SADCC		477 122					
Africa		2 966 477	5.5	0.6	26.4	23.5	44.0

[a]Arable land is land suitable for growing seasonal crops.
[b]Other land is land that does not fall into any of the four categories.
Source: SADCC Soil and Water Conservation and Land Utilisation Programme: Report No. 5, July 1986.

Table 11.2 Wildlife conservation areas in the SADCC Region (×1000 ha)

		National park		Other conservation areas[a]		Total conservation area	
Country	Land area	Area	% of total	Area	% of total	Area	% of total
Angola	124 670	5 466	4.4	2 765	2.2	8 231	6.6
Botswana	58 537	3 737	6.4	6 197	10.6	9 934	17.0
Lesotho	3 035	7	0.23	—	—	7	0.23[b]
Malawi	9 408	698	7.4	369	3.9	1 067	11.3
Mozambique	78 409	1 590	2.0	3 810	4.9	5 400	6.9
Swaziland	1 720	—	—	54	3.1	54	3.1
Tanzania	88 604	3 752	4.2	9 709	11.0	3 461	15.2
Zambia	74 072	6 359	8.6	16 049	21.7	22 408	30.0
Zimbabwe	38 667	270	7.0	2 290	5.9	4 990	12.9
SADCC	477 122	24 309	5.1	41 243	8.6	65 552	13.7

[a]Including game reserves, game management, safari and controlled hunting areas, recreational and botanical parks, and nature reserves, but excluding forest reserves.
[b]If recently established Wildlife Management areas are added, the figure is 37%.
Data for Namibia are not included.
Source: Protected Areas data Unit of the International Union for Conservation of Nature, Gland (1986).

utilisation of natural resources be integrated into the overall regional development strategy.

Land evaluation for land use planning attempts to look at the problem of land use from a SADC perspective. That each of the 10 countries have their own land use approaches is not in dispute, but as a region the community agrees that development of small economies is better sustained through regional integration. Land, being a finite resource, needs to be evaluated so that all relevant land use options are taken into consideration when final decisions on land use are made.

SADC's strategy seeks to ensure that the management of natural resources will contribute to improved agricultural productivity and increased incomes, especially for the rural population, while at the same time ensuring that agricultural and other forms of economic development do not undermine the diversity and richness of the region's natural resource base.

As for land issues in particular, there are two projects, i.e. the Land and Water Management Research Programme (L & WMRP) and the Environment and Land Management Sector (ELMS), which address specific land-related problems. Current land use in the various countries as well as the areas occupied by wildlife conservation practices are summarised in Tables 11.1 and 11.2.

LAND EVALUATION FOR LAND USE PLANNING

In addition to various other training workshops the SADC Land and Water Management Research Programme organised a workshop on land evaluation for land use planning, which was attended by all countries of SADC except Angola. Experiences from some of the countries are summarised below.

Swaziland

The land evaluation policy of Swaziland is determined by the Ministry of Agriculture, through the land use planning department. A land capability and soil map was completed by Murdoch in 1968 with a classification system suited only to broad planning, but the land use planning department presently prepares land suitability, capability and soil maps for districts at detailed levels.

Lesotho

In Lesotho the land use planning division was established in 1981 with the help of the Food and Agriculture Organization (FAO), after which the FAO land suitability rating was adopted to prepare national, district and village land use plans. The majority of work done presently is assisting communities to decide how best to use their lands. This is done by allowing farmers full participation in selecting the most suited land use options.

Tanzania

The Tanzania National soil service was established way back in 1974 and was charged with carrying out soil surveys for land evaluation, national resources and agro-ecological inventories and also with co-ordinating, correlating and standardising Soil Survey and land evaluation work in the country. A soil information system for Tanzania (SISTAN) has been developed and plans are underway to establish a soil correlation centre. Work on compiling soil survey and land evaluation guidelines for the country is ongoing.

Malawi

In Malawi the FAO project MLW 185/011 (Strengthening Land Resources Evaluation Capacity) has developed methodologies for a land resources survey and land suitability appraisal, and has applied these methods in nationwide surveys at a scale of 1:250 000. These surveys were carried out on soils, physiography, agro-climate, vegetation and present land use. An automated land evaluation system (ALES) has been applied which, together with the land resource database (LRDB), will allow rapid land evaluation, once the land use requirements model has been built and the LRDB has been established.

Zambia

The Zambian Land Evaluation System (ZAMLES) is a Software package designed to determine land suitability of areas targeted for rainfed agriculture. It matches climate and soil data with agro-ecological requirements of specific crops, indicating possible limitations. The package has already been tested using data from district soil surveys. Suitability classes obtained from the evaluation compare favourably with observed yields of crops in the districts.

General

Overall it is clear that several land use planning approaches have been used, i.e. land capability, land suitability and modelling or simulation.

The land capability classification (USDA) is being used in Zambia, Malawi and Zimbabwe, while the land suitability classification (FAO) is being used in all countries except Swaziland and Zimbabwe.

CONCLUSIONS

The SADC as a region has taken the problem of land and its future very seriously. The problems of population pressure require improved land management and related husbandry practices to ensure increased food production.

Apart from its importance for human nutrition, wildlife is the backbone of most SADC countries because of tourism, and recently the total land area allocated to wildlife management has increased (Table 11.2).

Individual countries' efforts in land use planning are supported by numerous projects at regional level, and several workshops have been organised to equip regional scientists with skills to deal with planning land and water management.

FURTHER WORK

(1) Because all SADC countries depend entirely on land for food production, land management should receive more emphasis in disciplinary and interdisciplinary projects.
(2) Overall coordination and information flow within the region is still poor. Some kind of networking should be developed to help those countries with little or no resources to carry out their own surveys and research and to more effectively benefit from results obtained elsewhere.
(3) The production of many maps, such as soils maps, erosion hazard maps, land suitability maps, etc., will not be useful unless put to good use. A lot of effort is still required to teach all those interested in their interpretation and use.
(4) SADC as an organisation depends on donor support to implement projects in the natural resources sector. Most of these have not as yet had any funding.

CHAPTER 12

Selected Case Studies at Supranational Level

MODELLING AND OPTIMISING RICE-BASED CROPPING SYSTEMS

B. A. M. Bouman,[a] F. W. T. Penning de Vries[a] and M. J. Kropff[b]
[a]Research Institute for Agrobiology and Soil Fertility, Wageningen, The Netherlands.
[b]International Rice Research Institute, Los Banos, Manilla, The Philippines

In many Third World countries, growing food demands from increasing populations and declining natural resources necessitate careful planning and optimisation of sustainable agricultural land use options. Tools have to be developed for efficient design and testing of new crop varieties, rotations and production systems in various agro-ecological environments. In the project Systems Analysis and Simulation for Rice Production (SARP, 1986–1995), such tools were developed in close cooperation with a number of National Agricultural Research Centres (NARCs) in Asia. Crop-growth simulation models are developed and validated on common and site-specific field experiments. The main crop of interest is rice (upland and lowland), but other crops of rice-based cropping systems are modelled as well, e.g. wheat, corn, soybean and cowpea. The simulation models are currently being used to explore the production potentials of newly developed varieties, to establish optimum sowing dates, to evaluate yield stability and risk in relation to weather variability and to estimate the effects of climate change on global rice production. In a next step, simulation models for individual crops will be linked so that the potentials of crop rotations and production systems can be explored too. The developed simulation tools focus on the biophysical parameters of production systems and their environments; the outcome can be used for socio-economic analyses, to guide further agricultural research, and to assist regional planners, policy-makers and extension services.

The SARP-project is supported by the Ministry of Foreign Affairs of the Netherlands, and executed by AB-DLO, IRRI, Theoretical Production Ecology (Wageningen Agricultural University) and 16 NARCs in south, east and south-east Asia.

The Future of the Land: Mobilising and Integrating Knowledge for Land Use Options
Edited by L. O. Fresco, L. Stroosnijder, J. Bouma and H. van Keulen.

A COUNTRYSIDE INFORMATION SYSTEM FOR EUROPE

R. G. H. Bunce and D. C. Howard
Institute of Terrestrial Ecology, Grange-over-Sands, Cumbria, UK

At regional, national and supranational levels the primary requirement for development of a computer system is to provide information for planning purposes in terms of an integrated database describing the rural environment. The system must allow easy manipulation and interpretation in response to diverse policy issues. One such system, the Countryside Information System (CIS), is currently being developed and bench-tested by policy advisers in Great Britain (GB). It holds a variety of consistent national and regional data sets describing the environment, but the majority of ecological information, important in rural planning, has had to be collected by sample surveys. Sample data in the CIS are structured according to a Land Classification resulting from statistical analysis of environmental data, giving 32 environmental strata in GB. This allows the efficient production of national estimates of land cover, habitats, vegetation and potential change. The approach adopted in GB could be adopted at the wider European scale. The same principles of objectivity and integrated databases apply. Such a project would require proper consideration of the definitions of categories used in different rural databases, and the problems of integrity and consistency. Although the CORINE project has the potential to provide excellent land-cover data for Europe, there are many other data sets that could be integrated in a product that would be of great benefit to policy-makers. A preliminary environmental land classification of Europe, based on climate variables alone, is being developed and could be used in a Rural Information System for Europe (RISE). Such a system should be developed to handle existing data sets and information. The feasibility of developing such a system, including the likely requirements of those with rural policy responsibilities, is being discussed.

A EUROPEAN ECOLOGICAL MAIN STRUCTURE: A BASIS FOR INTERNATIONAL SCIENTIFIC COOPERATION

R. H. G. Jongman
Wageningen Agricultural University, Wageningen, The Netherlands

A tentative ecological structure has been developed for Europe mainly based on (European) CORINE data and regional information. The objective was the enforcement of nature conservation and the halting of species decline by facilitating migration. The project shows that nature conservation cannot remain a national or regional policy but needs instead international cooperation. However, the design showed differences in strategies of planning for nature, in criteria in the available data on nature and in the state of knowledge between member states.

Moreover, a coherent European Main Structure cannot be restricted to the European Community but has to be expanded to include all Europe. In this way it can become a common base for a European policy and scientific cooperation.

For the development of a European strategy for nature conservation and planning, resulting in a European Ecological Main Structure, several steps have to be taken, such as the inventory of objectives of nature conservation and the status of nature in Europe. Important scientific problems are, on the one hand, the comparison between national nature conservation definitions and strategies and, on the other, the development of databases for (1) the planning of the network and (2) statistics on changes in nature.

Differences in nature conservation history and in definitions of conservation categories hamper international comparison and influence planning of nature conservation areas. That makes it important to exchange this information between countries and planning agencies. The development of existing European databases towards a statistically sound common database is possible, as can be shown by examples of national inventories.

FOCUS ON LAND USE: A COMPARATIVE STUDY OF THREE REGIONS IN EUROPE AS A BASIS FOR LAND USE PLANNING

H. N. van Lier
Wageningen Agricultural University, Wageningen, The Netherlands

The future of rural regions in Europe will change, most probably dramatically. The surplus production in agriculture, the environmental problems, and the need to protect and restore natural ecosystems demand careful planning of future land uses. Land use plans need detailed predictions of socio-economic developments as well as of physical potentials. Such predictions should be made on the level of local communities and on the regional scale.

The study presented here aimed at developing a method to make such predictions, based on the use of several scenarios regarding possible changes in EC-farm policies as well as on increasing environmental constraints. The method intends to predict the future amount of farmland, number of farmers and farm incomes for a specific region. The method itself is based upon a combination of EC-farm scenarios, environmental constraints, crop suitability mapping and a farmers' decision model. The last of these predicts the number of farmers that stops farming because of changing EC-farm policies and increasing environmental constraints by using a stepwise procedure in which such steps as increasing production, lowering costs, changing farm systems (other crops), finding an additional income source or selling the land are respectively taken into account. Applying the model for three different regions in Europe (Groningen, The Netherlands; south-west Ireland; Auvergne, Central Massive, France) showed that the same scenarios worked out differently for each of these regions. What is good for the reduction of farm surpluses and for the EC-farm budget, is bad for the number of farmers (and therefore, bad for the employment in agriculture) and for farm incomes.

Additional studies are in progress. They deal with potentials for other crops, the influence of non-farmland demands on the willingness of farmers to terminate farming, and the farmers' decision model itself. For the region of Groningen, the Netherlands, the amount of farm land sold for other land uses does not depend on the demand for land from outside agriculture but on the internal problems and forces of farming itself.

New research should focus on the way in which individual farmers react to changing conditions.

THE CONSTRUCTION OF A KNOWLEDGE BASE FOR WHEAT IN EUROPE

G. Russell
Institute of Ecology and Resource Management, The University of Edinburgh, Edinburgh, UK

The Joint Research Centre of the European Communities at Ispra coordinates a large project which has the aim of predicting crop production in the European Community. Models are being constructed to predict yield on a grid square basis. The aim of the present work was to compile information about wheat (*Triticum aestivum* and *T. durum*) which could be used to parameterise and provide inputs for these models. Information from a variety of sources was collected and compiled, durum and winter and spring common wheat being treated separately.

National agricultural statistics were first examined to identify the major causes of variation. Reliable models of yield prediction need to take into account these factors, which are not necessarily the same as those operating at the farm scale (consider, for example, the effect of rainfall on the production of wheat from a region in which half the soils are clay and the other half sandy).

Regional agronomic data, such as, *inter alia*, the earliest and latest dates of sowing, ear emergence and harvest, were obtained from a postal questionnaire sent to correspondents throughout Europe. The size of these administrative regions varied with country, but was typically about 5000 km^2. More than 100 responses were received and these sample data were then interpolated to produce a database covering the entire EC. In addition, climatological data from representative meteorological stations were used to estimate the thermal time from sowing to ear emergence and from ear emergence to harvest as a function of latitude and date of sowing.

Typical and maximum values of crop attributes, such as components of yield and thermal time from sowing to emergence, and process rates, such as grain growth rate, were abstracted from published field trial reports conforming to criteria such as the trial being conducted according to normal agricultural practice. These values can be used to flag unrealistic predictions as well as to parameterise the models.

Since many administrative regions are heterogeneous, rules based on data in a geographic information system (GIS) were developed from information both in the literature and in the questionnaire responses to predict where wheat is actually grown. The GIS should thus be able to provide weather and soil information appropriate for representative crops in each grid square.

The knowledge base approach is powerful but it is essential to specify precisely the use to which it is to be put and the scale at which it is valid. It also only works when experts are able to share their valuable knowledge. There are many traps for the unwary.

PART III

NATIONAL LEVEL LAND USE PLANNING

CHAPTER 13

Knowledge Transfer to Farmers and the Use of Information Systems for Land Use Planning in Thailand

H. Huizing,[a] M. C. Bronsveld,[a] S. Chandrapatya,[b] M. Latham,[b] M. Omakupt,[c] S. Panichapong,[d] S. Patinavin,[c] B. Saengwan[d] and A. Sajjapengse[b]

[a]ITC, Enschede, The Netherlands. [b]IBSRAM, Chatuchak, Bangkok, Thailand. [c]Department of Land Development, Land Use Planning Division, Chatuchak, Bangkok, Thailand. [d]TA&E Consultants Co., Ltd., Bangkok, Thailand

INTRODUCTION

Agricultural research supports land use planning. Land use planning in Thailand is officially the responsibility of the Department of Land Development of the Ministry of Agriculture and Cooperatives (DLD). Since 1983, this department has produced plans at different levels of detail ranging from the regional to the farm level. Besides DLD, consultancy firms are also involved in land use planning. Agricultural research has been conducted in Thailand for many years. Yet the question that often arises is: how much of this research has been accepted or implemented by farmers (Panpiemras, 1991)? The issue of effective communication between land management research and local land users was well identified by the then Deputy Minister of Agriculture and Cooperatives, Dr Kosit Panpiemras. The answer is unfortunately simple: communication is poor and transfer through the official channel (for example,the marketing organisation for farmers which is supposed to obtain information from the research) is still minimal. Most of the information on new varieties, fertilisers, pesticides and tillage equipment comes from the commercial circuit, which has little connection with the research and which is more profit-oriented than anything else (Trebuil, 1993).

Yet the ingredients are there: reasonably good research results, and scientists and farmers who are generally ready to accept innovations to solve acute land management

The Future of the Land: Mobilising and Integrating Knowledge for Land Use Options
Edited by L. O. Fresco, L. Stroosnijder, J. Bouma and H. van Keulen.

problems. Many motives may be advanced for this lack of communication. Fujisaka (1991) listed 13 reasons for the nonadoption of innovations intended to improve the sustainability of upland agriculture by farmers. These include inappropriate innovations, incorrect identification of adoption domains, nonadoption due to cost, and lack of extension or lack of security in land tenure. A fourteenth reason could be added in the case of Thailand: the fragmented and clustered way in which land management research and extension is conducted. Not less than four ministries (Agriculture and Cooperatives, Interior, University Affairs, and Science, Technology and Environment) collaborate with non-governmental organisations (NGOs) and other charitable foundations.

Obviously, communication is essential with this diversity of sources, and if proper procedures are not followed the application of innovative technologies may be hampered. In this context, what is the role of IBSRAM, the International Board for Soil Research and Management?

Land-resource data, socio-economic data and agricultural-statistics data are available for many areas of south-east Asia. However, these data are seldom integrated for evaluating the physical sustainability and economic viability of alternative land use options at the subregional level at present. The aggregated form in which socio-economic and statistical data are disseminated, incompatible data formats (e.g. tables, maps) and lack of geo-referencing of part of the data are some of the problems that make the processing and integration of the data difficult.

An information system can facilitate data processing and integrating. This chapter demonstrates the use of an integrated information system that combines GIS, database programs, various models and multiple goal analysis techniques. The information system was developed by ITC, The Netherlands, and the Department of Land Development (DLD), Thailand, as part of a joint research programme. The main objective of this research programme was to develop an operational, integrated information system, optimally using existing data, to support land use planning at the sub-regional (village, district, province) level.

This chapter will first review current land use planning activities in Thailand. Then, problems of communication between land management research and local land users will be analysed, emphasising the role that is being played by IBSRAM. Having demonstrated the difficulties of communicating research, a case study is presented covering an information system for evaluating physically sustainable and economically viable land use options. The authors do not pretend that use of information systems will solve all communication problems. However, they believe that information technology and land use planning help to improve this communication. All studies were made in close cooperation with the Department of Land Development in Bangkok, Thailand.

LAND USE PLANNING IN THAILAND

Since 1963, DLD has had four major tasks: (1) soil survey, (2) land classification, (3) soil and water conservation, and (4) land policy. In 1983, agricultural land use planning officially also became part of DLD's responsibility and a Land Use Planning Division was established within DLD. The task of this Division is to present to planners and decision-makers information that enables them to implement sustainable agricultural

Table 13.1 Information used for land use planning in Thailand

Level of land use planning	Purpose of plan	Information used
Regional land use planning, scale 1:500 000	Agricultural production policy and regional land use allocation for agriculture, forestry, urban and industrial development.	Reconnaissance soil map, 1:500 000; national policy and economic plans; laws and regulations; present land use; etc.
Provincial land use planning, scale 1:100 000	Provincial development plans	Topographic map; soil map 1:100 000; present land use map; forest and water resources maps; socio-economic information, etc.
Project and farm planning, scale 1:100 000 to 1:5000	To identify the best land use options in biophysical, socio-economic and environmental terms; feasibility studies, environmental impact assessment; etc.	Topographic map; semi-detailed soil map; land use map; water resource and forest maps; engineering design/infrastructure maps; socio-economic data, etc.

development policies and strategies. The Land Use Planning Division uses the FAO Land Evaluation Framework as a basis for its planning work.

The function of land use planning in Thailand is to guide decisions on future land use in such a way that land resources are put to the most beneficial use for man whilst at the same time conserving these resources for the future (FAO, 1983). Land use planning for agriculture requires integration of data on physical resources, socio-economic conditions, present land use, current or attainable levels of technology and management, infrastructure, and regional and national demands for agricultural products. It must take into account the competing demands for land, water, human resources and finance by both the agricultural and non-agricultural sectors.

Since the Land Development Department started its land use planning work, a large number of land use plans at different levels of detail has been produced (Table 13.1):

- Regional level (1:500 000): land use planning reports for the four regions of the Kingdom of Thailand.
- Provincial level (1:250 000): provincial land use plans for 73 provinces (66 have been published already); land use Master Plans for 18 provinces; and coastal zone development operation plans for 10 provinces.
- Project level (1:100 000 to 1:5000): reports with maps at various scales depending on the objectives of the project.

Land use plans that adequately take into account the potentials of an area for development should be based on available data on land management research. Biophysical as well as socio-economic data are utilised to formulate land use plans for particular areas (Table 13.1). What data are used depends on the purpose of the land use plan and the (quality of) available data. Land use plans must be based on productive, stable and sustainable ecosystems.

Besides government organisations, many consultant companies that deal with natural resources are now using a GIS in Thailand. Problems of establishing a GIS are (Saengwan, 1992):

- lack of appropriate computer hardware and software;
- data of various government agencies are presented in different formats—this hampers the exchange of data and makes data integration for land use planning difficult;
- there is not yet a clear policy on promotion of data exchange between government agencies—this is necessary to make a GIS effective.

A land use planner is interested in the storage, retrieval and processing of large sets of spatially as well as logically related data. He needs a database management system that can handle these data. However, methodologies have to be developed for such a system and case studies are needed for testing.

Consultant firms are engaged in land use planning in Thailand, for example, in feasibility studies for irrigation development, resettlement programmes and environmental management (OEPP, 1992; RID, 1992, 1993; TA&E, 1993). The Royal Irrigation Department used a GIS for land use planning of an irrigation development project in eastern Thailand (RID, 1992). This Department has recently extended the application of a GIS to other development projects (RID, 1993).

The private and public sectors differ in working environment; for example, in scope and objectives of studies, levels of planning, budget and multidisciplinary expertise. The FAO framework for land evaluation can be a tool for both sectors. The objectives of feasibility studies can, therefore, be systematic. These firms can hire specialists in the different fields that are needed for such studies. For government agencies, objectives are generally less clear and possibilities to employ multi-disciplinary expertise are more limited.

LAND MANAGEMENT RESEARCH AND KNOWLEDGE TRANSFER TO FARMERS

Research strengthening on priority land management issues

One of IBSRAM's first principles is to work on and promote efforts on national priorities and not to focus on those of its own. Whereas there are still some uncertainties about prioritisation of land management issues in Thailand, the following list of issues appears to be the closest to what it should be:

- conservation farming on sloping lands to make hillside and mountain agriculture profitable and at the same time to combat further forest encroachment and to preserve soil and water resources;
- alternative utilisation of the soil complex (sandy Ultisols or saline soils) in north-east Thailand from the present cassava–rice cropping system;
- intensification of existing cropping systems on upland acid soils through inputs and improved cropping systems;

- management of lowland soils, especially clayey soils, for upland crops after rice;
- environmental consequences of shrimp and fish farming in mangroves and coastal areas and ways to alleviate them;
- pollution of water by pesticides, fertilisers and other chemicals due to intensive farming (horticulture or orchards); and particularly land-management issues in the vicinities of the great cities and ways to alleviate them.

Since its establishment in 1985, IBSRAM has conducted, in Thailand, workshops and training activities on four of the six above-mentioned priorities: sloping lands (IBSRAM, 1989), sandy soils (IBSRAM, 1986); lowland clayey soils (IBSRAM, 1990); and upland acid soils (IBSRAM, 1991a). At the same time, IBSRAM has promoted research activities on these priorities within two of its networks: ASIALAND management of sloping lands for sustainable agriculture, and ASIALAND management of acid soils. Five experimental sites are actively engaged in this research—three for sloping lands (Chiang Dao, Chiang Rai and Nan) and two on acid soils (Hat Yai and Rayong).

Role of IBSRAM towards Thai management research organisations

Within its involvement in agricultural research in Thailand, IBSRAM tends to create awareness, foster cooperation within and outside Thailand, promote innovations, disseminate information, and train researchers on land management issues (IBSRAM, 1991b).

The awareness is created hand in hand with the outcome of the IBSRAM research networks through workshops/visits by decisions-makers and farmers, exhibitions, articles in the local newspapers, and presentations on the TV. These means have been used in the past seven years and IBSRAM has contributed to the increased awareness of some of Thailand's land-management problems such as steepland agriculture in the north or intercropping of annual crops with rubber in the south. These efforts are being pursued to focus researchers' and policy-makers' ideas on the major priorities in land management.

IBSRAM has made an effort to promote adaptive research projects of a sufficient magnitude by fostering collaboration between Thai and other institutions.

Most of these cooperative projects are on farms, covering an area of one hectare or more. They necessitate a costly infrastructure and equipment investment but they are centred on farmers, which encourages technology transfer.

Another role of IBSRAM in these projects is to promote innovative technologies such as the use of pigeon pea in soil conservation hedgerows or the proper management of crop residues. These are not new techniques as such but they are novel in these agro-environments and they have proved to be inexpensive and efficient. Their adoption by scientists on a testing basis can come from IBSRAM, but they can also be promoted by visits and contacts with scientists from neighbouring countries. One of the most popular events at the annual meetings of the networks is always the field trip where partners in the network observe what their neighbours are doing.

Innovation or broadening of knowledge is also included in the methodological training activities conducted within the IBSRAM network. Experience has shown that training activities give good results as most of the subjects tackled have immediately been adopted by the network collaborators. They also demonstrate that the application of new methods and analytical techniques need to be monitored afterwards by the coordinator as they may not be fully understood at initiation. Training has also concerned a few graduate students who have used the network as a base for their MS or PhD theses. Innovations are also spread through specialised journals (*Soils Management Abstracts* (SMA), *IBSRAM Newsletter*, *Thai Newsletter*, and network reports), which provide scientists with literature reviews on specialised articles. Unfortunately the dispersion of the land management research group makes the distribution of and the accessibility of the information provided in such journals, i.e. SMA, difficult. Panichapong (1991) conducted a survey in Thailand on the impact of SMA and found that only 78 persons had seen the journal when 61 copies were distributed. This is a serious problem which IBSRAM has tried, with mixed success, to address through better collaboration between libraries and scientists. IBSRAM's role in land management research is, therefore, to catalyse ideas on priorities and the efforts of the national system to tackle them.

Roles of national research organisations

The role of Thai research and training organisations collaborating in the networks is threefold: (1) implementing research and contributing to national knowledge as well as to the networks' aims, (2) training, and (3) being involved in dissemination of information to policy-makers.

The research component within a national project is fully conducted by the national partner. IBSRAM provides coordination and assistance when needed, but the research from planning to interpretation of the results is a national responsibility. IBSRAM's collaborators not only present their achievements and results to their peers during annual review meetings (and publish these results, with proper acknowledgement of the collaboration at IBSRAM meetings), but also at international workshops and conferences or in journals.

Unfortunately these publications by IBSRAM collaborators are still too restricted and their scientific and extension impact is limited. There were only four Thai papers out of 113 at the international soil conservation conference held in Bangkok in 1988 (Rimwanich, 1988). This is certainly not enough in a country where soil conservation is a priority and where research and implementation on the subject is widely practised. Research cannot be well conducted without publication support. This is a major issue in which IBSRAM and other similar institutions can play a major role.

Thai research and training organisations can and do play a major role in local training and the training of others in new land management techniques. The Department of Technical and Economic Cooperation (DTEC) sponsors regular regional training courses conducted in the English language at universities like Kasetsart for personnel from developing countries. IBSRAM has organised, with the support of IDRC (International Development Research Centre), study tours to Thai universities and the departments of the Ministry of Agriculture, for Vietnamese and Chinese collaborators. According to Khun Priya Osthananda, Director General of DTEC, Thailand now needs to share its experience with its neighbours and to expand its role as a leader in Indochina. IBSRAM may be able to help materialise this expectation in this respect.

A final role that Thai research and training organisations should play more forcefully is to influence policy-makers on the decisions to be taken for the management of natural resources. This role has partially been adopted by the Thailand Development Research Institute (TDRI) which has organised regular forums in policy-making such as the 1990 conference on 'Industrializing Thailand and its Impact on the Environment' (TDRI, 1990). The role has also been taken by environmental NGOs so that academics, researchers, and users of the land can voice their concerns to the Government (Nutalaya et al., 1993). It seems that the official research and training organisations should play a more active role in this area to provide the necessary guidance and promote sustainable land management concepts among policy-makers.

Transfer of technologies to farmers

The process of technology transfer in land management is conceptually well known (Denning, 1984) but seldom well applied: 'There is insufficient cooperation between extension and research despite their common ultimate goal' (Cerna et al., 1984). Basically, between research, extension and farmers there should be two-way communication to ensure accurate mutual understanding. IBSRAM's Thai collaborators do conduct field days, demonstration activities, and training of farmers. They also provide services and facilities to farmers for improved land management techniques.

Yet the approach remains very much top-down and the efficiency of these activities remains limited. In the northern province of Chiang Mai, the Department of Land Development provides services to farmers to establish soil conservation infrastructures on some 3000 ha per year, which is hardly significant when the cultivated areas on sloping lands can be estimated to be around one million hectares. And not all the few hundred farmers who have visited the demonstration sites will immediately implement the techniques demonstrated.

The effort needs to be implemented more vigorously and requires the involvement of all the parties including the Royal Forestry Department (when the land is registered as forestland after clearance) and the land department to provide secure land-tenure titles to farmers on a permanent basis to encourage sustainable land management. Regrettably, unlike new varieties which can be very quickly adopted by farmers, fertilisers or pesticides, or proper land management practices require a middle- to long-term effort. According to experiments in Chiang Mai, conservation-farming methods on a maize/red kidney bean system show an advantage only after two to three years. In the meantime, they require an initial investment which the farmer can hardly afford due to insufficient capital, or is unwilling to adopt due to no proper land title and the immediacy of the treatment's response. This situation raises four separate issues:

(1) Land degradation due to erosion not only affects the upland farmers; it also affects the whole community due to siltation of the waterways and of the lowlands. Siltation and pollution from the uplands raises further controversies from lowland farmers (Ekachai, 1990) and, therefore, merits serious consideration by the policy-makers to create initial incentives for farmers to apply soil conservation measures.

(2) Land users without land titles have investment difficulties and are unable to borrow from banks; land users willing to adopt soil conservation measures should be provided with appropriate land titles to be able to make the necessary investment.
(3) Conservation-farming methods only impact after three to four years. There should, therefore, be bank schemes especially designed for conservation farming with a three-to-four-year grace period.
(4) This being said, conservation-farming practices will only be sustainable if they are profitable; therefore land use plans should carefully review areas and techniques that can respond to this criterion.

Other techniques designed to solve the land management problems indicated earlier, face similar constraints; therefore, the transfer of land-management technologies cannot reach its target without a concerted effort by all the parties involved: researchers (to provide appropriately profitable and sustainable land management techniques), policy-makers (to design policies ranging from the land attrition issue to the extension of techniques and their implementation), bankers (to organise credit for farmers), extension officers (to promote appropriate innovations and helpful information), and farmers (to accept them).

The following case study will demonstrate the use of information technology to support land use planning and to improve communication of research results. The authors do not consider this approach to represent a definite solution to the communication problem, but they wish to demonstrate its potential usefulness.

THE USE OF AN INFORMATION SYSTEM FOR LAND USE PLANNING: A CASE STUDY IN PHETCHABUN PROVINCE

The case study area is part of Lom Kao District, Phetchabun Province, Thailand. The area consists of lowland and adjoining hills. The lowland is traditionally used for rice growing. Where irrigation water is available, crops such as mungbean and tobacco are grown in the dry season. The hilly land has been deforested in the last 20 years and is now mainly used for growing maize, a cash crop that is exported as animal feed. Land preparation in the hills is mainly by downslope ploughing. Surface soil losses are very high and yields are declining (Table 13.2). The cultivation of maize in the hills leads to a high runoff in the wet season and reduces dry season base flows.

Table 13.2 Yield decline of maize in study area in relation to period of cultivation (on soils derived from shale, siltstone and mudstone)

Years since reclamation	Average maize yield (kg/ha)	Standard deviation (kg/ha)	Relative yield (%)
0–3	3500	625	100
4–8	2750	?	80
11–15	2100	440	60
16–30	1750	500	50

Source: Farmers interviews (Funnpheng, 1988; Adebanjo, 1989; Palmer, 1989). Total number of interviews is 62.

Table 13.3 Main physical and economic characteristics of the selected villages

Village number	Land type (% of village area)[a]			Average annual income per household (US$)
	Flat	Sloping	Steep or rocky	
1	5	30	65	240
2	50	45	5	680
4	30	55	15	450
8	15	55	30	300
13	60	40	0	750

[a]Based on 1:50 000 soil map of DLD.
Source: Thammasart University (1990).

The case study area comprises five villages in Hin Hoa Subdistrict that differ with respect to (1) their dependence on sloping and hilly land for crop growing and (2) average income (Table 13.3).

Most of the households in the villages are solely dependent on agriculture for their income: more than 85% in villages 2 and 13, and 93–99% in the other villages (Thammasart University, 1990). Table 13.3 shows that differences in income between the villages are related to the amount of flat land (lowland) in the village. The income of villages with only small areas of lowland depends for a large part on maize grown on hill slopes.

Information needs for land management

Evaluating alternative land use and land management options requires information on the effects of these options on:

- food production
- income
- capital requirements
- employment
- environmental degradation

For this evaluation, data are needed on:

- local development needs and goals
- present land use
- alternative land use and land management options
- the suitability of the available land resources for these options
- instruments (e.g. extension, credit) available to the Government to promote these options
- population characteristics, including size and location of administrative units
- farming systems

Existing data

Except for mountainous and steep areas, soils and land use maps at detailed reconnaissance scale (1:100 000) are available for the whole of Thailand. For smaller areas, more detailed soils and land use information is also available at scales of 1:25 000 to 1:50 000. A digital database ('*Kho Cho Cho*' or KCC database) with socio-economic and land use data at the village level covers the whole country. This database is updated every two years and is used for the selection of priority areas, project planning and budget allocation purposes.

For the case study area, the following data are available:

- a semi-detailed soil map at 1:50 000 scale (DLD, 1990)
- a land use map at 1:100 000 scale (DLD, 1986; based on aerial photographs of 1974 and field data of 1985)
- KCC data of 1989 (Thammasart University, 1990)

The soil mapping units are characterised by individual soil series or associations of soil series. The soil map provides an adequate basis for assessing the suitability of land for various land uses. The land use map, on the other hand, does not contain sufficient detail for the purpose of the case study. Its information is, at least partly, out of date because land use changes in the case study area are rapid.

Surveys by ITC students show that KCC data generally give a good picture of the socio-economic situation of villages. There are some problems, however, when land use data contained in the KCC database are used because:

- village areas and boundaries are not properly registered, and
- crop areas refer to areas cultivated by villagers and *not* to areas cultivated within a village. A village, for example, may report a cultivated area larger than the village area itself when many farmers grow crops on land in other villages.

Existing information is lacking with respect to the following essential subjects:

- farming systems,
- farmers' preferences and priorities with regard to crops,
- crop yields and income from crops in relation to landforms, soils and land management, and
- rates of soil loss in relation to slope, soils, land use and land-management practices

Supplemental data collection

Because of the partial inadequacy and/or absence of existing data, supplemental data collection was essential. Updating of existing data and collection of new data included the following:

- The 1:100 000 scale land use map was updated with information from Landsat Thematic Mapper (TM) images of December 1988 and March 1989.

- Data on crops, land management and yields was collected from staff of local Agricultural Extension Offices.
- Present land use and farming systems data was obtained through surveys by DLD and ITC postgraduate and MSc students (DLD, 1990/1991; ITC, 1989/1990).
- Detailed information on boundaries of the five villages included in the case study was obtained from village heads and farmers using aerial photographs as a reference.

Land use and land management options

Twelve land use types (LUTs) were selected. These LUTs are crops or cropping patterns that already occur in the area. Most LUTs are based on the management practices and yields of 'average' farmers in 1989/1990 (data of Anaman, 1990; Krishnamra, 1991; DLD 1990/1991):

- maize as a sole crop; maize followed by mungbean
- rice as a sole crop; rice followed by mungbean or tobacco
- wet season mungbean, dry season mungbean and tobacco as sole crops
- sweet tamarind

In addition, three 'improved' land use and land use management options are considered:

- soybean as a sole crop
- rice followed by soybean (in areas with dry season irrigation)
- sweet tamarind with improved land management (intermediate inputs; cover crops to provide an effective protection of the soil)

The improved land use options were already practised by some farmers in 1989/1990.

Soybean is a promising dry season crop for lowland areas that could increase farm income considerably. Sweet tamarind was selected as an alternative to maize. Sweet tamarind plantations, with proper management, protect the land against erosion. The expansion of sweet tamarind growing is supported by extension services and improved plant material is available. Market prices are high and high gross margins per hectare can be obtained (Table 13.4). Prospects for export are favourable.

Indicators selected for assessing sustainability and viability

To assess the effects of alternative land use and land management combinations on food production, income, capital requirements, employment and environmental degradation at the sub-regional level, the following indicators are chosen:

- total annual food (rice) production
- total annual gross margin from crops
- total annual employment that is generated
- total annual soil loss

Table 13.4 Gross margins of maize and sweet tamarind in US$/ha (1989/1990 situation)

		Sweet tamarind	
Land suitability	Maize	Low level management[a]	Intermediate level management[b]
High	200–250	250–400	900–1300
Moderate/marginal	100–200	100–250	400–900

[a]Management by the majority of sweet tamarind growers in the study area.
[b]Management by 'better' sweet tamarind farmers.
Note: high levels of management as practised by few owners of large sweet tamarind plantations are not considered in the case study.
Source: Anaman (1990); ITC (1990).

These indicators are calculated per village and/or group of villages subject to a number of constraints:

- availability of suitable land
- availability of labour to meet monthly peak labour demands
- total available annual capital for recurrent and investment costs

Assumptions and goals

The following assumptions were made with respect to desired (goal) values of the indicators mentioned above:

(1) Average gross margin from crop production per household should be at least US$800 per year. This value is based on the minimum income target for rural households of the Lom Kao District Office in 1992.
(2) Average annual input costs do not exceed US$240 per household. This value assumes that annual capital inputs for crop growing do not exceed the maximum amount of credit that can be obtained from formal sources in one year.
(3) Average soil loss rates will not exceed 1.5 mm surface layer per year in sloping and hilly land.
(4) About 200 kg of rough rice per person is needed annually to meet the food requirements of the population.
(5) Full employment of all persons actively involved in crop growing in the five villages implies that the selected land uses should create work for 235 000 person days per year.

The information system

The structure of the information system used in the case study is shown in Figure 13.1. The information system consists of several modules. In each module, available data are combined and integrated using various operations or functions. Outputs produced in a module are, after checking and verification, used as inputs in one of the other

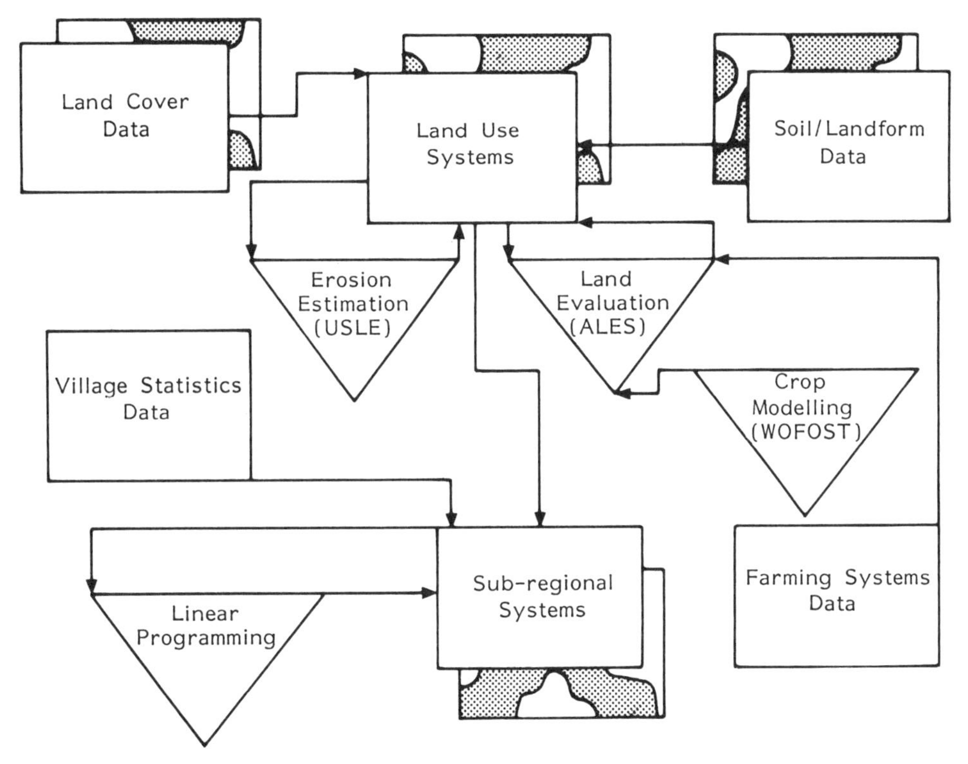

Figure 13.1 Structure of the information system

modules. All outputs are finally combined in the 'sub-regional (village, district, province) system'. The information system is flexible and makes it possible to answer a wide range of questions of administrators, planners and land users. The system is implemented on MS-DOS based personal computers. It uses the following software algorithms:

- the ALES land-evaluation expert system (Rossiter and Van Wambeke, 1989) to assess biophysical land suitability, proportional crop yields and gross margins based on procedures of the FAO Framework for Land Evaluation (FAO, 1976);
- the WOFOST crop-growth simulation model (Van Keulen and Wolf, 1986) to assess yield reductions due to water shortages;
- adapted Universal Soil Loss Equations (USDA, 1978) for different slope ranges to estimate soil losses; adaptations are based on field measurements;
- a relational database;
- multiple goal analysis techniques based on linear programming; and
- GIS (ILWIS) (i) to extract information from remote sensing data; (ii) to combine spatial data for further analysis, for example, for linear programming; and (iii) to produce maps showing present and alternative land use and land management situations

The WOFOST model was validated through crop experiments with maize in Phetchabun Province in 1989 (Kuneepong, 1990). WOFOST calculates yield reductions for (groups of) land mapping units on the basis of (i) rainfall and (ii) land characteristics such as soil depth, soil texture, soil stoniness, and slope. Yield reductions predicted by WOFOST are used to set proportional yields for the land quality 'available water' in ALES. Existing ('expert') knowledge and local knowledge specific for a study area on relations between land, land management and crop yields is used to assess proportional yields for other land qualities. Modified Universal Soil Loss Equations (USLE) are used to estimate soil losses by water erosion (Harper, 1988; Meijerink, 1989, Srikhajon et al., 1991). The selection of the equations was made in such a way that the calculated soil losses were in agreement with field estimates of the magnitude of soil losses on different soils and slopes. These field estimates are based on a combination of rill erosion (rill volume) measurements and estimations of inter-rill erosion by measurement of root exposure of selected plants (ITC, 1989/1990; Palmer, 1989; Huizing and Bronsveld, 1992).

The number of land use options that can be analysed in a GIS is very large. To restrict the analysis to options (scenarios) that are realistic with respect to the physical and socio-economic conditions of a study area, multiple goal analysis techniques are applied. These techniques, based on linear programming, make it possible to limit the number of land use options to those that are feasible when specific local constraints or development targets are taken into account. Constraints may be the availability of labour, credit or suitable land for different land uses. Targets may be minimum income (or gross margin) levels or soil loss rates below a certain, acceptable level.

Discussion of results

The effects of several future land use situations were assessed. These future land use situations are examples of what can be achieved in terms of food production, income, employment, etc. They are based on (i) the assumptions and goals mentioned above, and (ii) constraints (e.g. availability of suitable land, availability of labour and capital). Any other land use situation, based on preferences of different users, can also be evaluated by the information system using the same basic data. Evaluations can show the extent to which preferences expressed by individual land users, the community, planners and decision-makers can be met.

The assessment of future land use situations, based on the selected land use options, shows that:

- it is not possible to meet the goal of providing full employment due to the limited employment opportunities in the dry season (perennial crops do not require much labour in this season; and the size of areas with irrigation facilities is limited);
- labour availability during peak labour demand periods is not a problem in nearly all the land use situations considered in the analysis; only slight problems may occur in December because of concurrent rice and sweet tamarind harvesting; and
- a surplus of rice can be produced above the subsistence needs in all the villages; growing surplus rice for sale on flat land is favourable for the village income because gross margins for other land uses on this land are generally low.

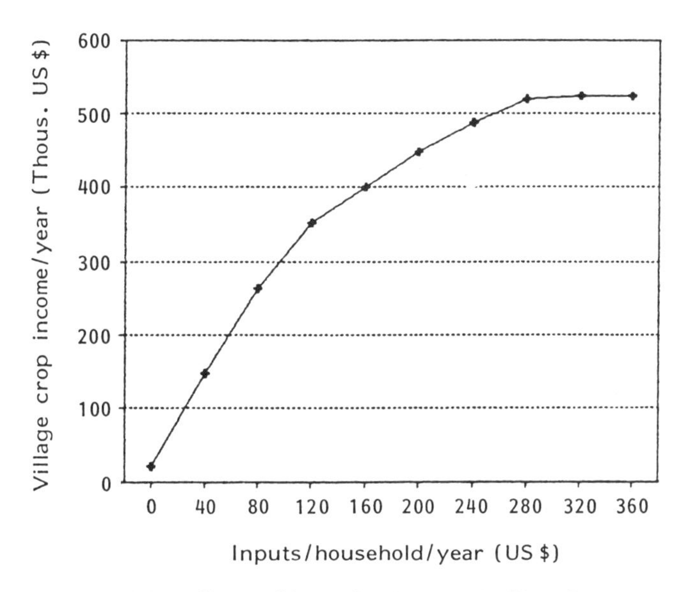

Figure 13.2 Effects of input level on total village income

In the case study area, the most sensitive factor in achieving the assumed goals is the amount of money available for recurrent inputs and investment. The sensitivity of village income to capital inputs is shown in Figure 13.2.

In Figures 13.3 and 13.4, a comparison is shown between two alternative, future land use situations. Figure 13.3(a) shows the current land use. Figure 13.3(b) is based on the assumptions mentioned above, except for soil loss rates for which no restriction is made; additional assumptions are (i) that the area planted to sweet tamarind is the same as in 1989, and (ii) that maize is not planted on slopes steeper than 35%.

The maize area in Figure 13.3(b) is much smaller than the area presently planted to maize because of the exclusion of steeply sloping areas. Figure 13.3(c) shows a second alternative in which soybean growing and an expansion of the sweet tamarind area is considered; however, this area is subjected to the following constraints: (i) the maximum average capital input cannot exceed US$240 per household per year and (ii) the average annual soil loss on sloping and hilly land should be less than 1.5 mm surface soil layer. In this alternative, the area planted to maize is less than in the first alternative. Figure 13.4 shows the income, total erosion, food and employment created by the two alternatives for the combined area of the five villages included in the study. A comparison of the two alternative land use situations shows that target incomes at acceptable erosion rates can be achieved (i) when soybean and sweet tamarind growing is expanded and (ii) when agricultural activities are concentrated on the most suitable land.

Figure 13.5 shows that these target incomes can be achieved in four of the five villages.

A field trip in 1993 to the case study area showed a land use pattern similar to the second alternative above. The area planted to maize had decreased drastically since 1989, mainly due to low maize prices. On the other hand, there was a substantial increase

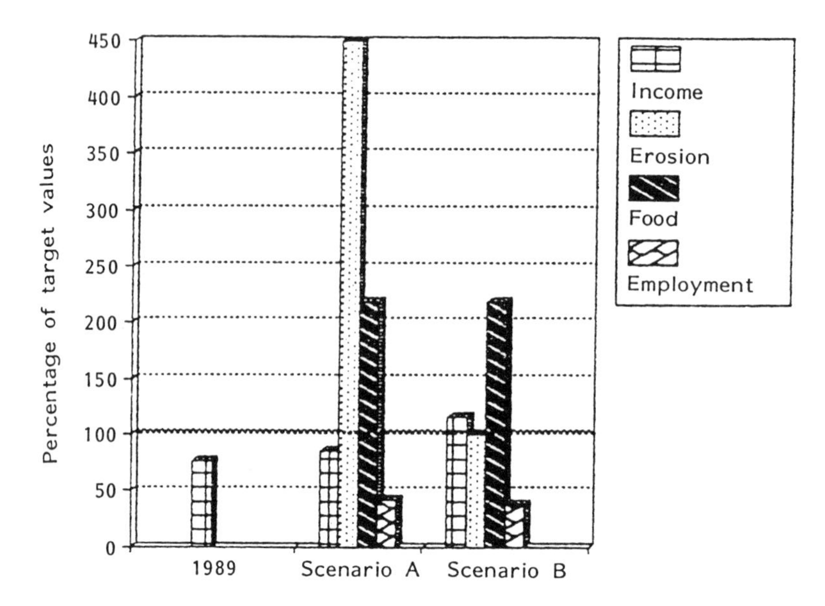

Figure 13.4 Income for 1989 and two alternative future land use scenarios. The 1989 income is based on KCC database; Scenario A is based on the sweet tamarind area of 1989; and Scenario B is based on the expansion of soybean and sweet tamarind

of the sweet tamarind area, particularly in sloping areas close to settlements. This strongly resembles the second land use situation described above (Figure 13.3(c)).

Biophysical and economic effects of interventions can be assessed by the information system in an interactive way using assumptions or goals of different users. They show biophysical and economic potentials that can be reached. Social aspects such as land tenure and unequal distribution of resources and unequal access to services, however, may provide additional constraints that are not covered by the information system in this case study.

Implementation of the information system in local planning and extension procedures

The information system was used with success in several project areas (district, subdistricts, watersheds) by national, Bangkok-based, Government staff. One of the objectives of the DLD/ITC research, however, was to use the information system as an operational tool in sub-regional planning and extension procedures. This objective is difficult to achieve, however. The main reasons are:

(1) the centralised planning structure in Thailand in which influence of the local administration (at province, subdistrict and district level) on decisions regarding budget allocation for programmes and projects is restricted;
(2) the 'vertical' organisation of the Government which results in the assignment of a Department in Bangkok to implement a project: a Department, however, is generally responsible for one aspect of natural resource management only

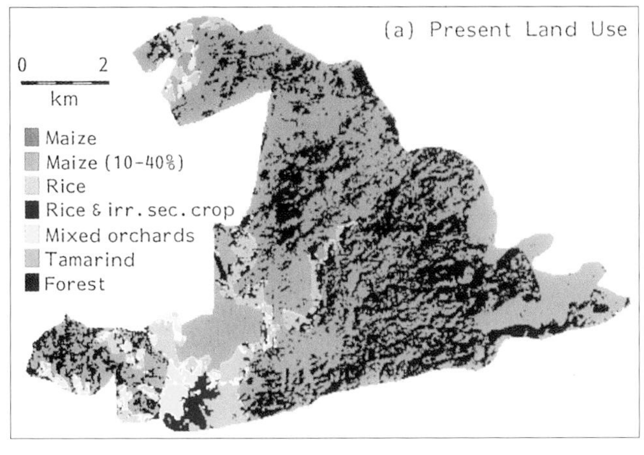

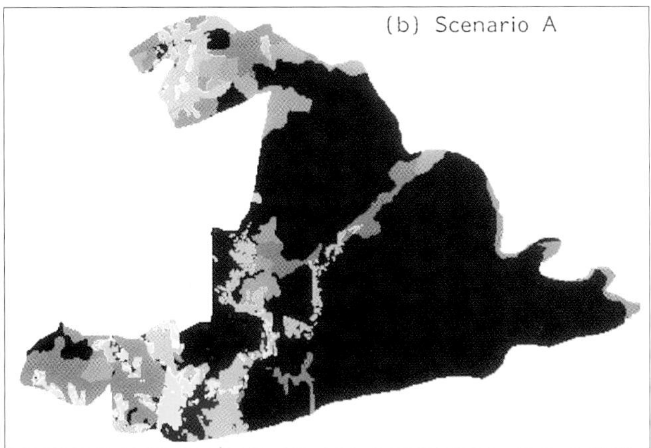

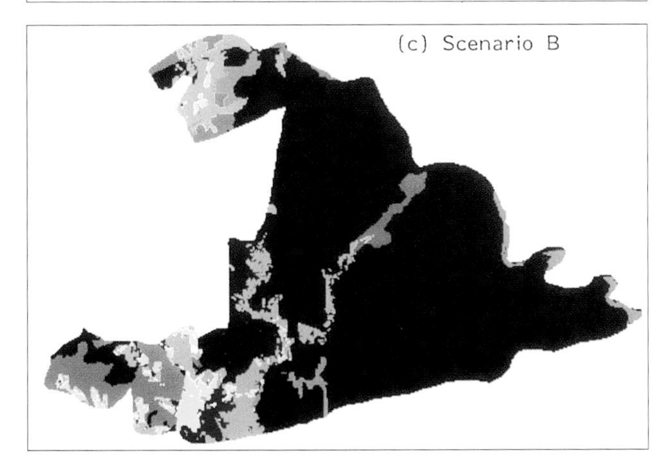

Figure 13.3 (a) Current land use; (b), (c) two scenarios for future land use (see text)

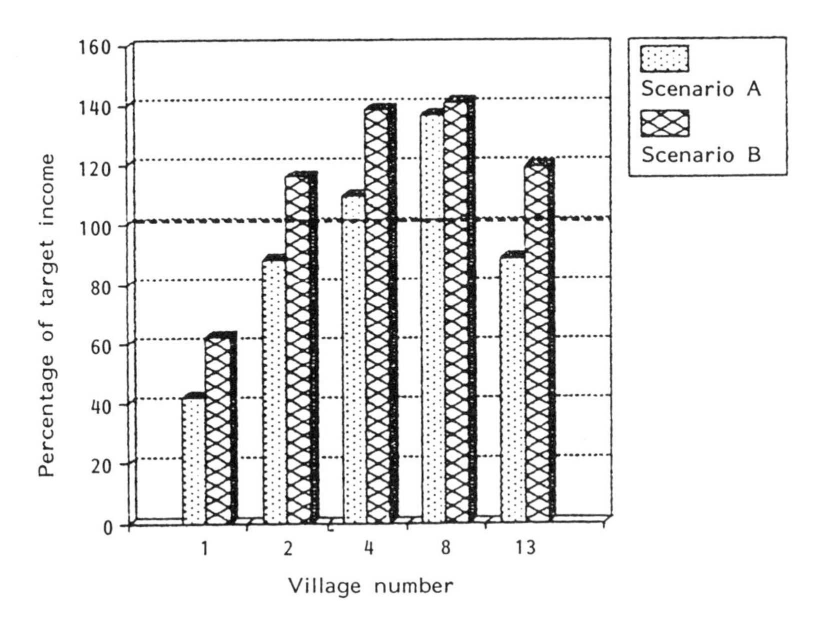

Figure 13.5 Percentage of target income achieved in the five villages studied in the two alternative land use scenarios. Scenario A is based on the sweet tamarind region of 1989 and Scenario B is based on the expansion of soybean and sweet tamarind growing

(i.e. irrigation, soil conservation, extension, livestock, management of forest land, etc.);

(3) the lack of funds, particularly at the district and subdistrict level, to implement a PC-based information system.

Effective use of an information system as described above at the sub-regional level will only be possible when local interdisciplinary teams (or task forces) are created in which all relevant departments are represented.

CONCLUSIONS

(1) Integration of landform, soils, land cover, land use, farming systems and agricultural-statistics data is essential for land use planning. Except for farming systems data, these data are available for most areas in Thailand.

(2) A model of real conditions can be obtained by developing an information system based on GIS (ILWIS), database programs, models (USLE, ALES and WOFOST) and linear programming techniques. The case study shows that the information system makes it possible to assess biophysical and economic effects of the introduction of alternative land use options at the sub-regional level. The information system facilitates decisions as to the best combinations of land uses in a particular biophysical and socio-economic environment and is, therefore, a powerful tool for land use planning and environmental management.

(3) Implementation of the information system in sub-regional planning and extension procedures in Thailand will be difficult, however. Main reasons are the centralised planning structure, the vertical organisation of the Government and lack of funds at the sub-regional level. Communication of research results to the sub-regional and farmer level is, therefore, bound to remain a serious problem. Iterative discussions are needed between researchers, policy-makers, extension agents and farmers. The authors believe that the systematic and flexible presentation of data, as illustrated in the case study, can be helpful in improving communication in the future.

ACKNOWLEDGEMENTS

The authors wish to thank Mrs Chalinee Niamskul, Director of International Relations in IBSRAM, and Dr Sumalee Sutthipradit, Professor of Soil Science at the Prince of Songkhla University, for their contributions to this chapter.

REFERENCES

Adebanjo, O., 1989. Farmers' first approaches to soil conservation programmes. A case study of parts of the Upper Pa Sak river basin, Phetchabun Province, Thailand. Unpublished MSc thesis. ITC, Enschede, The Netherlands.

Anaman, T., 1990. Data base development and use as a tool for farming systems analysis: A case study in tambon Lom Kao and tambon Na Saeng. Upper Pa Sak watershed area. Phetchabun Province, Thailand. MSc thesis, ITC, Enschede, The Netherlands.

Cerna, M. M., Coulter, J. K., Russel, J. F. A., 1984. Building the research–extension–farmer continuum: some current issues. In: *Research–Extension–Farmers: A Two Way Continuum for Agriculture Development*, pp. 3–10. The World Bank, Washington, DC.

Denning, G. L., 1984. Integrating agricultural extension programs with farming systems research. In: *Research–Extension–Farmers: A Two Way Continuum for Agriculture Development*, pp. 113–135. The World Bank, Washington, DC.

DLD (Department of Land Development), 1986. Land use map of Phetchabun Province. DLD, Bangkok, Thailand.

DLD (Department of Land Development), 1990. Semi-detailed soil map of Amphoe Lom San, Lom Kao and Khao Kor. DLD, Bangkok, Thailand.

DLD (Department of Land Development), 1990/1991. Farm and land use interview data. DLD, Bangkok, Thailand.

Ekachai, S., 1990. *Behind the Smile: Voices of Thailand*. Thai Development Support Committee, Bangkok, Thailand.

FAO, 1976. *A Framework for Land Evaluation*. FAO Soils Bulletin No. 32, FAO, Rome.

FAO, 1983. *Guidelines: Land Evaluation for Rainfed Agriculture*. Soils Bulletin 52, FAO, Rome.

Fujisaka, S., 1991. Thirteen reasons why farmers do no adopt innovations intended to improve the sustainability of upland agriculture. In: *Evaluation for Sustainable Land Management in the Developing World*, pp. 509–522. IBSRAM Proceedings No. 12(1), IBSRAM, Bangkok, Thailand.

Funnpheng, P., 1988. Contribution to erosion hazard assessment in land evaluation for conservation. Ban Sila area, Pa Sak valley, Thailand. Unpublished MSc thesis. ITC, Enschede, The Netherlands.

Harper, D., 1988. Improving the accuracy of the Universal Soil Loss Equation in Thailand. In: Samarn Rimwanich (ed.), *Proceedings of the Fifth International Soil Conservation Conference*,

18-29 January 1988, Vol. 1, pp. 531-540. Department of Land Development, Bangkok, Thailand.

Huizing, H. and Bronsveld, M. C., 1992. The use of geo-information systems and remote sensing for evaluating the sustainability of land use systems. In: *Proceedings of International Workshop on 'Evaluation for Sustainable Land Management in the Developing World'*, pp. 545-562. IBSRAM Proceedings No. 12, Vol. II, IBSRAM, Bangkok, Thailand.

IBSRAM (International Board for Soil Research and Management), 1986. *Soil Management Under Humid Conditions in Asia and Pacific*. IBSRAM Proceedings No. 5, IBSRAM, Bangkok, Thailand.

IBSRAM (International Board for Soil Research and Management), 1989. *The Establishment of Soil Management Experiments on Sloping Lands*. IBSRAM Technical Notes no. 3. IBSRAM, Bangkok, Thailand.

IBSRAM (International Board for Soil Research and Management), 1990. *Management of Lowland Clayey Soils for Upland Crops after Rice Workshop*. Report no. 8, IBSRAM, Bangkok, Thailand.

IBSRAM (International Board for Soil Research and Management), 1991a. *The Establishment of Experiments for the Management of Acid Soils*. IBSRAM Technical Notes no. 5, IBSRAM, Bangkok, Thailand.

IBSRAM (International Board for Soil Research and Management), 1991b. *IBSRAM's Strategy 1991-2001*. IBSRAM, Bangkok, Thailand.

ITC, 1989/1990. Results of fieldwork of post-graduate ITC students in the Upper Pa Sak watershed, Phetchabun Province, Thailand. Courses in Rural and Land Ecology Surveys. Unpublished reports. ITC Enschede, The Netherlands.

Krishnamra, S., 1991. Land evaluation and farming systems analysis for land use planning at the village level using geo-referenced data. A case study in the Upper Pa Sak watershed, Phetchabun Province, Thailand. MSc thesis, ITC, Enschede, The Netherlands.

Kuneepong, P., 1990. Crop modelling of maize as a tool on the data base and as an input to land evaluation for the Upper Pa Sak watershed, Phetchabun, Thailand. MSc thesis, ITC, Enschede, The Netherlands.

Meijerink, A. M. J., 1989. Modelling in the land and water domain with a versatile GIS (ILWIS); experiences from a large tropical catchment. In: Bouma, J. and Bregt, A. K. (eds), *Land Qualities in Space and Time*, pp. 73-87. Proceedings of a Symposium organized by the International Society of Soil Science (ISSS), Wageningen, 22-26 August 1988. PUDOC, Wageningen, The Netherlands.

Nutalaya, P., Pansawat, T., Sophon Sakunrat, W. (eds), 1993. *Natural Resource and Environment Conservation in Thailand. Environment Year 1992*. Thailand Development Research Institute, Bangkok, Thailand.

OEPP, 1992. Interim report of the feasibility study of central metropolitan park (Bang Krachao). Office of Environmental Policy and Planning (OEPP), Ministry of Science, Technology and Energy, Bangkok, Thailand.

Palmer, M., 1989. The relationship between land cover/land use and erosion/land degradation. A case study in the Upper Pa Sak watershed, Phetchabun Province, Thailand. MSc thesis. ITC, Enschede, The Netherlands.

Panichapong, S., 1991. IBSRAM Information Service (IBSRIS). *IBSRAM Newsletter* **19**: 8.

Panpiemras, K., 1991. Ensuring farmers acceptance of sustainable land management technologies. In: *Evaluation for Sustainable Land Management in the Developing World*, pp. 615-620. IBSRAM Proceedings No. 12(1), IBSRAM, Bangkok, Thailand.

RID, 1992. *Prasae Irrigation Development Project*. Royal Irrigation Department, Ministry of Agriculture and Cooperatives, Bangkok, Thailand.

RID, 1993. Interim report of the feasibility study and EIA of Lower Nam Kam Irrigation Project. Royal Irrigation Department, Ministry of Agriculture and Cooperatives, Bangkok, Thailand.

Rimwanich, S. (ed.), 1988. *Land Conservation for Future Generation*. 2 Vol. Department of Land Development, Bangkok, Thailand.

Rossiter, D. G. and Van Wambeke, A. R., 1989. *Automated Land Evaluation System (ALES). Version 2 User's Manual*. Department of Agronomy, Cornell University, Ithaca, NY

Saengwan, B., 1992. *ILWIS: From ITC to DLD*. Proceedings of a Workshop of the Department of Land Development, Thailand and ITC, the Netherlands on *GIS and Remote Sensing for Natural Resource Management by ILWIS*. Pattaya, Thailand, 25–27 November, 1992. Department of Land Development, Bangkok, Thailand.

Srikhajon, M., Chomchan, S., Pramojanee, P., Liangskul, M., Meekangwan, S., Somrang, A., Petchsangsai, C., and Pongsamart, A., 1991. Soil erosion in Thailand. Third printed paper, Department of Land Development, Bangkok, Thailand.

TA&E, 1993. Soil survey of the proposed area for developing citrus plantation in the People's Republic of China. TA&E Consultants Co., Ltd., Bangkok, Thailand.

TDRI (Thailand Development Research Institute), 1990. *Industrializing Thailand and Its Impact on the Environment*. TDRI, Bangkok, Thailand.

Thammasart University, 1990. National 'Kho Cho Cho' (socio-economic) digital data base at village level for 1989. Thammatsart University, Bangkok, Thailand.

Trebuil, G., 1993. Agriculture pionnière, révolution verte et degradation de l'environnement en Thailande: Le cinquième dragon ne sera pas vert. *Revue Tiers Monde* **134**: 365–383.

USDA, 1978. *Agriculture Handbook No. 537. Predicting Rainfall Erosion Losses. A Guide to Conservation Planning*. US Department of Agriculture, Washington, DC.

Van Keulen, H. and Wolf, J. (eds), 1986. *Modelling of Agricultural Production: Weather, Soil and Crops*. Simulation Monograph no. 25, PUDOC, Wageningen.

CHAPTER 14

Nutritional Consequences of Different Land Utilisation Types and Different Income Sources

P. van der Molen[a] and J. W. Schultink[b]

[a]Wageningen Agricultural University, Wageningen, The Netherlands.
[b]SEAMEO-TROPMED/GTZ, University of Indonesia, Jakarta, Indonesia

INTRODUCTION

In view of the still large problems concerning food and nutrition in developing countries, the International Conference on Nutrition (ICN) organised by FAO/WHO in December 1992 in Rome, strongly recommended the improvement of the interface between agricultural production and human nutrition at household level (FAO/WHO, 1992). The interface between nutrition and agricultural production comprises the interrelationships between nutritional status and production quantities (Deolalikar, 1988). Crop choice can influence nutritional status as was found in cases of agricultural commercialisation (Von Braun and Kennedy, 1986). Land utilisation types (LUTs) are described by the selection of a crop or variety and a set of management/technology attributes of land use (FAO, 1976). To reduce complexity, in this chapter land utilisation types are defined by the main crop cultivated at household level. The best way to investigate the effects of land utilisation types on nutritional status is a longitudinal study (parameter developments are followed over a period of several years). However, during such a study many developments might take place which confound the relationships studied. Therefore, in fast developing countries a cross-sectional study (parameters are measured once) may give more insight than a longitudinal study. The objective of the cross-sectional study described here was to explore the relation between nutritional status, land utilisation types and socio-economic conditions in a fast developing country.

The Future of the Land: Mobilising and Integrating Knowledge for Land Use Options
Edited by L. O. Fresco, L. Stroosnijder, J. Bouma and H. van Keulen.

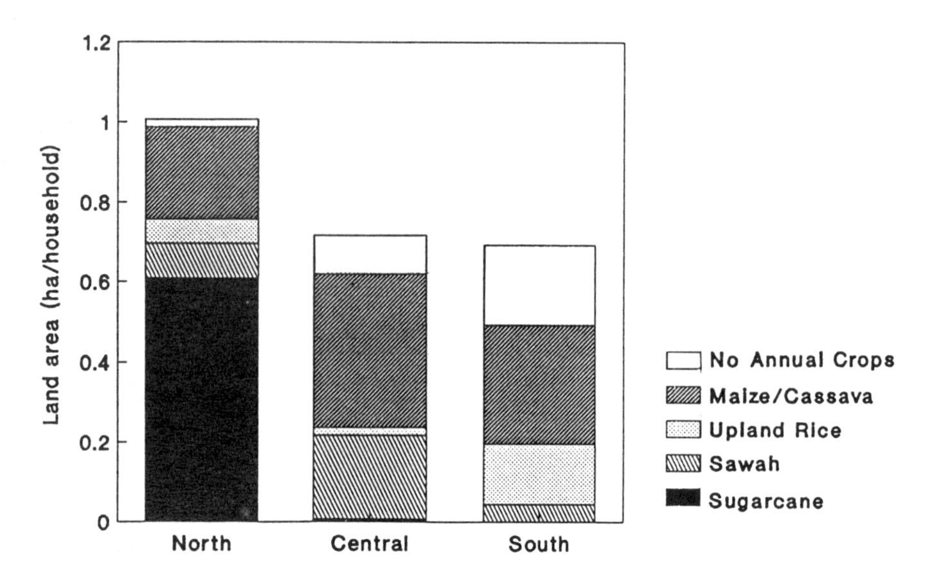

Figure 14.1 Average land utilisation type area per household in the three regions

STUDY AREA

The study was carried out in South Malang district in an upland limestone area that is situated along the south coast of East Java. The area has a tropical monsoon climate. The rainy season lasts from November until May. Altitude varies between 200 and 500 m above sea level. Due to water erosion, steep hill slopes are less fertile than valley bottoms and flat hill tops. Many terraces have been constructed on the hill sides to stop erosion and to give better crop yields. Typically, all annual crops are grown in combination with perennials like coconut, fruit trees (banana, jackfruit, etc.), timber trees (teak, mahogany etc.) and glyricidia. The dominant staple crops in the area are maize and cassava, which are mainly intercropped. Sugar cane is grown in the northern part of the limestone area and often intercropped with maize. Other crops cultivated are upland rice, groundnut and soya. In addition to crop cultivation, many farmers also rear ruminants.

Within the study area three different regions can be distinguished on the basis of land utilisation type and income generating activities: North, Central and South. Land utilisation types in each of these regions is shown in Figure 14.1. The land utilisation type 'no annual crops' comprises land without annuals but it may have perennials. This type of land is often very stony, with shallow soils and steep slopes.

Figure 14.1 shows that the areas cultivated with rice (sawahs or flooded rice and upland rice) in the Central and Southern region are almost equal. This may result from a general desire on the part of the farmers to provide their households with a minimum amount of rice. In the Northern region less rice is grown, presumably because sugar cane provides enough money to buy rice.

Sugar cane was introduced recently and has become an important cash crop. This expansion of sugar cane cultivation into the limestone area has been part of a general shift of sugar cane cultivation from the lowlands to the uplands as a result of a law that was passed by the Indonesian Government as part of the third national five-year plan, stipulating that all irrigated lowlands had to be cultivated with rice. This measure was made to support Indonesia's drive to become self-sufficient in rice production (Edmundson and Edmundson, 1983). Better land quality and suitability, larger parcels, larger farms, a shorter distance to the sugar factory, better infrastructure and a better developed market structure, all possibly explain why sugar cane cultivation has remained restricted to the North.

So, sugar cane dominates in the Northern region, the Central region is dominated by rice cultivation on sawahs, and in the Southern region many soils are at present not suitable for annual crop growing but only for wood production. Given these differences in land utilisation types, the study area lends itself quite well to studying the nutritional effects of different land utilisation types.

METHOD

In the limestone area in the southern part of the Malang district, two villages were selected: Putukrejo and Kedungsalam. Putukrejo is situated in the northern region, the northern part of Kedungsalam in the Central region, and the southern part of Kedungsalam in the Southern region. Twenty-one RT (wards), of on average 25 households each, were selected in these villages, in such a way that they were evenly distributed in space and that all socioeconomic and ethnic groups, Javanese and Madurese, were proportionally represented in the sample. Within the selected RT all households were visited during January to March 1992 for an interview, resulting in a study population of 556 households. All the members of the same households were asked to participate in a weight and height measurement session in their RT in April and May 1992. During the interview information was gathered on household composition, employment, farm size, crops cultivated, animal husbandry, housing facilities, hygienic conditions, health services and the habitual intake frequency of 20 food items.

The nutritional consequences of the present agricultural situation are investigated. The agricultural situation is assumed to be constant over the last several years. Therefore, the nutritional patterns of the last several years are crucial in this chapter. Nutritional patterns are reflected in the nutritional status. The nutritional status of children is used because children are most vulnerable to nutritional insufficiency. The parameter HFA (height-for-age) was chosen because it provides the best information about the nutritional situation over the last several years. The nutritional status of children below 10 years is determined by comparing the measured height of a child with international reference values (NCHS, 1977). For each child a HFA standard deviation score (Z-score) was calculated (WHO, 1983). Where the Z-score of a child is below -2 the child can be classified as being chronically malnourished. The average HFA Z-score for a group of households is used to indicate the average nutritional status of that group.

To study the nutritional consequences of different land utilisation types and different income sources, three steps were taken. First categories of households having different land utilisation types and different income sources were defined. Then, the dietary

Table 14.1 Number of households per income source category, for the three regions

	Income source[a]	North	Central	South	Total
A	dom	21	9	0	30
B	farm + dom	39	51	14	104
C1	farm + ofl	71	1	0	72
C2	farm + lsb	3	19	98	120
D	farm	72	111	47	230
	Total	206	191	159	556

[a]Farm = farming, dom = dominant non-farming income source, ofl = off-farm labour, and lsb = limestone burning.

patterns of these groups were analysed and, lastly, the differences in nutritional status between these groups were examined.

RESULTS

Household categories

Households can be categorised in three ways: on the basis of the region in which they live, source of income and land utilisation type. Farming is the most prevalent source of income. Other sources of income can be dominant income sources or supporting income sources. Examples of dominant income occupations are teacher, shopkeeper, trader, employee and chauffeur. Supporting income sources are off-farm labour (ofl) and limestone burning (lsb). The breakdown of the categories over the regions is shown in Table 14.1.

Nearly all households (95%) were active in farming. However, for 24% of the households (A + B) farming was not the most important source of income. Within the subsample of households without dominant non-farming income sources, the percentages of households with supporting activities for the three regions are respectively 51%, 15% and 68%. The supporting activities off-farm labour and limestone burning are highly area-specific. This is a direct result of the different land utilisation types: off-farm labourers are working on sugar cane fields of large farmers in the Northern region, while limestone burning requires wood and labour which are abundant in the Southern region. Limestone burning is a labour intensive activity in which limestone rocks are gathered and then burned with wood produced or gathered by the household. The lime industry is so lucrative that at the time this study was being conducted wood was being imported into the Southern region. The small scale kilns provide the households with cash. Some people have benefited from hiring out their kilns, enabling them to build larger kilns and employ more people.

Groups A and B have stronger socio-economic positions. In these groups the dominant income sources tend to determine nutritional conditions. Therefore, the groups of special interest for this chapter are C and D. The farmers of groups C and D can be divided on the basis of their land utilisation type. Three land utilisation type groups have been

Table 14.2 Number of households without dominant income source per land utilisation type group and income source

	North		Central		South		
LUT group[a]	fa	fa + ofl	fa	fa + lsb	fa	fa + lsb	Total
1 SUG	33	21	1	0	0	0	55
2 SAW	3	4	47	6	6	11	77
3 MCU	36	49	63	14	41	87	290
Total	72	74	111	20	47	98	422

[a]SUG = sugar cane, SAW = sawah, MCU = maize/cassava and/or upland rice, fa = farming, ofl = off-farm labour, lsb = limestone burning.

formed as is shown in Table 14.2. A farmer was placed in the sugar cane cultivating group if the area cultivated with sugar cane exceeded 0.20 ha. The same area threshold was used for the sawah owning group. When a farmer could be placed in more than one category ($n = 22$), the first category in the below sequence was used. Table 14.2 shows that sugar cane cultivating and sawah owning farmers are less involved in supporting activities than farmers growing maize/cassava and/or upland rice (32% against 52%).

Dietary pattern

Rice is a high prestige food in Indonesia. It is thought to be the best staple food and people aim to include rice in their diets as often as possible. In the study area people mix rice with maize and/or cassava in times of budget stress. Many people have to buy their rice because they produce no or too little rice themselves. Almost all households cultivate maize and cassava to obtain better food security. The staple food consumption pattern is highly seasonal. Households were asked how many months per year they ate a certain 'mix' of staple foods. Although the answers always included rice as a component of the staple food mix, visual checks proved that at times the quantity of rice was not more than a few grains. On average, the rice portion was one-third of the staple food consumption. Table 14.3 shows the staple food consumption pattern by presenting the number of months during which rice and mixtures with rice were eaten. Because of the strong relation of rice consumption with cash availability, the trends shown are as expected. A clearly visible trend is that rice consumption is replaced by rice/cassava consumption for the lower socio-economic groups, while the rice/maize consumption is constant. Staple food consumption patterns of the groups SAW and MCU are almost the same, although a higher rice consumption of the sawah-cultivating households was expected.

Besides staple food, people in the study area consumed a wide range of side dishes. Table 14.4 shows relative consumption frequencies for some selected side dishes (total population mean = 100). The first three columns (tahu/tempe, egg and fish) are foods providing protein, banana is the fruit eaten most frequently, and amaranth the most frequently consumed vegetable. Carrot consumption is shown for its distinct relation of its consumption frequency with the region. The side dish consumption pattern showed

Table 14.3 Number of months that the staple food 'mixes' are consumed for groups in different regions, with different income sources and with different land utilisation type

	Rice	Rice + maize	Rice + cassava
North	5.4	4.9	1.7
Central	3.0	4.1	4.9
South	1.8	4.6	5.6
A dom	8.1	2.9	1.0
B farm + dom	4.0	4.4	3.6
C1 farm + ofl	4.5	5.5	2.0
C2 farm + lsb	1.8	4.6	5.6
D farm	3.4	4.5	4.1
1 SUG	5.1	5.1	1.8
2 SAW	3.3	4.1	4.6
3 MCU	2.7	4.8	4.5

Table 14.4 Relative consumption frequency of several side dishes for groups in different regions, with different income sources and with different land utilisation type. Total population mean = 100. Frequency is presented as average number of consumptions per week

	Tahu/tempe	Egg	Fish	Banana	Carrot	Amaranth
North	110	150	125	130	130	115
Central	100	85	90	75	120	80
South	85	55	80	90	35	105
dom	170	280	90	205	275	115
farm + dom	125	150	120	115	135	95
farm + ofl	70	80	115	95	65	100
farm + lsb	90	50	85	85	55	120
farm	95	85	95	90	95	90
SUG	105	135	125	115	120	115
SAW	95	95	90	125	95	100
MCU	85	60	90	75	75	95
Consumptions/week	2.3	0.7	2.1	0.7	0.5	3.4

significant differences between the groups (Mann-Whitney rank correlation test). Foods that have to be bought like tahu, tempe (soy products) and carrots are eaten more frequently by the upper socio-economic class (groups A and B). Also some locally produced food items (like egg and banana) were consumed more regularly by the group of households with dominant income sources. The consumption frequency of 'common' foods like fish and amaranth was constant over all groups. Noteworthy are the low consumption frequency of tahu/tempe for off-farm labourers and the low consumption frequency of egg and banana for farmers active in limestone burning and those growing maize/cassava and/or upland rice.

Table 14.5 Average nutritional status for income source groups in each region. Values are mean HFA Z-scores of children under 10

Income source	Region			
	North	Central	South	Total mean
A dom	−0.79	−0.90	—	−0.84
B farm + dom	−0.77	−1.20	−2.17	−1.15
C farm + sup	−1.66	−2.00	−1.92	−1.83
D farm	−1.35	−1.60	−2.08	−1.63
Mean per region	−1.35	−1.51	−1.98	−1.59

Farm = farming, dom = dominant non-farming income source, sup = supporting income source.

Table 14.6 Average nutritional status for land use groups in each region. Values are mean HFA Z-scores of children under 10

LUT group	Region						
	North		Central		South		
	fa	fa + ofl	fa	fa + lsb	fa	fa + lsb	Total mean
1 SUG	−1.38	−1.74	—	—	—	—	−1.56
2 SAW	—	—	−1.54	−2.37	−1.98	−1.30	−1.62
3 MCU	−1.32	−1.62	−1.61	−1.81	−2.11	−2.02	−1.79
Region mean	−1.35	−1.66	−1.60	−2.00	−2.08	−1.92	−1.72

—, No data available.

Nutritional status

The differences in dietary patterns are reflected in the nutritional status of the population. Table 14.5 shows the average nutritional status per region and income source. When a dominant income source is present the nutritional status is significantly better (note that category B/South is a small group). The nutritional situation in the South is significantly worse (t-test, $p < 0.01$). The practice of supporting activities apparently does not improve nutritional status. In the North and Central regions the practice of supporting activities is a sign of the need for extra income, while in the South the practice of limestone burning seems to give some relief in nutritional stress generally.

Note that the average socio-economic position in North and Central are better, mainly because of the higher percentage of households with dominant income sources. Supporting activities are practised by households with marginal socio-economic positions. Therefore, the group practising supporting activities in the North and Central regions has a worse nutritional status compared to the average in the same region, while the group practising supporting activities in the South has a somewhat better nutritional status than the average in the South region. Table 14.6 shows that this holds for all land utilisation type groups. This is probably because the profits of limestone burning are on average somewhat larger than of off-farm labour. Income calculations should support these assertions.

Table 14.6 also shows that there is a region dependent difference in nutritional status between land utilisation type groups. For the North and the Central regions the nutritional status is not related to land utilisation type. In the Southern region, however, this relation is clearly visible. When the farmers growing maize/cassava and/or upland rice are divided in small (<0.25 ha) and large (>0.26 ha) farmers, a further distinct difference in nutritional status is found. For the small farmers in the South not owning any sawah the average HFA Z-score was −2.45. For the North and Central regions such a relation with operated area did not occur.

DISCUSSION

The different land utilisation types as found in the small study area with three regions offer the possibility to evaluate the nutritional consequences of different land utilisation types. From a nutritional point of view it can be seen that the best diets are found in the North. This might be a spin-off of sugar cane cultivation. Another possible spin-off of sugar cane cultivation is the higher percentage of households with dominant income sources, who showed by far the best diets, resulting in the highest nutritional status. Among the farmers without dominant income sources, those growing sugar cane showed the best nutritional status.

For farmers without dominant income sources the average HFA Z-score was −1.60, except for the group growing maize/cassava and/or upland rice in the South, with an average score of −2.04. Presumably the farming systems in the North and Central regions provide the household with enough means to have a level of food sufficiency that results in an average nutritional status of -1.60, which is also reached by sawah owning farmers in the South but not by the maize/cassava and/or upland rice growing farmers in the South. Especially, the small farmers of this group are far below this level. Reasons can be a lesser average soil quality and lower use of fertiliser resulting in lower yields, and connected with these, lack of possibilities for profitable cash crops and off-farm activities.

Therefore, only in the South are sawah ownership and the owning of more than 0.25 ha of land determinants of nutritional status. This implies that only a division of households, on the basis of their land utilisation type, is not relevant from a nutritional point of view. Soil quality or, in other words, yield potentials, should also be considered. This means that land use systems (FAO, 1976) would be a better characteristic in relation to nutritional status.

It is difficult to find ways for improvement of the nutritional status of the population. Limestone burning gives some relief for the farmers in the South. Sustainable wood production and better limestone rock transportation facilities need special interest in this area. Because a strong relation exists between the nutritional status and the consumption frequency of tahu, tempe and eggs, the importance of the consumption of these products should be stressed, if cash is available.

ACKNOWLEDGEMENTS

The authors wish to thank Prof Dr J. G. A. J. Hautvast and Dr R. Gross for their guidance and support, Dr J. W. Nibbering for his valuable comments and the INRES-project staff members of the Brawijaya University in Malang for their cooperation, especially during the data collection.

REFERENCES

Deolalikar, A. B., 1988. Do health and nutrition influence labor productivity in agriculture? Econometric results for rural South India. *Review of Economics and Statistics* **70**: 2.

Edmundson, W. C. and Edmundson, S. A., 1983. A decade in village development. *Bulletin of Indonesian Economic Studies* **29**(2):46–59.

FAO, 1976. *A Framework for Land Evaluation*. Soils Bulletin No. 32, FAO, Rome.

FAO/WHO, 1992. *Improving Household Food Security*. International Conference on Nutrition (ICN), Theme paper no. 1, December 1992, FAO, Rome.

NCHS, 1977. National Centre for Health Statistics growth curves for children, birth–18 years, United States. US Department of Health, Education and Welfare, Washington DC.

Von Braun, J. and Kennedy, E., 1986. Commercialisation of subsistence agriculture: income and nutritional effects in developing countries. Working Paper no. 1, IFPRI, Washington, DC.

WHO, 1983. *Measuring Change in Nutritional Status*. WHO, Geneva.

CHAPTER 15

Selected Case Studies at National Level

CHARACTERISATION OF RICE-GROWING AGRO-ECOSYSTEMS IN WEST AFRICA

W. Andriesse,[a] N. Van Duivenbooden,[b] L. O. Fresco,[b] and P. N. Windmeijer[a]
[a]The Winand Staring Centre for Integrated Land, Soil and Water Research, Department of International Cooperation, Wageningen, The Netherlands.
[b]Wageningen Agricultural University, Wageningen, The Netherlands.

Inland valleys are among the most promising rice-growing agro-ecosystems in West Africa for increased wetland rice production. These valleys are the upstream parts of the drainage systems and are characterised by the submerging of their valley bottoms during the rainy season. They occur all over West Africa and their hydromorphic sections cover a total area of about 35 million hectares.

The agro-ecology of West Africa is very heterogenous. Therefore, agro-ecological characterisation of inland valleys is an important tool for the development of improved management technologies and integrated land use planning. For the agro-ecological characterisation of inland valleys, a three-step approach with increasing detail is used: macro-level, semi-detailed and detailed characterisation. Using these steps safeguards the representativity of selected key areas and inland valleys for further research.

The macro-level characterisation has resulted in the description of three broad agro-ecological zones, with different lengths of growing period (FAO definitions): the Equatorial Forest (>270 days), Guinea Savanna (165–270 days) and Sudan Savanna Zone (90–165 days). Rainfall, landscape, soils, population density, income/caput and rice production data are used to subdivide these zones into smaller units.

Representative key areas have been selected in several West African countries, including Côte d'Ivoire, for the semi-detailed characterisation. The latter focuses on the comprehensive description of biophysical and socio-economic parameters and their interrelation. The first results of the characterisation in Côte d'Ivoire show large differences between and within the different agro-ecological zones. In the semi-humid Guinea Savanna Zone the valley bottoms are very narrow (10–50 m) and moist or submerged for only 5–6 months, while in the humid Equatorial Forest Zone the valley bottoms are much broader (65–160 m), and are moist throughout the year. Although

The Future of the Land: Mobilising and Integrating Knowledge for Land Use Options
Edited by L. O. Fresco, L. Stroosnijder, J. Bouma and H. van Keulen.

in the Guinea Savanna Zone the area under cultivation (LUR) per valley is smaller (20–35%) than in the Equatorial Forest Zone (45–80%), the total actual cropping intensity (ACI) shows the same values (25–65%). This is also valid for the individual land subelements. The bottoms of the valleys formed in schists have in the Guinea Savanna Zone a LUR of 30% and an ACI of 40%. In the Equatorial Forest Zone these values are 50% and 30% respectively.

Within the Guinea Savanna Zone the valleys formed in schists are broad (1600–2200 m) and the valley bottoms dry up in the course of the dry period. On the contrary, the valleys formed in granite are less wide (100–1650 m) and the valley bottoms dry up in a later stage, if at all. The main land use parameters are the same for valleys formed in schist or granite.

Further research will focus on hydrology, nutrient cycles, effects of cropping patterns and techniques in representative inland valleys.

ENVIRONMENT VERSUS DEVELOPMENT: A CRITICAL LAND MANAGEMENT ISSUE IN A SMALL LAND ECOSYSTEM OF THE HUMID TROPICS

A. K. Bandyopadhyay
Central Agricultural Research Institute, Port Blair, Andamans, India

Andaman and the Nicobar Group of Islands are a group of 572 islands in the Bay of Bengal. The average rainfall is about 3000 mm. Data collected at a red oil palm plantation at Little Andaman showed loss of soil ranging from 3.5 t/ha/year to 45.78 t/ha/year. If this is allowed to continue, a time will come when the soil will cease to produce any crop. It is estimated that 840, 150, 840, 310 and 325 kg/ha each of N, P, K, Ca and Mg, respectively, are immobilised in an oil palm plantation. Productivity of these land areas is reduced due to acid development because of increased deforestation, neglect of deforested lands, and poor management practices.

Because of high rainfall, chemical pesticides are less useful and runoff to the sea occurs within a few hours after application, causing pollution and resulting in mortality of sea animals including fish. Since there are no pastures or grasslands on these islands and because the cattle and goat population is very high, overgrazing of agricultural lands and forest areas enhances soil erosion which ranges from 5 to 15 t/ha/year of soil.

SUGGESTED LAND USE PLAN

All lands in land capability classes IV to VIII should be kept for the growing of trees and grasses, adopting appropriate technologies and restricting agricultural land to classes III and II and multi-storey plantation agriculture to classes III and IV. Intercropping of forest land as well as coconut plantations with spices, incorporating grass/leguminous crops for soil cover, will help in maintaining soil fertility. Also, loss of soil and nutrients due to erosion and runoff will decrease. Bio-control methods along with integrated pest management practices may be used to save crops as well as the environment from insect pests. Sea farming and development of coastal aquaculture on a large scale will divert the attention of farmers from land-based cropping systems, thus reducing pressures on natural resources.

USING COUNTRYSIDE SURVEY DATA IN AN INFORMATION SYSTEM FOR PLANNING PURPOSES

C. J. Barr, R. G. H. Bunce, D. C. Howard and T. W. Parr
Institute of Terrestrial Ecology, Grange-over-Sands, Cumbria, UK

In 1990, the Institute of Terrestrial Ecology completed the third in a series of countryside surveys. With funding from the United Kingdom Department of the Environment, the objectives of the 1990 countryside survey were: (a) to establish the stock of landscape features and habitats of the countryside of Great Britain (GB) in 1990; (b) by reference to earlier data, identify the change in these resources over the past 10 years and in the previous two decades of 1970 and pre-1960; and (c) to ensure that all survey methodology can be accurately repeated in future years to provide reliable estimates of future change. To meet these objectives, an integrated survey approach was used which brought together data from a variety of sources, all of which could yield data at the same 1 km scale of resolution. The data included land cover for every 1 km square in GB, deduced from satellite imagery, detailed land cover, habitat types and vegetation from field surveys (using a stratified sampling approach), freshwater biota and soils information. The results of the survey are being made available as contract reports and journal publications, but also within a Countryside Information System. This computer-based system uses windows, icons, mice and pull-down menus to allow the user to obtain land use information for any part of GB. Results (including stock and change) of any of the surveyed features can be given for data sets that are based on a 1 km grid and have great potential for use by countryside planners; it is currently being bench-tested within several UK Government Departments and Agencies.

OPERATIONALISING SUSTAINABILITY IN MGP-MODELS FOR LAND USE STUDIES

J. J. E. Bessembinder
Atlantic Zone Programme, CATIE-WAU-MAG, Guápiles, Costa Rica

The possibilities of including aspects of ecological sustainability in multiple goal programming (MGP) models were investigated by studying the literature of some explorative land use studies. The results are to be used in the development of an exploratory MGP model with emphasis on sustainable land use. Three approaches to including sustainability aspects were found:

(1) assuming production with the 'best technical means', the application of technical developments for a higher efficiency of inputs or the use of less harmful chemicals, more sustainable land use systems with a certain technology level (LUSTs) can be formulated as inputs to a MGP model;

(2) restrictions can be imposed on inputs and outputs, the selection of certain LUSTs and the use of certain areas for agricultural purposes; and
(3) sustainability objectives may be formulated relative to the minimisation of certain inputs or outputs, the maximisation of efficiency or the selection of certain LUSTs.

Two examples illustrate these possibilities. In the MALI5 model only LUSTs with a balance between the input and output of nutrients were included. This balance was obtained either by the use of fertilisers or the application of fallow years (Veeneklaas et al., 1991). In the study 'Rural areas in the European Community' LUSTs were defined using the concept of 'best technical means'. Minimisation of the amount of nitrogen losses per hectare and per unit product were included as objectives, and while optimising other objective functions the highest acceptable amounts of nitrogen loss served as restrictions (Netherlands Scientific Council for Government Policy, 1992). By means of objective functions the possibilities for a more sustainable land use can be explored within a MGP model. This exploration takes place within the limits imposed by the LUSTs and the restrictions. Without sustainability objectives these limits will be respected; however, the most sustainable scenario will not be generated automatically. Therefore, to promote the overall sustainability of the resulting land use scenario, concrete objectives leading to sustainability should also be included.

Sustainability depends on time-scales (Fresco and Kroonenberg, 1992). Yet many MGP-models present static scenarios suggesting stable situations. Besides this, interannual variation in, for example nutrient leaching might be unacceptable, although the average leaching is acceptable. Therefore, taking the dynamic aspects into account implies the use of short-term as well as long-term restrictions and/or objectives in multi-period MGP-models.

Until now the range of sustainability aspects included in MGP-models has been very limited. Most attention has been given to nutrients, since nutrient flows are relatively easy to quantify. However, aspects like water use efficiency, physical soil qualities and pollution by chemicals also deserve attention in further research.

REFERENCES

Fresco, L. and Kroonenberg, S., 1992. Time and spatial scales in ecological sustainability. *Land Use Policy* **9**:155–168.

Netherlands Scientific Council for Government Policy, 1992. *Ground for Choices*. SDU, Den Haag, The Netherlands.

Veeneklaas, F. R., Cissé, S., Gosseye, P. A., Van Duivenbooden, N. and Van Keulen, H., 1991. *Development Scenarios*. CABO-DLO/ESPR, Wageningen.

AGRICULTURE AND SPATIAL ORGANISATION IN THE NETHERLANDS

H. Hetsen and M. Hidding
Wageningen Agricultural University, Wageningen, The Netherlands

The main objective of this study is to explore future options for spatial development of rural areas in the Netherlands. It especially focuses on regional differences in agricultural development related to differences in urban and regional economic development and physical potentials for nature conservation and drinking water supply.

The study includes an analysis of developments in the above-mentioned fields since 1950 and a reconnaissance towards 2010. The analysis shows that three types of problems can be distinguished:

- urban–agrarian congestion, which especially occurs in the central zone;
- problems in the relationship between agriculture and regional economic development, which become most manifest in the peripheral zone; and
- degradation of environmental quality as well as of potentials for different forms of land use, which is most serious in the higher parts of the Netherlands.

Within the frame of the reconnaissance, three options for future spatial development have been elaborated, with specific attention to the northern part of the Netherlands:

(1) reallocation of the (growth of) agricultural production to the north;
(2) the north as an experimental area for a clean and integrated agriculture;
(3) reallocation of agricultural production, safeguarding nature conservation and drinking water resources.

From the reconnaissance it can be concluded that solutions for the above-mentioned problems are conflicting in different respects. The conflicts become acute with respect to the reallocation of production to the north. It is questionable if reallocation is beneficial with respect to environmental quality, nature conservation and the drinking water supply in the north.

LAND TREATMENT APPROACH FOR SUSTAINABLE AGRICULTURAL DEVELOPMENT IN ST LUCIA

K. M. Severin,[a] L. Chitoli,[a] Z. Alikhani[b] and C. A. Madramootoo[b]
[a]Mabouya Valley Development Project, Ministry of Agriculture, Land, Fisheries and Forestry, St Lucia. [b]McGill University, Montreal, Canada

St Lucia is a small island country (616 km^2) in the West Indies with an agriculturally based economy. Banana is the principal crop and is mainly produced by farmers owning less than 4 ha, predominantly on steep slopes. The high level of landlessness, exacerbated by growing tourism, construction and manufacturing industries, has resulted in extensive squatting on reserve forest lands.

It was the Government of St Lucia's concern that squatters could cause irreparable damage to the natural environment and thus the Mabouya Valley Development Project (MVDP) was founded. The MVDP, financed jointly by the Government of St Lucia and the European Development Fund, has the following agricultural development objectives:

(1) to provide awareness among farmers of the fragility of the soil resource in the tropics; to introduce them to a regime based on land characteristics and adopt best management practices for hillside farming;

(2) to employ an integrated approach to rural development by involving farmers in the decision-making process;
(3) to encourage farmers to participate in a crop diversification programme to ensure sustainable agricultural development;
(4) to facilitate rural settlement by augmenting community development services, such as more reliable water, roads, and solid waste management.

Since its inception in 1989, the MVDP has been instrumental in eliminating slash and burn practice in Mabouya Valley and has helped 121 families with permanent settlement. In the second phase of this project, as part of the whole effort to protect the natural resources of the Valley, the activities will extend beyond the government-controlled land to include the entire water catchment area.

IN SEARCH OF THE OPTIMUM SCALE OF LAND RESOURCE MAPS

J. J. Stoorvogel and A. Nieuwenhuyse
Atlantic Zone Programme, CATIE-WAU-MAG, Guápiles, Costa Rica

Land use planning projects normally collect information on a number of land properties which are presented on one scale. However, land properties may not be equally suitable for a certain map scale. Quantitative thematic information, the semivariogram, is often used to get a more detailed insight into the workability of different map scales. For thematic and qualitative data one may consider using the spatial-difference-probability function for a similar analysis. Both methods of analysis are suitable for thematic information (e.g. drainage and rainfall). For the determination of the optimal map scale for a soil map with its complete set of soil variables these techniques are less useful, and it tends to be the funds, time and future use of the map that determine the scale. Nevertheless, detailed scales do not always provide more accurate information. Often large mapping units can be found, even on detailed scales, simply due to the occurrences of complexes or associations which at that specific scale are not mappable. Two independently produced soil maps of parts of the Atlantic Zone of Costa Rica were compared. The northern part of the Atlantic Zone (5400 km^2) has been mapped at a scale of 1:150 000. A smaller pilot area (11.5 km^2) has been mapped at a scale of 1:10 000. The two maps differ especially in the number of associations and the level of detail in the legend. On the basis of these two maps, two other maps on intermediate scales were generated and evaluated using the above-mentioned criteria of production costs (time and funds) and possible use. The spatial-difference-probability function for soil type which has been derived on the basis of the 1:10 000 soil map results in an optimum but, nevertheless, different map scale. An index of maximum reduction was calculated on the basis of the average size of the delineated areas on the map.

Land resource maps are often made on standard map scales. It is concluded that several techniques are available to improve the choice of map scale. Small pilot areas of available detailed soil surveys may form the basis for such analyses.

SOCOX: A SOIL CONSERVATION EXPERT SYSTEM

A. C. Vlaanderen
TAUW Infra Consult, Lochem, The Netherlands

INTRODUCTION

The collection and dissemination of knowledge and information is one of the key activities of the Food and Agriculture Organization of the United Nations (FAO). The Land and Water Development Division of FAO is making a wealth of information and data available through, for instance, its Soils and Irrigation Bulletins, the Agro Ecological Zones Database, its Soil Conservation Notes etc. One of the more recent initiatives of the Division is the development of a soil conservation expert system (SOCOX) to support field staff in the selection of appropriate soil conservation measures.

Expert systems, also called knowledge-based systems, are user-friendly computer applications in which general problem-solving knowledge of experts has been stored. When a new problem occurs, the system is able to apply the expert knowledge and to present suggestions as to how that particular problem may be tackled. The Land and Water Development Division decided to develop such an expert system, because it realised that knowledge and experience gained in field projects is not systematically made available outside the project and that with the increasing availability of microcomputers in the field, new ways of supporting field projects are being created. The reason for concentrating on the selection of soil conservation measures and not, for instance, on the analysis of soil erosion problems was that knowledge and experience about soil conservation measures are thought to be more concrete and easier to identify. It was decided to concentrate the expert system on measures at the farmer and village level and to start with problems in the humid tropics. The primary end-users of the system are supposed to be district-level subject matter specialists, who may consult the expert system a few times a year to find answers for the problems encountered by themselves or by more junior staff they are advising.

Although the Land and Water Development Division already has a considerable experience in developing databases, the development of an expert system is a new undertaking. Because expert systems are concerned with (subjective) knowledge, rather than with (objective) data and because an expert system has to be used by and be useful to many different soil conservation experts in a wide range of situations, it was felt that a thorough preparation period would be necessary before the software could be written. During this preparation period a great number of discussions between FAO staff, soil conservation experts, field staff and expert system specialists was held to identify the objectives and requirements for the envisaged expert system. To show the various software possibilities, a demonstration model was developed based on an expert system shell and a description of the envisaged project was written. The discussions on the system's objectives and the development of the demonstration model enabled FAO staff and potential end-users to get a thorough understanding of the advantages and disadvantages of the envisaged expert system and to identify the requirements and concepts on the basis of which software can be selected and designed.

CONCEPTUAL DESIGN OF SOCOX

SOCOX will consist of four complementary facilities: a database with information on soil management and conservation measures; an indexed collection of field experiences; a diagnostic knowledge model with criteria for decision-making; and a decision support procedure providing advice on the appropriateness of soil management and conservation measures under given circumstances.

An important feature of the expert system will be its interaction with the end-user, who will not only be able to comment on the advice given, but will also be able to add specifications that are meaningful in his situation, thus calibrating the system to his specific needs.

The selection of appropriate measures proceeds through three steps:

(1) The system requests information from the user about the erosion problem he is confronted with and the soil conservation objectives he is aiming for (for instance, whether he is looking for measures to improve soil cover, or rather for measures to improve soil retention). Based on this information, the knowledge model will then select soil conservation measures that are capable of achieving the defined objective.
(2) The system requests information from the user about site conditions, to determine whether selected measures are technically feasible in the particular circumstances. As the expert system contains information about minimum requirements of the conservation measures, the system will select only those measures that are feasible at that particular site.
(3) The user is asked to indicate the major farmer or village constraints and potentials. The expert system is able to select those conservation measures that are likely to be acceptable.

FEEDBACK FROM THE END-USER

SOCOX is designed to facilitate the collection and exchange of experiences among soil conservation experts. Therefore, it is considered essential to enable the user to give feedback on the advice that is being offered. SOCOX shows the user which measures have been rejected or selected, and how its advice was arrived at. The user is given the opportunity to indicate whether he agrees or does not agree with the advice or the selection steps given, and to indicate what measures would, in his opinion be more appropriate.

The user has several options to do this. First, he is offered the opportunity to go back into the reasoning process and change the information he provided earlier. This may change the end result and may better fit his situation. Secondly, the user may, according to a format provided by the program, indicate where he would have departed from decisions made by the expert system. In this way, the user is able to indicate that, for his situation, changes or additions should be made to the arguments the expert system has been using. Thirdly, the user may describe, in his own words, his particular problem and his solution. This case study will be stored separately from the expert system in a special file, but will be saved together with the expert system, so that it becomes available to all users of SOCOX.

TECHNICAL DESIGN

The SOCOX expert system will consist of a knowledge base (containing the diagnostic model); a relational database (with descriptions of soil conservation measures and field experiences); an inference mechanism (to consult the knowledge base and provide the advice); and a user-friendly interface. Hypertext files will be used to offer additional flexibility in information retrieval. Hypertext links related concepts, logic or procedures in different files. Thus, hypertext offers the possibility to store large quantities of data, while letting the end-user decide what information is relevant and what has to be consulted in his particular case. The expert system will be developed for IBM-compatible microcomputers. Future use of CD-ROM may be envisaged. The expert system will be built in separate modules, to facilitate maintenance and updating.

ENVISAGED CONTRIBUTORS TO SOCOX

The core knowledge in the knowledge base will be acquired through interviews with one or two renowned soil conservation experts. A technical expert group will expand this knowledge, providing research results, personal knowledge and project experiences. In addition, a selected group of field technicians, with experience in a variety of erosion problems and soil conservation solutions, will make their experiences and knowledge available. Under the umbrella of FAO, who would be expected to be responsible for development and maintenance of the expert system, SOCOX will draw upon the knowledge and experiences of a large number of individuals, projects and programmes all over the world. Contacts have been made with the National Soil and Water Conservation Service of Costa Rica, the Department of Agriculture, Bureau of Soils and Water Management of the Philippines and the Wageningen Agricultural University. The SOCOX expert system will be linked with the World Overview of Conservation Approaches and Technologies (WOCAT) program that is currently being set up by the World Association of Soil and Water Conservation (USA).

THE COST OF AGRICULTURAL LAND PRESERVATION

W. van Vuuren and S. Sappong
University of Guelph, Guelph, Ontario, Canada

Urbanisation in North America is absorbing a disproportionately high percentage of prime agricultural land. Food security concern over this loss of high quality land has resulted in most states and provinces adopting some form of agricultural land preservation programme. Economic analysis has been scant in the preservation debate and programmes.

Prime land preservation usually implies diverting urbanisation from prime to lower grade agricultural soils. In most cases such diversion entails costs. Urbanisation of either site is associated with sacrificing net revenues from current uses and sacrificing possible environmental and aesthetic amenities as well as conversion costs. Moreover, both sites may be located at different distances from the Central Business District, implying a difference in urban transportation costs. Various costs are site specific, such as aesthetic and environmental sacrifices as well as urban transportation costs. Other costs associated with urbanisation do depend on soil quality such as agricultural rent foregone and the cost of converting the land from agricultural use to making it ripe for building purposes. Quality-specific costs loom large in land conversion and are prominent in determining the siting of urban development under free market conditions.

The quality of agricultural land in Canada is rated into seven classes: class I being prime agricultural land and class VII unsuitable for agriculture. The classification is based on a variety of soil conditions such as topography and stoniness. Each land class is determined by the degree of total limitations and hazards to agriculture. Topography, stoniness, drainage and depth to bedrock not only affect agricultural production but also conversion costs. The higher the soil class, the higher the cost for levelling, filling, draining, excavating, and installing water and sewage pipes.

In a case study for a municipality in southern Ontario, soil quality specific costs associated with urbanisation were considerably higher on low quality soils than on prime quality soils. Diverting urbanisation from class I to class IV land resulted in approximately \$150 000 per ha additional costs, all related to soil quality specific costs. This shows that preservation can be highly costly, unless offset by site-specific benefits. Whether or not the land to be preserved is needed for future food production is highly uncertain under currently foreseeable conditions. Preservation retains the option to use the land for future food production, if needed. However, providing this option entails a cost. If the cost is reasonable, preservation might be a worthwhile objective to pursue. However, in the above case study the price appears to be excessive. Even if food becomes scarce in the future, it is highly unlikely that prime agricultural land would fetch a price of \$150 000 per ha in agricultural use.

PART IV

REGIONAL LEVEL LAND USE PLANNING

CHAPTER 16

Sustainable Land Use Planning in Costa Rica: A Methodological Case Study on Farm and Regional Level

R. Alfaro,[a] J. Bouma,[b] L. O. Fresco,[b] D. M. Jansen,[a] S. B. Kroonenberg,[b] A. C. J. van Leeuwen,[a] R. A. Schipper,[b] R. J. Sevenhuysen,[a] J. J. Stoorvogel[a] and V. Watson[c]

[a]Programa Zona Atlántica, Estación Experimental Los Diamantes, Guápiles, Costa Rica.
[b]Wageningen Agricultural University, Wageningen, The Netherlands.
[c]Centro Científico Tropical, San José, Costa Rica

INTRODUCTION

Land use planning coming of age

Different methods have been used to derive land use plans. Two related methods are land evaluation and farming system analysis. *Land evaluation* (LE) originated within the soil science discipline. Based on a series of selected land qualities, land suitabilities are defined for well-described land utilisation types including information on social and economic conditions affecting production. Land evaluation is based on characteristic land units, distinguished on soil maps, and climatic data. Defined suitabilities for actual and potential conditions are rather rigid and do not reflect the range of possibilities presented by alternative forms of soil management to be selected by the land user. Another method with an agronomic and economic background is *farming system analysis* (FSA) which includes a detailed analysis of the social and economic structure of farmers' households, but pays little attention to the natural variability of the land and geo-referenced information. Moreover, land evaluation is often used on a regional scale and farming system analysis on a farm level.

While both approaches have merits of their own and are to some extent complementary, there is little integration because each belongs to a different school of thought. Integration by a multi-disciplinary team resulted in a new framework, called

The Future of the Land: Mobilising and Integrating Knowledge for Land Use Options
Edited by L. O. Fresco, L. Stroosnijder, J. Bouma and H. van Keulen.

LEFSA (Fresco et al., 1990). Realisation of the LEFSA approach has been boosted by the development of new tools, of which the following are the most important:

(1) Development of operational simulation models for crop growth, allowing predictions of both actual and potential yields that solely depend on clearly defined environmental conditions. This eliminates the confounding and often obscure effects of farmer-specific management on actual crop production, as found through surveys. Moreover, yield data from farms were hard to obtain in the past, and conditions under which crops were grown were often poorly defined. Properly validated models allow answers to the 'what if' questions that are common in modern land use planning. For each land unit a series of options can be defined in terms of yields under different forms of management.
(2) Refining of remote sensing techniques to characterise land use patterns and crop conditions. Patterns obtained tend to vary in space and time, and provide a more realistic and dynamic image for land behaviour than was possible using static soil and crop maps which were by necessity based on gross generalisations.
(3) Use of interactive multiple goal linear programming techniques (MGLP) which allow formulation of different land use scenarios on the basis of selected socio-economic boundary conditions. For each land unit, different land use systems and associated production levels are defined as a function of different forms of management. Land use patterns in an area, be it a farm or a region, depend on priorities being formulated in the MGLP. Thus, different interests can be balanced in a quantitative manner providing a rational basis for land use planning.
(4) Development of geographic information systems (GISs) which can rapidly produce maps of land uses as a function of management or economic conditions, derived by modelling and application of MGLP models. Geo-referencing has proved to be important for communicating research results to users.

Sustainability

The need for an integrated approach is pressing, as it is impossible to envisage sustainable land use without encompassing all the different factors that influence sustainability. The term 'sustainability' is generally used to indicate the limits placed on the use of ecosystems by humans, or more specifically the way in which resources can be used to meet changing future needs without undermining the natural resource base (cf. Fresco and Kroonenberg, 1992). Sustainable land use is related to the concept of sustainable development. The research programme adopted a working definition of sustainable development by Pearce and Turner (1990): 'it involves maximising the net benefits of economic development, subject to maintaining the services and quality of natural resources over time'. According to Pearce and Turner, maintaining the services and quality of the stock of resources over time implies acceptance of the following rules: '(a) utilise renewable resources at rates less than or equal to the natural rate at which they regenerate, and keep waste flows to the environment at or below the assimilative capacity of the environment'; and '(b) optimise the efficiency with which non-renewable resources are used, subject to substitutability between resources and technical progress'.

Sustainability is dependent on time and spatial scales (Fresco and Kroonenberg, 1992). Therefore, different sustainability criteria have to be envisaged for different scales of observation (farm, regional and global levels).

The Atlantic Zone Programme in Costa Rica

The Costa Rican case study presented here is based on a unique combination and integration of the four techniques discussed above, enabling portrayal of different land use scenarios on computer-generated maps for users at farm and sub-regional levels. The Atlantic Zone Programme in Costa Rica aims to develop a methodology for analysis and planning of sustainable land use to support policy-making at regional and farm levels. In the programme the Wageningen Agricultural University cooperates with CATIE, the Centre for Research and Education of Tropical Agriculture in Central America, and MAG, the Costa Rican Ministry for Agriculture and Livestock. The methodology integrates the ecological and economic dimensions of sustainability in a three-level approach:

(1) the plant–soil level, in which the demands of the main crops of the area are analysed in relation to the soil resources in terms of actual, limited and potential production;
(2) the farm system level, in which the economic and agronomic decisions of farm management are evaluated;
(3) the regional level, analysing boundary conditions beyond the farm level, such as employment and the marketing system of products, but also ecological events like flooding risks.

THE COSTA RICAN SETTING

Costa Rica

Costa Rica, the second smallest country of Central America, has an area of 51 000 km^2 and is bordered by Nicaragua, Panama, the Pacific Ocean and the Caribbean Sea (Figure 16.1). It has great potential, but at the same time the country presents complex limitations with respect to land use planning.

Costa Rica has more ecological variety than all of Western Europe (Halpin et al., 1991). It possesses 12 life zones (Tosi, 1969) against about 10 in Western Europe. Climate ranges from humid tropical via temperate mountain to semi-arid within a distance of about 120 km. This climatic diversity is related to the high mountain range (up to 3820 m above sea level) that divides the country into two zones: the Atlantic and the Pacific. Nearly 69% of the country's soils are on lands of more than 30% slope, 17% between 8 and 30% slope, and only 14% on lands with less than 8% slope, which do not have a major risk of erosion (Vásquez, 1989). However, many of the mountain soils (approximately 30%) are recent soils, derived from volcanic ashes, with a moderate to high fertility.

The economy of the country depends to a large extent on agriculture, tourism and small industries. The present population amounts to about three million and the

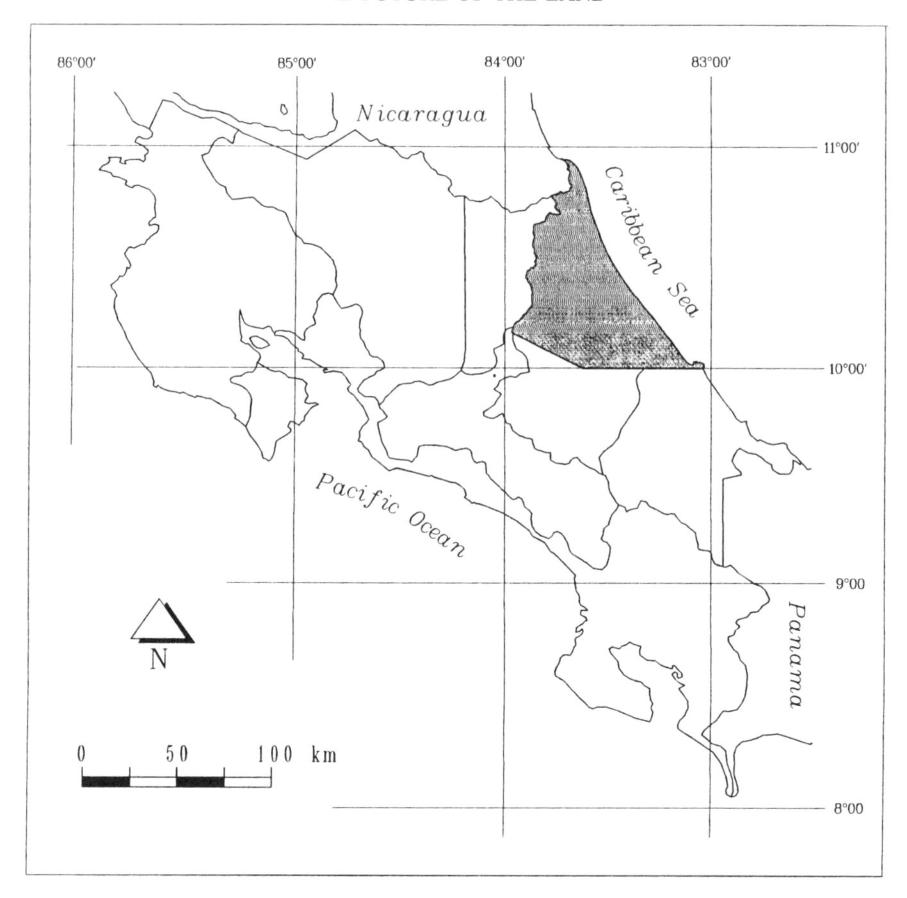

Figure 16.1 The location of the study area (shaded region) in Costa Rica

demographic growth rate is 2.6%. Indices for education, health and economy are high for a developing country. The schooling rate is 94%, life expectancy at birth is 75 years, similar to that of many developed countries, and per capita gross income is $2600 per year. Structural Adjustment Plans were established to make industries more aware of markets, to stimulate producers to be more efficient and at the same time produce more valuable products. Ecological aspects are increasingly considered at all levels of society and eco-tourism is developing into an important source of income.

Agricultural lands are comparatively scarce in view of population growth. While only 30% of its land surface has an agricultural land use potential (WRI-TSC, 1991), the country must feed a population of more than twice the current size within the next 25 years. This will force producers to increase agricultural land productivity.

The discrepancy between actual land use and land capability has led to considerable land degradation. According to WRI-TSC (1991), more than a million hectares of originally forested lands have lost their original vegetation cover. Here, approximately 85% of Costa Rican soil erosion occurs. The economic cost of deforestation, expressed in loss of timber, nutrients and fertile soil, varied between 3 and 7% of the gross national product for the period 1970–1989. Furthermore, in many areas low income from agricultural production, commonly some $100 per ha per year, precludes implementation of soil conservation measures. On the other hand, horticulture on high volcanic lands can yield up to $8000 per ha per year (all expressions are given in US$).

The Atlantic Zone

With the increase of population in recent decades, a strong decrease of natural forest occurred: 67% of the country was covered with forest in 1940; in 1983 only 17% remained (Sader and Joyce, 1988). The same trend occurred in the northern part of the Atlantic Zone, the area where the present case study has been carried out. In this area, which covers 530 000 ha, the natural forest diminished by nearly 70% in the same period.

The present land use in the Atlantic Zone can be characterised as follows: colonisation (including urbanisation) by small- and medium-size farmers covers 45% of the surface area of the Atlantic Zone, large rangelands 15%, estates for export products 10% (bananas, ornamentals), and the remaining 30% forest. Recent settling took place mainly along the roads and railways made for the production of bananas and wood logging (Veldkamp et al., 1992), and was inspired by the decline of agricultural productivity in other parts of the country, especially in the semi-arid Guanacaste province. Indeed, the Atlantic Zone offers comparative advantages for agricultural production: a humid tropical climate with a well-distributed rainfall surplus over evapotranspiration in all months of the year and large areas of fertile soils of recent volcanic origin, situated in a rather flat alluvial plain. These conditions enable the production of a large variety of crops ranging from food crops to cash crops. Main food crops in the area are maize, cassava, plantain and pineapple. The highly commercialised cash crops include bananas, palm heart and ornamental plants. In addition, silvopastoral systems for beef production and forestry for timber are found. There are, however, also inherent problems on the ecological side: high groundwater-tables, rapid leaching of nutrients and fertiliser, and high requirements for weed and disease control. A considerable area in the Atlantic Zone (20%) is set aside as national parks or forest reserves in which eco-tourism is being developed (Wielemaker and Kroonenberg, 1992).

On the economic side, high costs for inputs and investment and a limited demand for agricultural products in the (small) national market result in specific marketing problems. Also, the absence of a well-developed social structure in the new settlements hampers continuity and efficiency of production, leading to over-exploitation of resources and widespread land speculation. The Costa Rican Government is well aware of these problems and has issued several regulating measures, including a land settlement law and a (de)forestation law, which are implemented by specialised institutes. At the same time there is a growing concern, mainly among the urban population, regarding loss of biodiversity and sustainable use of natural resources in general. Since 1986, land use has been regulated on the basis of a classification of land capabilities, and sustainability has become an important aspect of agricultural policy-making.

THE USTED SYSTEM

Introduction

USTED (Uso Sostenible de Tierras En el Desarrollo; Sustainable Use of Land under Development) is a methodology for analysing and planning sustainable land use. The methodology comprises the collection, processing and analysis of relevant land use information, as well as the incorporation of this information in an integrated multi-disciplinary model. In its present state, USTED may serve as an aid in land use planning at a sub-regional level, which in Costa Rica refers to settlements or small districts. If data are available, they might also be applied to higher scale levels. So far, we have studied the Neguev settlement (4675 ha) in the southern part of the study area (Figure 16.3).

LUSTs (Land Use System and Technology) are the basic units of analysis. A LUST is defined as a specific combination of a land unit with a land utilisation type together with a well-defined technology. (This terminology differs from the FAO and Land Evaluation procedures, where a LUT also specifies the technology.) Each LUST describes a unique, quantitative relation between physical inputs and outputs. This relation is determined by various factors like soil type, weather, effect of management practices on crop/animal performance, etc. Given objectives and available resources, farm decision-makers will select LUSTs for actual implementation. Farms differ significantly in their potential for agricultural production, which is reflected in the farmers' decisions. To account for these differences, farms were classified according to their size and soil types.

Sustainability is defined on the basis of a number of theoretical considerations and current and location-specific land use problems. Using the working definition for sustainability given in the Introduction (p. 184) (optimal efficiency in the use of non-renewable resources and the use of renewable resources at a rate lower than or equal to the natural replenishment), the nutrient balance and the use of biocides are considered sustainability criteria for the Atlantic Zone of Costa Rica.

Figure 16.2 gives an overview of the methodology, starting with data collection. Basic information necessary for the LUST descriptions, farm typology and the validation of the different models is obtained from a number of surveys and literature. Geographic data are stored and manipulated via a GIS. For the storage of LUSTs and the calculation of input parameters, *LUSTPZA* (LUSTs at the 'Programa Zona Atlántica') has been developed. LUSTPZA calculates the sustainability parameters and other attributes for the different LUSTs. It then creates input files for the linear programming model. The definition of each scenario is related to a number of coefficients in the linear programming model. Changes in attributes (like prices), boundary conditions or goals will each result in separate scenarios. LUSTPZA functions as a connection between data collection and the linear programming model. The results of the linear programming model will return subsequently to the GIS, where they can be displayed and interpreted.

The methodology described in this section is operationalised using several software packages; among others, PC Arc/Info version 3.4D (ESRI, 1990) for the GIS operations

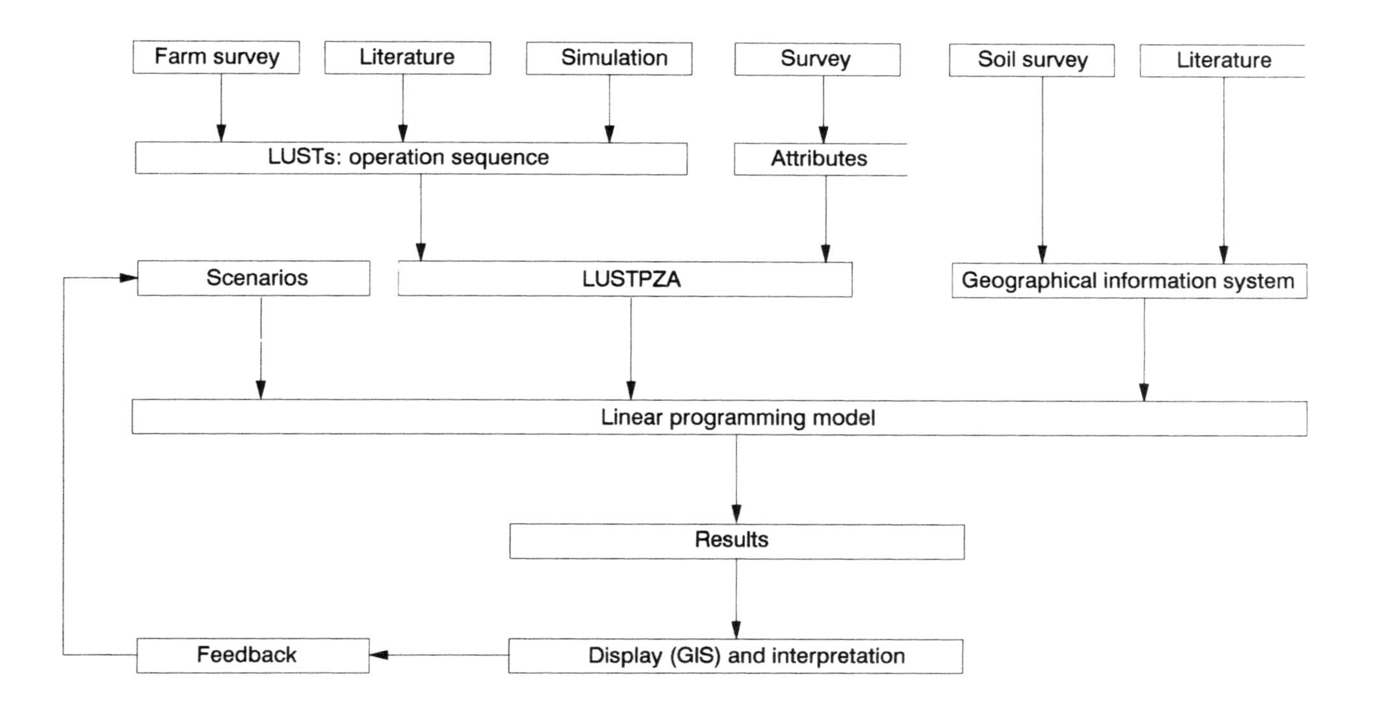

Figure 16.2 The set-up of USTED

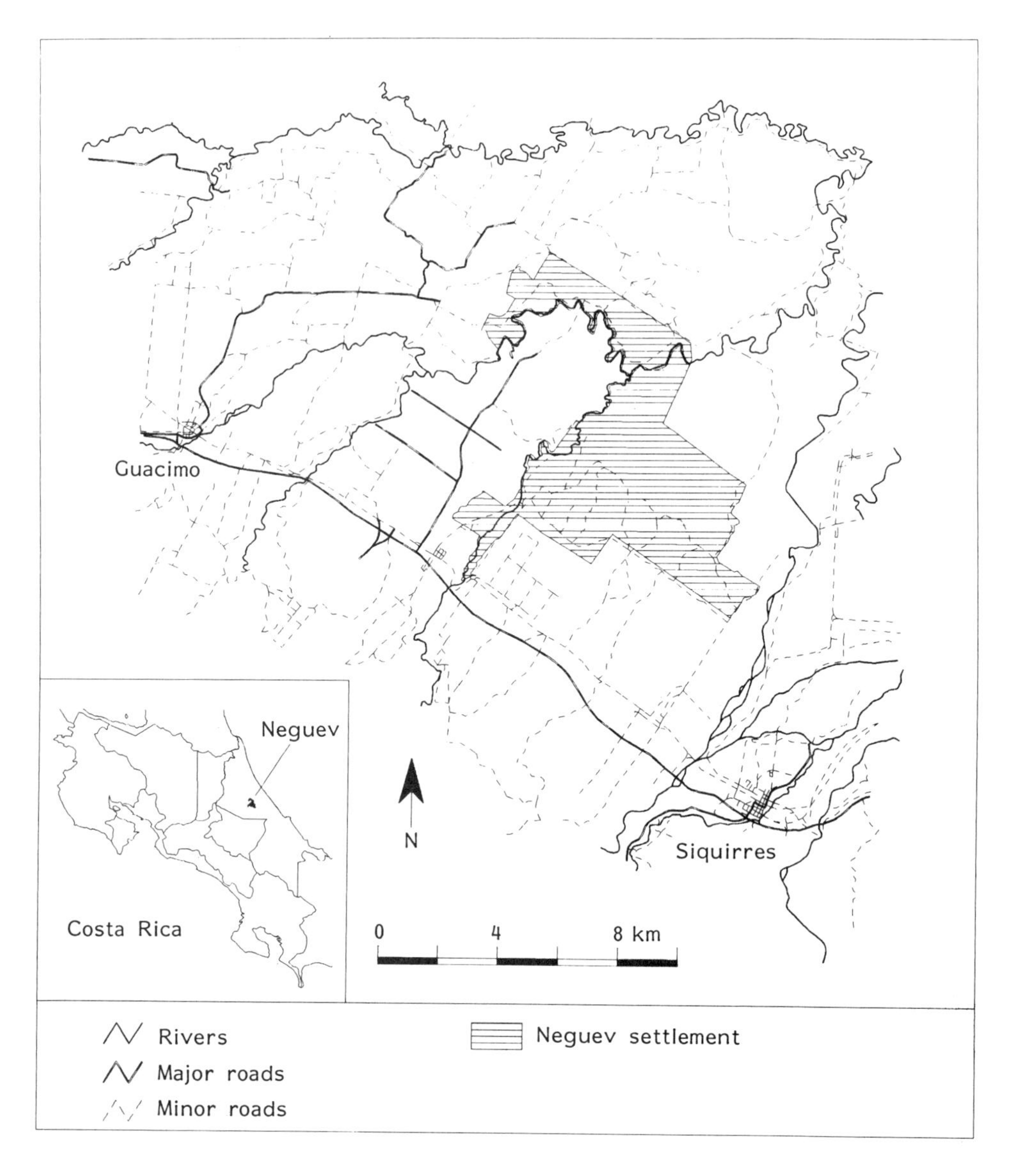

Figure 16.3 The Neguev settlement

and OMP Version 4 (Beijers and Partners, 1992) for the linear programming models. However, the link between these software packages is based on LUSTPZA, a unique set of computer programs, developed at the Atlantic Zone Programme.

LUST description

Farmers have different options to use a particular soil type on their farm. The linear programming model indicates which combination of these options results in optimal

Table 16.1 Combinations of soil groups and land utilisation types distinguished in the Atlantic Zone Programme

	Major soil groups		
	Fertile well drained	Fertile poorly drained	Unfertile well drained
Maize	X		
Cassava	X		
Pineapple			X
Palm-heart	X		X
Plantain	X		
Pasture	X	X	X
Forestry	X	X	X

use of the soils on a farm, taking into account the goals to be achieved and the constraints under which a farm has to operate. In order to enable use of the linear programming model under different conditions, the LUST description needs to include the items in which the goals and the constraints are expressed. One of the essential points in the methodology is the link between the different, mostly disciplinary, databases. Each LUST is identified by a land unit (soil type), land utilisation type and technology level. In the database for the Neguev, 120 different soil types and eight different land utilisation types are identified. Several technology levels for each land use system (LUS) are needed to provide alternatives to the linear programming model. In order to facilitate methodology development, the 25 soil types are grouped into three large soil groups, which are relatively homogeneous. Only the relevant combinations of land use and soil type are taken into consideration, as shown in Table 16.1.

This consideration leads to LUST descriptions that include a chronological operation sequence (Stomph and Fresco, 1991), such as soil preparation, sowing/planting, care and harvesting. For each operation, the amount of inputs is indicated, e.g. labour, biocides, equipment, other materials and, when applicable, the amount of output produced. A simplified LUST is shown in Table 16.2. Different LUSTs may follow the same operations and use the same type but different quantities of inputs, or may produce the same type but different quantities of output.

Attributes, such as prices, nutrient contents, or toxicity, are not included in these LUST descriptions. Instead they are stored in attribute files, that can be invoked by the LUSTPZA program and the GIS. To facilitate cross-referencing between LUST and attribute files, unique identifiers are used for each operation or input/output item.

Field experiments and expert knowledge, combined with crop-growth simulation models, yield LUSTs that are not *a priori* influenced by the current goals and constraints. Instead they indicate crop production potentials. Here three levels are distinguished: potential production when no physical limitations are present; water-limited production, as determined by limited availability or excess of water, which is related to weather and soil physical characteristics; and nutrient-limited production, as determined by soil chemical characteristics. Description of these 'potential' LUSTs, enables evaluation of scenarios that incorporate future technical developments.

Table 16.2 A simplified LUST for the cultivation of maize

SFWLZM01
A LUST of maize on a fertile well-drained soil

030693									(Date of sowing)
1									(Soil type)
1									(Slope class)
Year	Operation	Date	Duration	Traction	Equipment	Materials code	Quantity	Unit	
1	1	− 10	35	6013	3456				Pre-planting
1	2	0	40			5542	8	7	Sowing
1	4	15	20			3452	9	7	Fertilising
1	8	120	60			6210	5000	7	Harvest

Data sources

The Land and Soil Geographical Information System (SIESTA)

A reconnaissance survey of land and soil units was carried out using interpretation of aerial photographs (IR 1 : 80 000, 1984; B&W 1 : 35 000, 1981) followed by extensive field surveys. Photographs (1 : 10 000, 1989) were used for detailed surveys in pilot areas. Standard soil physical and chemical laboratory analyses have been carried out for all major soil types. The data are stored in an Arc/Info-based Geographical Information System called SIESTA (Sistema de Información y Evaluación de Suelos y Tierras del Atlántico).

Mapping units are described by specific combinations of terrain units, the smallest survey units. Each terrain unit is characterised by 10 terrain properties like geology, physiography, landform, parent material, slope gradient, subsurface stoniness and soil. The database structure is such that thematic maps, interpretations and links with other databases can be made easily (Wielemaker and Oosterom, 1992; Oosterom et al., 1992; Krabbe, 1993). The land properties define the constraints for crop growth, incorporated in simulation models. The natural boundaries between the mapping units form the boundaries of the different land use systems. Therefore, these boundaries, together with the limits of farms, also appear in the ultimate scenario maps (Figure 16.4).

Crop and animal production

Crop and animal production are described in terms of LUSTs. Data collection focused on two types of LUSTs: LUSTs currently found in the Atlantic Zone of Costa Rica and potential LUSTs, defined on the basis of expert systems and crop-growth simulation. To define actual LUSTs, data were collected in two steps. First, a general farm survey in three pilot areas was carried out. Secondly, an intensive two-year survey on a limited number of farms was carried out for the definition of operation sequences for the different LUSTS. In addition, specific crop-oriented studies were carried out to obtain quantitative data on, for example, nutrient concentrations and management. For the definition of potential LUSTs fertiliser experiments were carried out for maize, palm heart and pasture. On the basis of these experiments and expert knowledge, potential

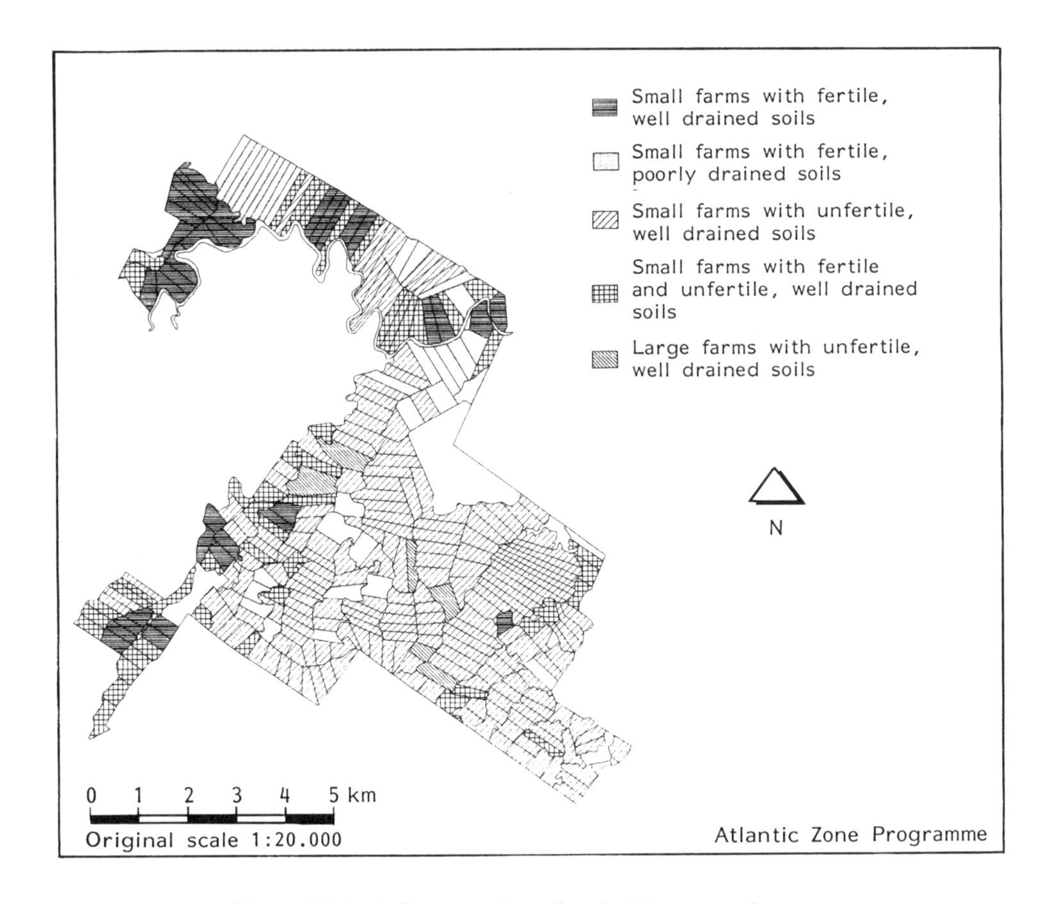

Figure 16.4 A farm typology for the Neguev settlement

LUSTs were defined. At that stage a validated crop-growth simulation model was available only for maize (MACROS, Penning de Vries et al., 1989), which was used to define the potential LUSTs. For the other crops the operation sequences were based on expert knowledge.

A large variation in yield-determining factors is found at the (sub-)regional level. The present LUST descriptions are valid for average regional conditions and do not take into account this variation. For the use of these LUSTs at larger scales, one may need a more detailed LUST description to deal with this variability: LUSTPZA analyses the selected LUSTs and exports information on costs of inputs, production and prices of products, labour demand, nutrient balance and biocide use to the LP model.

Farm typology

Two considerations were important for farm typology ('farm' refers to farm household and its resources): the geographical reference and the stability of the farm types. Although at first sight land use may seem a logical criterion for the typology, it is not included, since it forms the output of the linear programming model. Instead, the classification is based on the potential for agricultural production of the farm, as defined by farm size and soil types (Figure 16.4). In combination with the assumption of a similar household size and thus labour resources, the resulting classes have a similar proportion of land and labour availability. The information on farm size and soil types was derived from a 1:20 000 soil map (De Bruin, 1992). A map of the farm limits, prepared by IDA (Instituto de Desarrollo Agropecuario), was digitised and stored in the GIS. It should be clear that farm types solely based on the quantity and quality of one resource (i.e. land) cannot be called a proper farm household classification. Other aspects, like present land use, capital and labour availability and allocation, as well as—most importantly—objectives and strategies should be taken into account (see the next section; p. 196).

Linear programming model

The linear programming model selects and distributes a number of LUSTs for each farm type. The models for each farm type are incorporated in a sub-regional linear programming model, encompassing regional constraints such as the availability of off-farm employment. The linear programming model is formulated in such a way that the Neguev settlement is considered to consist of five farm types (in reality the Neguev consists of about 300 farms, each between 10 and 17 ha). Each farm type may select from all the available land use activities or LUSTs.

Up to now, the model contains variables for the main crops (cassava, maize, palm-heart, papaya, pine-apple, plantain and pumpkin), pasture and cattle. To produce these, the farm types select LUSTs. Next, the model has monthly labour variables (according to labour type—own, hired and off-farm work on other Neguev farm types or plantations). Furthermore, the model contains equations for costs of fertiliser (N, P and K) use. To account for sustainability, nutrient (also N, P and K) depletion and biocide use are included as constraints.

The constraints are related to product and costs balances, to land availability per farm and per soil type on a monthly basis, and to labour requirements, balanced by labour availability, specified per labour type for each month. The first step involves maximising the net benefits, i.e. the difference between benefits and cost, measured at 1991 product, input and factor prices. The second step is to evaluate the other objectives related to the sustainability parameters.

The program restricts the analysis of sustainable land use to two parameters: nutrient depletion and biocide use. For all the LUSTs the nutrient balance can be modelled (Stoorvogel, 1993). Initially, in USTED, nutrient depletion is included as a boundary condition, with various limiting values. Separate solutions of the linear programming model (e.g. maximising the farmer's income versus minimising nutrient depletion) permit the evaluation of the reciprocal opportunity costs of these objectives. The thresholds for nutrient depletion can be set as the result of an iterative process.

The amount of biocides used in all LUSTs is known. An index value for biocides is created on the basis of the amount of active ingredients, their toxicity (according to the WHO classification) and their half-life time. Like nutrient depletion, biocides form

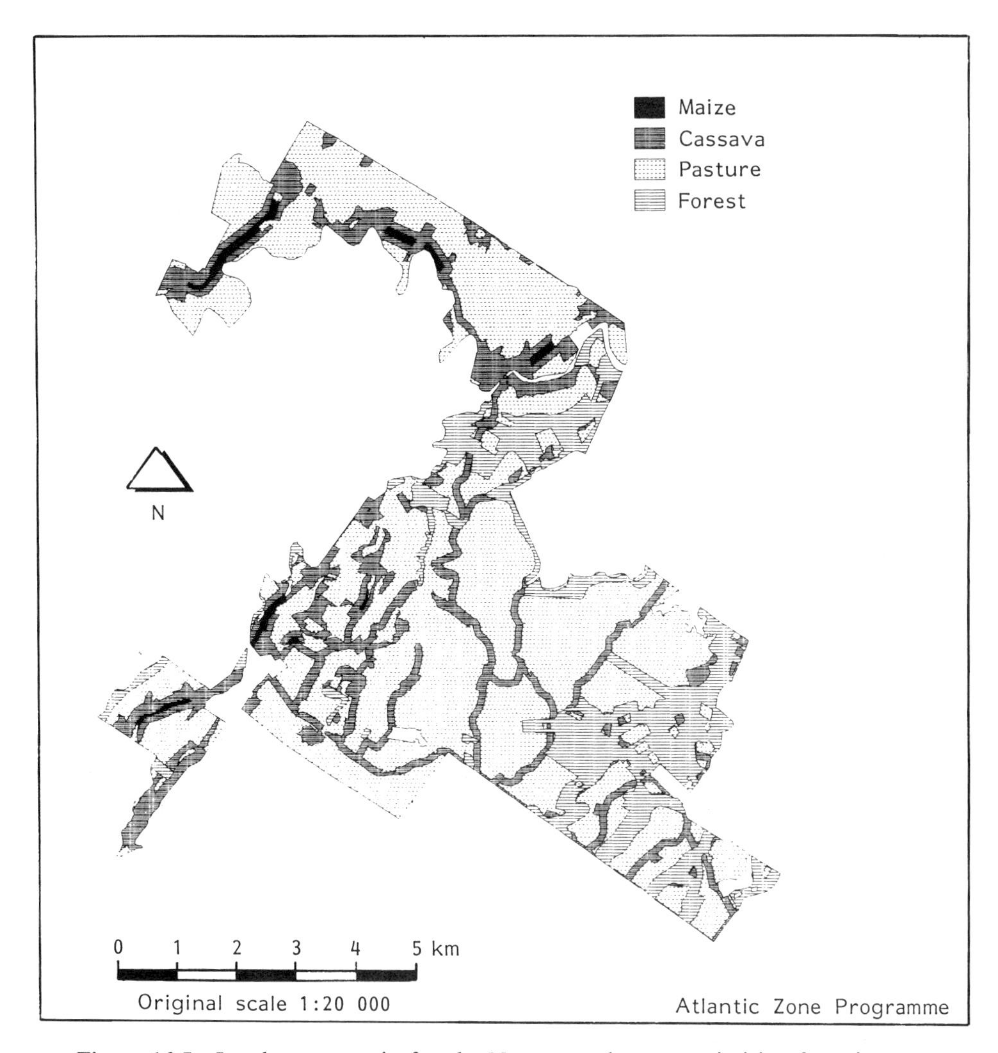

Figure 16.5 Land use scenario for the Neguev settlement optimising farm income

a separate boundary condition in the linear programming model. However, threshold values for the biocide index remain tentative.

Scenarios of alternative land use

Policy-makers have a large number of measures at their disposal such as incentives and regulations to influence land use (Lutz and Daly, 1991). In many cases the effects of these measures and other major land use determinants (e.g. population growth, demand prices, sustainability issues) are unknown. Scenarios, indicating possible trends in major land use determinants and/or policy measures, may be evaluated through USTED with regard to their influence on the use of land.

Table 16.3 Land use (ha) and biocide use in farm type 5 for two scenarios, without constraints on biocide use and with a reduction of 50% of the total biocide index

	No biocide restrictions	Biocides restricted with 50%
Fertile well-drained soil		
Cassava	68	6
Pasture	—	60
Fertile poorly-drained soil		
Pasture	125	125
Unfertile well-drained soil		
Pasture	1460	1487
Cassava	313	287
Plantation work (10^3 hours)	191	217
Biocide use[a]		
Paraquat (litres)	3048	2269
Dimetoate (kg)	1009	908
Biocide index[b]	189	142

[a]The amounts are given in commercial formulations (Gramoxome for Paraquat and Perfektion for Dimetoate).
[b]The restriction on the biocide index has been set at 50% for the settlement as a whole.

Figure 16.5 presents a scenario for the Neguev after optimising farmers' income. A total of 120 different LUSTs for maize, pineapple, plantain, palm heart, cassava and pasture is provided, through LUSTPZA, to the linear programming model. The LP model selected different land utilisation types under a selected number of technologies for the five farm types (Figure 16.4). Table 16.3 shows the results with special attention to the use of biocides: an alternative scenario in which total biocide use in the settlement had been limited to 50% of the biocide index (as defined above) resulting in a different land use pattern and a reduction of 20% of farm income. The reduction of biocide use has been established by selection of other crops and different technologies (Table 16.3 and Figure 16.6).

The output of USTED for different scenarios of land use for the Neguev settlement also functions as feedback to new scenarios. This iterative process allows the determination of the sensitivity of the sustainability parameters. With the selection of LUSTs for each of the separate farm types, a number of results with regard to farm income and labour requirements can be calculated. Both land use and other results can be graphically displayed by the GIS as shown in Figures 16.4 and 16.5 and facilitate the interpretation of the outcome of the scenarios. This includes comparisons with other scenarios or with current land use for which a map (Overtoom et al., 1987; Mücher, 1992) is available in the GIS.

FARM CLASSIFICATION BASED ON FARMER CHARACTERISTICS

Introduction

Because USTED/LUSTPZA uses only a limited set of resource indicators such as land qualities and farm size (see p. 194), it reflects insufficiently the complexity of

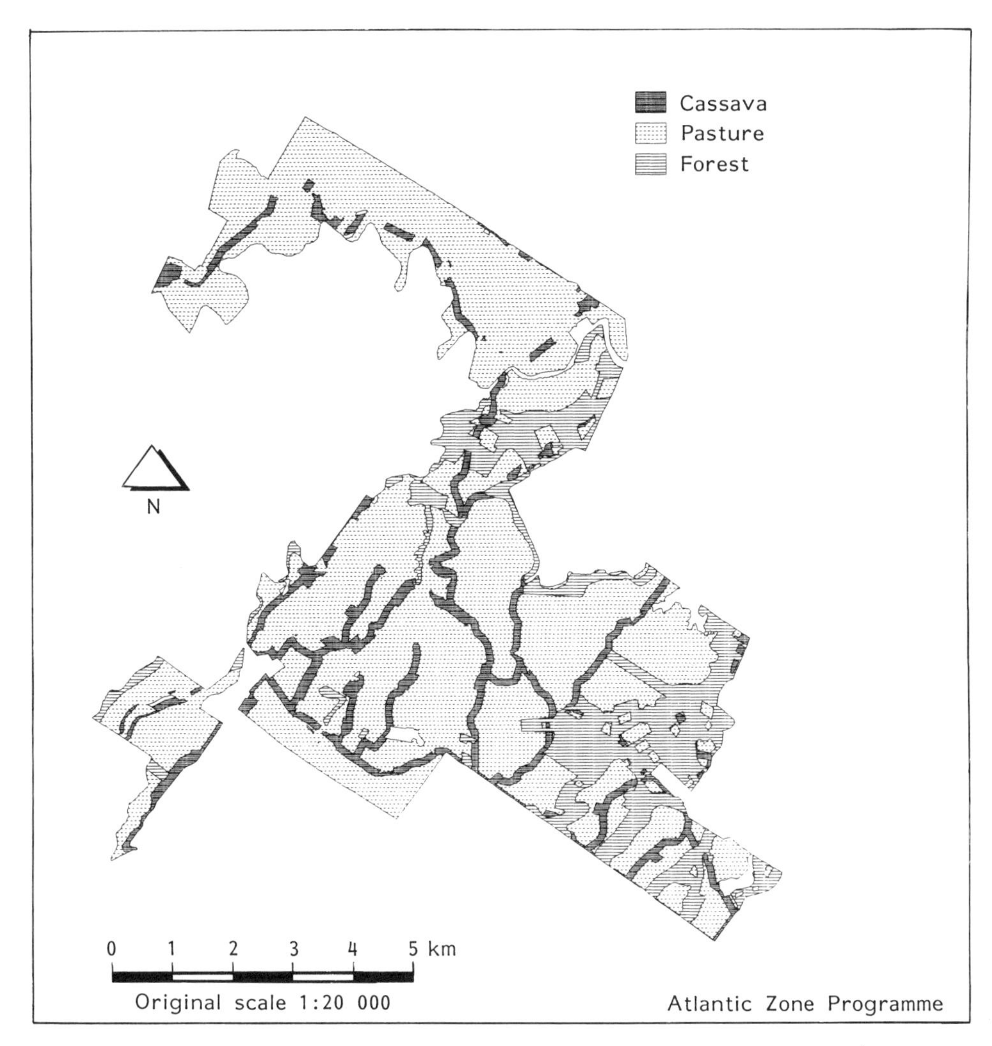

Figure 16.6 Land use scenario for the Neguev settlement optimising farm income and restricting the biocide index with 50%

socio-economic variables and farmer attitudes. The Atlantic Zone Programme, therefore, developed an additional farm classification based on research on farmers' decision-making processes. It combines qualitative (e.g. farmers' objectives and strategies) as well as quantitative criteria for the establishment of the farm classes.

Farm classification

A farm classification constitutes a grouping of farms with the objective to obtain homogeneous classes with regard to particular characteristics. The reasons to reduce the number of farms can be numerous but they determine the type of farm classification to be developed. Therefore, before carrying out a farm classification the purpose of the task has to be clearly defined. Basic questions such as, 'Why are the farms to be grouped?' and 'How will the farm classification be used?' have to be answered.

A farm classification to be applied for land use planning should aim at defining groups of farms which are more or less homogeneous in land use decisions. Land use decisions of farmers are determined by aims and goals of the farm family and by their resources, and are influenced by natural, socio-economic and political circumstances in a region as well as by planned interventions as a result of land use planning.

In the Atlantic Zone of Costa Rica most of the land is used for cattle ranging, which is not directly related to soil type. Also, farms in a settlement area in the Atlantic Zone, with minor differences in farm size still show considerable differences in land use patterns. These examples corroborate the assumption that in the Atlantic Zone the socio-economic situation of a particular farmer is the major determining factor in land use decisions (Akkermans, 1993; Alfaro, 1993). However, it is also recognised that at the same time the physical possibilities of the farm family will also determine land use. The biophysical and socio-economic situations of a farmer are not independent, but are interrelated and should, therefore, both be taken into account in a farm classification. The following farm classification is being developed at an experimental scale in a small peasant community of the Atlantic Zone of Costa Rica, called Agrimaga.

Methodology of farm classification

The methodology for this type of farm classification basically consists of two parts: (1) development of a farmer typology, the mainly qualitative part; and (2) a grouping of farms per farmer type, the mainly quantitative part.

The farmer typology

The farmer typology is a qualitative classification which defines representative socio-economic rationalities of the farm household regarding land use and farm management (the farmer types). First, the different farmer types present in the region have to be identified and described. This should be done by a sociological study of a (random) sample of farm families. In such a study the differentiating criteria between farmer types should be identified. Then, on a selected number of differentiating criteria, a survey will be carried out in the region to allow attribution of each farmer to a particular farmer type.

In the Agrimaga case, the farmer types were distinguished on the base of: (1) their level of economic dependency on farm production; (2) their perception of soil suitability for cropping; (3) type of agriculture practised (traditional or modern); and (4) integration in the labour market. As a result, six types of farmers were defined:

- investor
- investor/farmer
- seller of services (other than agricultural)
- full-time farmer who occasionally hires labour

- farmer/day labourer
- permanent worker/farmer

A grouping of farms per farmer type

Each farmer type is representative for a number of farms which are different in biophysical production factors like soil type, farm size, machinery, etc. These factors also influence a farmer's land use decisions. Therefore, for each farmer type a subdivision of farms will be made in order to arrive at homogeneous farm classes with respect to farmers' objectives and strategies as well as their physical possibilities. The differences in biophysical possibilities in the region have to be identified and can then be included in the regional survey.

The final classification, then, consists of farmer types, which represent different objectives and strategies of farm families, subdivided by farm types, which represent different availabilities of biophysical production factors. The final classification results in a limited number of classes as a basis for modelling. The structure of this farm classification is similar to the structure of a linear programming model: the farmer types represent and provide the different objective functions for the model; the farm types represent and supply the constraints of the model.

DISCUSSION

The USTED methodology: strengths and weaknesses

The USTED system is a tool to evaluate land use scenarios for a certain region. It provides possibilities for land use planners to examine the effects of certain incentives. The system is operational, using various computer programs for the links between the different components. It provides an excellent integration of socio-economic and biophysical data. A major asset is its flexibility, enabling incorporation of other crops, soils or sustainability criteria and extension into other areas and at other scales. However, expert knowledge is crucial to interpret the results of the scenarios and perform the feedback to the input data.

Although the system is operational, it still requires enormous data-collection efforts. Surveys of soils and land use should provide the necessary geographic reference of the different farm types. A farm survey and fertiliser experiments should form the basis for the description of actual LUSTs and potential LUSTs. For the step from fertiliser experiments to potential LUSTs, validated crop-growth simulation models or expert systems are indispensable, and even then the results are greatly simplified. Relevant sustainability criteria have to be identified and the underlying processes studied. The methodology requires a multidisciplinary team to account for the different components of the system. For a good functioning of the system, links between the various, mostly disciplinary databases are essential. The links are accomplished by the different attribute files of LUSTPZA.

At this moment USTED operates at a sub-regional level. On a regional level such as, for example, the Atlantic Zone, the assumption of stable attribute files will not be valid. Especially, prices of inputs and products will differ significantly. To deal with this variation different sub-regions will be selected in such a way that the assumption of stable attribute files is justified. The LP model will then perform the calculations for each sub-region with its specific set of attribute files. In addition, other sustainability criteria, like a decrease in infiltration through soil compaction, may become relevant.

Potential applications

Land use planning plays a major role in the decisions the country has to take, such as the final designation of wildlife protection areas; designation of watersheds for eco-tourism thus excluding power generation; designation of areas for reforestation or for the stimulation of natural regeneration; designation of areas for intensive and sustainable agricultural production, including appropriate soil conservation techniques.

The land use analysis initiated in the Atlantic Zone of Costa Rica comes at an appropriate moment. The data collection efforts described in the section 'Data Sources' (p. 192) are on the way for other parts of the country as well. Many institutes in Costa Rica are working with GIS technology and their number is still increasing. With the introduction of GIS the generation of digital data is increasing. For example, information at a 1:50 000 scale of land use capability for the Central Pacific Region; a map of land use capability of forestry lands in Costa Rica is being developed; and the World Bank is expected to finance the development of a similar map for the rest of the agricultural area of the country, with the intention of identifying crops and production systems appropriate for the areas in question. In addition, the cadastral maps are being digitised, as is the infrastructure and the information from various censuses carried out primarily in 1973 and 1984.

At the non-governmental level, one of the preliminary projects elaborated by a subcommission of NGOs formed as a results of the Bilateral Sustainable Agreement between the governments of Costa Rica and the Netherlands aims at creation of a multi-purpose 'Catastro', to promote a territorial ordering of the country. Each particular farm of this multi-purpose 'Catastro' that eventually would cover the whole country, will be stored together with its characteristics, the actual land use and potential land use.

Moreover, the methodology might help traditional producers faced with the new government-imposed Structural Adjustment Plans to solve questions such as what to produce, by whom, how and where. USTED may be widely tested in different environments, by governmental agencies like IDA, MAG, National Bank System, Universities and also at a non-governmental level, such as the Tropical Science Centre.

Future research: validation and extrapolation to higher scale levels

In its present state USTED functions only up to the (sub-)regional level. The study of regional aspects of land use planning in the context of the whole Atlantic Zone (400 000 ha) is only in its starting phase. Also, aggregation of data from the farm level to the regional level is still difficult. This means that in the coming years still considerable effort is necessary to accomplish the regional model in the Atlantic Zone. A useful concept for a regional geographical grouping of land use is the Land Use Zone (LUZ:

Huising, 1993). Furthermore, regional criteria for sustainability will have to be elaborated in more detail. At the same time, however, efforts will be made to study by what policies a chosen regional scenario could be implemented at farm level. The regional model will also be the starting point for land use planning at the national level.

Another line of research to be tackled is a rapid low data-input validation of the methodology in a subhumid/semi-arid zone in Costa Rica (Guanacaste). Data collection in the Atlantic Zone has taken about five years. If the methodology has any practical use at all, it must be possible to run it also in cases with only limited data and in a short time.

A third objective is to extend the methodology to a national and subcontinental scale of observation. First, a comparative study of two ecologically and economically contrasting areas within Costa Rica, the Atlantic Zone and the Guanacaste area will be made. From the point of view of a national planner, land use changes in one region may affect land use in others, e.g. through migration of labour. So land use planning on a national scale requires consideration of processes outside the scope of the regional planner. Finally, an attempt will be made to link our regional models with global models of world vegetation and land use on the basis of similar climatic and biophysical constraints used in our smaller scale farm and (sub-)regional simulations, to predict the impact of global changes on land use.

REFERENCES

Akkermans, J., 1993. The 'Why' of decisions taken by farmers in the AGRIMAGA settlement. Report No. 50 (Phase 2), CATIE-UAW-MAG, Turrialba, Costa Rica.

Alfaro, R., 1993. Análisis de inventario en una comunidad campesina de la Zona Atlántica de Costa Rica: El caso de AGRIMAGA. Report No. 43 (Phase 2), CATIE-UAW-MAG, Turrialba, Costa Rica.

Beijers and Partners, 1992. *OMP Manual*. Beijers and Partners NV, Brasschaat, Belgium.

De Bruin, S., 1992. Estudio detallado de los suelos del asentamiento Neguev. Report No. 25 (Phase 2), CATIE-WAU-MAG, Turrialba, Costa Rica.

ESRI, 1990. PC Arc/Info 3.4D, Environmental Systems Research Institute Inc., Redlands, CA.

Fresco, L. O. and Kroonenberg, S. B., 1992. Time and spatial scales in ecological sustainability. *Land Use Policy* **9**: 155–182.

Fresco, L., Huizing, H., Van Keulen, H., Luning, H. and Schipper, R., 1990. Land evaluation and farming systems analysis for land use planning. FAO guidelines: Working document, FAO, Rome.

Halpin, P. N., Kelly, P. M., Secrett, C. M. and Smith, T. M., 1991. Climate change and Central American Forest Systems, Costa Rica Pilot Project. Background Information, Department of Environmental Science, The University of Virginia.

Huising, J., 1993. Land use zones and land use patterns in the Atlantic Zone of Costa Rica. A pattern recognition approach to land use inventory at the subregional scale, using remote sensing and GIS, applying an object-oriented and data-driven strategy. PhD Thesis, Wageningen Agricultural University, Wageningen.

Krabbe, W. K. 1993. *The SIESTA Geographic Database: Instruction for Its Use and Maintenance*. Atlantic Zone Programme, Guápiles, Costa Rica.

Lutz, E. and Daly, H., 1991. Incentives, regulations and sustainable land use in Costa Rica. *Environmental and Resource Economics* **1**: 179–194.

Mücher, C. A., 1992. A study on the spatial distribution of land use in the settlement Neguev. Report No. 9 (Phase 2), CATIE-UAW-MAG, Turrialba, Costa Rica.

Oosterom, A. P., Stuiver, M. J., Krabbe, W. K. and Hootsmans, R. M., 1992. Geographical Information Techniques and photogrammetry in soil and landscape survey of the Atlantic Zone in Costa Rica. In: Wielemaker, W. G. and Kroonenberg, S. B. (eds), *Generación y Aplicación de la Información de Suelos de la Zona Atlántica de Costa Rica* (Proceedings Workshop), pp. 23–32. Actas del Taller de Información de Suelos, Guápiles, 1990. Exposiciones y Guia de Excursión. Programme Paper No. 13, 2nd Phase Report No. 17. CATIE Serie Técnica, Informe Técnico No. 170. CATIE-UAW-MAG, Turrialba, Costa Rica.

Overtoom, T., Mudde, H. and Koffeman, I., 1987. Land use map of the Neguev settlement (1:20 000). The Atlantic Zone Programme, CATIE-WAU-MAG, Turrialba, Costa Rica.

Pearce, D. W. and Turner, R. K., 1990. *Economics of Natural Resources and the Environment*. Harvester Wheatsheaf, New York.

Penning de Vries, F. W. T., Jansen, D. M., Ten Berge, H. F. M. and Bakema, A., 1989. *Simulation of Ecophysiological Processes of Growth in Several Annual Crops*. Simulation Monographs 29, Pudoc, Wageningen, The Netherlands and IRRI, Los Baños, The Philippines.

Sader, S. A. and Joyce, A. T., 1988. Deforestation rates and trends in Costa Rica, 1940 to 1983. *Biotropica* **20**: 11–19.

Stomph, T. J. and Fresco, L. O. 1991. *Describing Agricultural Land Use*. FAO/ITC/WAU, Rome/Enschede/Wageningen.

Stoorvogel, J. J., 1993. Optimizing land use distribution to minimize nutrient depletion: a case study for the Atlantic Zone of Costa Rica. *Geoderma* **60**: 277–292.

Tosi, J. A. Jr., 1969. Mapa Ecológico de Costa Rica, Scale 1:750 000. Centro Científico Tropical, San José, Costa Rica.

Vásquez, A., 1989. Mapa de Sub-grupos de Suelos de Costa Rica, Scale 1:200 000. Project No. GCP-Cos-009-ITA, FAO, Rome.

Veldkamp, E., Weitz, A. M., Staritsky, I. G. and Huising, E. J., 1992. Deforestation trends in the Atlantic Zone of Costa Rica: a case study. *Land Degradation and Rehabilitation* **3**: 71–84.

Wielemaker, W. G. and Kroonenberg, S. B. (eds), 1992. *Generación y Aplicación de la Información de Suelos de la Zona Atlántica de Costa Rica* (Proceedings Workshop). Actas del Taller de Información de Suelos, Guápiles, 1990. Exposiciones y Guia de Excursión. Programme Paper no. 13, 2nd Phase Report No. 17. CATIE Serie Técnica, Informe Técnico No. 170. CATIE-UAW-MAG, Turrialba, Costa Rica.

WRI-TSC, 1991. *Accounts Overdue*. World Resources Institute, Washington, DC.

CHAPTER 17

Scenarios for the Peatland Reclamation District

W. J. M. Heijman,[a] J. M. L. Jansen,[b] G. H. A. Te Braake,[c] S. Meerman,[d]
P. C. van den Noort,[a] J. H. van Niejenhuis,[a] A. B. Smit[a] and S. Thijsen[e]

[a]Wageningen Agricultural University, Wageningen, The Netherlands. [b]The Winand Staring Centre, Wageningen, The Netherlands. [c]Reconstruction Committee, Groningen, The Netherlands. [d]Dir. Raw Materials, AVEBE, Veendam, The Netherlands. [e]Grontmij, Haren, The Netherlands

An open mind is the window to the future.

INTRODUCTION

In the Peatland Reclamation District (PRD) in the North of the Netherlands (Figure 17.1) about 90% of the cultivated land is used for arable farming. Since prices for arable products have decreased substantially, many farmers in the area face difficulties in gaining a sufficient income. The decline in agriculture implies a decrease in employment and has also a negative impact on other sectors of the regional economy, resulting in a general negative influence on the whole region.

In 1990 the estimated average net farm income on arable farms in the PRD was about DF 20 000 (based on 70% equity of the total assets). This is low compared to the average Dutch wage in agriculture of DF 60 000 in the same year. Therefore, on many farms there is an income problem (this is not so for all farms, for there is a large variation in average farm income). On 25% of the farms, additional income is generated by starting a second branch of farming like vegetables or dairying, and on 15% of farms through jobs outside agriculture. Nevertheless, the number of farms declines by 3% a year, generally because older farmers sell their land to existing farmers. The question now is whether opportunities exist that will enable the remaining farms to survive and what uses there are for land outside agriculture.

In our attempt to find answers to this regional question, we have to take into account some important characteristics of the planning process:

The Future of the Land: Mobilising and Integrating Knowledge for Land Use Options
Edited by L. O. Fresco, L. Stroosnijder, J. Bouma and H. van Keulen.

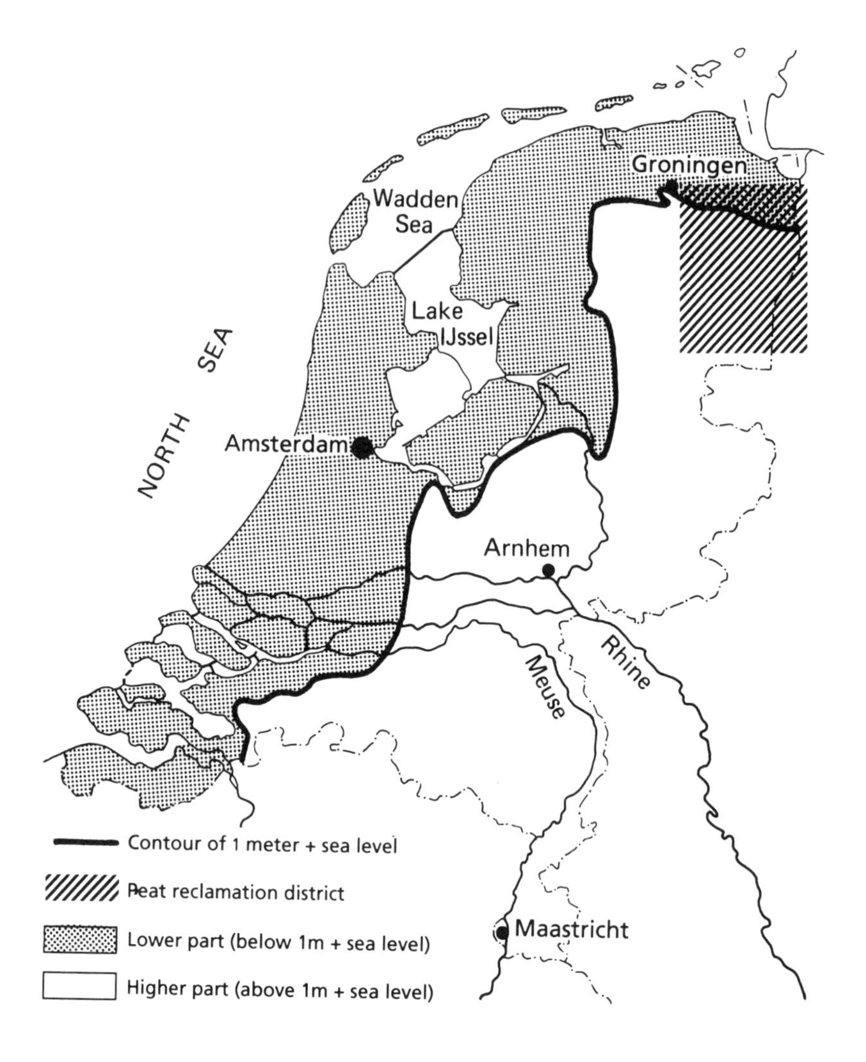

Figure 17.1 Location of the Peatland Reclamation District in the North of the Netherlands

(1) planning is a continuous process on different levels, carried out by parties involved;
(2) by involving the regional population as much as possible, the planning process and the execution of the plans take a considerable amount of time. The estimated time necessary for the reconstruction of the PRD is 25 years (1980–2005) (see the section on rural development, p. 214).

The sequence of this multi-authored chapter is as follows: in the second section (p. 205), P. C van den Noort describes and analyses agricultural change and regional development in the PRD, using the concept of management style as a tool for the explanation of the important role arable farming is playing in the region. In the third section (p. 206), J. H. van Niejenhuis and A. B. Smit explore regional land use options within agriculture, showing that increasing efficiency, part-time farming, alternative arable crops and livestock farming are all options that can contribute to the solution of the income and employment problem. However, even if all these options are selected extensively, the number of farms will decline drastically in the near future (by an estimated 20–50%). Land use options outside agriculture such as outdoor recreation, forestry and nature conservation are treated by S. Thijsen in the fourth section (p. 210). He concludes that, together with the agricultural options, these alternatives can play a role. The implementation of the planning process is sketched and analysed in the fifth section (p. 214) by G. H. A. te Braake. The complexity of the planning process in the region resulted in the special Reconstruction Act of 1979. The estimated costs of the reconstruction are DF 2600 million. In the final section (p. 221), S. Meerman deals with the consequences of reconstruction scenarios for the farmers, expressing the opinion that options outside agriculture can, at best, be supplementary to agriculture.

The main conclusion is that total employment in agriculture will decline in the near future. Therefore, an urgent need exists for jobs outside agriculture. Further, it can be concluded that land use options outside agriculture can contribute to the creation of jobs only in a modest way, hence, more employment has to be created in industry and services. This is a major consideration to be taken into account in the planning of the region.

PEATLAND RECLAMATION DISTRICT: REGIONAL ECONOMIC CHANGE

Geographical differences

The map of the Netherlands can be divided into two parts by a line running SW–NE (Figure 17.1). West of this line is the 'polder' area of the country, often below sea level; to the east is the higher part with mainly sandy soils. Depending on the drainage situation, the valleys in the eastern part were sometimes filled with peat, characteristic of a wild and inaccessible moorland. To the north there were several such regions, situated around the borders of the PRD. These areas were gradually reclaimed. Local farmers reclaimed the edges to get peat and more land. There were also large-scale operations, like that of the city of Groningen: digging canals for drainage and transport, and digging peat for domestic fires. The result was more agricultural land, which was used for arable farming. These farms were the envy of the farmers on the 'old land'. The peat operations and the agricultural development were referred to as the 'colonisation' of the peat and moorland. Despite many technical and economic problems, agriculture in these areas is still dominated by arable farming, whereas the area is suitable for dairy and pig farming. The reasons may be the high social status of arable farming in this province as well as the high grain prices and cattle diseases in the nineteenth century. The north was once the leader of Dutch agriculture, but now, due to structural problems, has lost that position to the south.

Differences in regional change

The use of land in the Netherlands has changed. In all regions land used for agriculture has decreased, but between 1975 and 1991, the PRD had one of the highest rates of decrease (0.5% per year). The number of farms in the region decreased by 2.6% per year in the same period; this was the highest percentage in the country (Kamminga et al., 1993). The highest increase in land per farm in the country during 1975–1991 was also in the PRD (2.1% per year), but this does not imply a lack of structural problems there. The concentration of animals in this area is far below the Dutch average. Farm products are grain and starch potatoes. As the latter are processed in the region, it has become rather dependent on this commodity (Wennekes, 1993).

Causes of regional change

The natural conditions or physical environment were once considered the dominating factor for the structure of agriculture. At present, however, it is believed that other factors are often more important, such as status and the influence of 'relevant others'. This is especially seen where a change from arable farming to dairying or horticulture would have been profitable. Regional differences exist in the adoption of new techniques and processes ('*bedrijfsstijlen*' or 'management strategies')(Van Der Ploeg, 1993), in the distance to the markets which keeps changing because of changes in means of transport and road systems and marketing, e.g. 'tele-auctions'. Regional differences also exist in costs per unit.

General economic development in the various regions of the country is not equal. The northern area lags behind the west. The PRD is not so attractive because nature is underdeveloped, wildlife and forest being scarce. Social life is sometimes difficult because of the structure of the villages and towns, and the unattractive landscape (of course, this is subjective). High unemployment and low incomes mean that many young people try to leave.

LAND USE OPTIONS WITHIN AGRICULTURE

Material and method

Options can be analysed for individual farms. A distinct farming situation has then to be defined as the 'average' farm, and the results are only valid for that farming situation. In reality, however, there are as many farming situations as there are farms. Moreover, many aspects of the near future cannot be foreseen in the current unsure situation of political developments and environmental regulations. Therefore, the present analysis is restricted to some qualitative reflections on the most promising options for agriculture in the PRD without answering the question of which solution is most suitable for any particular farm.

The chances for agriculture are determined by a large number of location-specific factors. In general these are physical, economic, social and political factors. The physical fertility of soils in the PRD is lower than that of the light clay soils in the Netherlands.

In some years drought causes damage to the crops and in spring there is a chance of wind erosion. Compared to other parts of the country, the PRD is less densely populated. The farms are specialised in arable farming, the main crop being starch potato for industrial processing, grown on about 45% of the area. This is only possible with soil fumigation which, however, has negative environmental effects. Farms are larger than average in the country, but too small to provide work for the farmer the whole year round. The Government is trying to improve the farming and non-farming situation in this region.

In discussing options for improvement, a distinction has to be made between options for existing farmers within the present farming system and options that require a completely new farming system. The latter often require investments in equipment and knowledge not available to the present farmer. Thus some changes can only be implemented by farmers from outside the PRD. This view is based on expert knowledge, existing literature, data of research stations and budgets for a limited number of typical farm situations (Ippel and Noordam, 1990).

Options within the existing farming system

Given his goals, a farm manager has the task of continuously adapting the farm's organisation to changing circumstances. His own experience, his learning ability, the existing farm situation and the risk associated with the establishment of a completely new farming system are the constraints in his space for decision-making. Many farmers are in this position and adaptation is often restricted to minor changes in the current farming system. Important opportunities include improvements in efficiency, other arable crops, part-time farming and an increase in area.

Efficiency

Given the cropping plan, income can be improved by an increase in the gross margin of the enterprise and/or a decrease in fixed costs. Gross margin will be increased by efficient use of fertiliser and chemicals; lower fixed costs require efficient use of machinery and labour. Continuous education is necessary if the farmer is to be made aware of new technological options in these fields. Registration systems provide him with up-to-date and relevant data about the production process, and evaluation of these data will result in insight in the strong and weak points of farm management and show opportunities for improvement. A special point for the PRD is the control of soil-borne diseases caused by the intensive potato cropping rotation.

Other arable crops

A long list of alternative crops can be given. For financial results the central function of cereals is important. As most other crops have similar or lower results they will not solve an income problem. For a limited number of farmers it is possible to produce, for example, wheat for special bread or malting barley, for small segments of the market at a higher price. The economic results of crops for industrial processing will either be of the same magnitude as cereals or less. Changes in the Common Agricultural Policy (McSharry proposals) may lead to even lower financial results (Struik et al., 1991).

Part-time farming

If income from farming activities is insufficient, the farmer has to look for other sources of income either on the farm or outside. Many small farmers cannot use all their labour on the farm for productive activities and need to look for part-time jobs elsewhere. The combination of farm and non-farm activities is complicated because of labour peaks, especially in the sowing and harvesting seasons and the surplus in other periods of the year. In many cases it is necessary to adapt the farming system to the off-farm work, resulting in lower farm results. Part-time jobs of interest to the farmer are not always available in the region. Part-time farmers are excluded from some government assistance programmes, as such farming is regarded as a transitory situation towards giving up farming. It has a negative influence on efficiency and on the development in farming structure which is regarded as undesirable. Nevertheless, there is an increasing number of part-time farmers.

Increase in farm size

In general an increase in farm size creates a higher income capacity. Such an increase is only possible by buying land. The Dutch tenancy law is so restrictive that private landlords are hardly willing to lease land if they have the opportunity to sell it freehold. The marginal value of land and land prices is high because many small farmers have some free capacity in equipment and labour. Given the supply on the land market, only a few arable farmers, with the best farm results, are able to buy land at these prices. Land prices in other parts of the country, with more intensive farming systems, are even higher. There is enough demand for land from these other regions to keep land prices in the PRD high.

Many farmers use all these options as far as possible. The economic climate seems to change so fast that the opportunities are insufficient to generate enough income for the average farmer.

Options outside the existing farming system

In this section some land use options are discussed outside the existing farming system that may contribute to the creation of productive work on the farm, specifically livestock production and horticulture. The number of animals is restricted by legislation for the protection of the environment, but the number in the PRD is so small that some expansion is possible. In some fields markets seem to be limited, but the PRD may have a comparative advantage that makes it possible to obtain a market share from traditional but less favourable areas.

Dairy farming

Dairy farming is only possible if the farmer has a milk quotum. This can be bought at high cost, or can be transferred to the PRD by a farmer who has left a dairy farm elsewhere in the country. The latter provides good options for dairy farms in the region, as land prices are much lower than in traditional dairy farming areas, so that the farmer can meet the environmental standards by increasing farm size, given the same number of cows.

Field production of vegetables and other horticultural crops like trees

The sandy soil in the PRD is suitable for a number of horticultural crops. Water for irrigation has to be available, which is not always the case. Markets for these crops do not allow expansion, but the farm structure in the PRD seems favourable. Promising commodities are several species of flower bulbs, ornamental trees and some species of beans and leek.

Greenhouse horticulture

Expansion is taking place in recently established centres such as Emmen. The low land prices, the support of local authorities and the availability of labour are favourable factors. The low temperature and the lower light level are less favourable, compared to the traditional production centres in the western part of the Netherlands. Since greenhouse horticulture uses relatively small plots of land, but is labour intensive, it is more a solution for the employment problem than for the land use problem.

Meat production

In the past, pork and broiler production were concentrated on specialised farms. Environmental legislation has re-established to some extent the relation between land and the production of manure. In the PRD the farmer can either utilise the manure quota by livestock production on the farm itself or utilise manure from other farms against payment. Mixed farming systems consisting of arable farming and meat production are also favoured by changes in the Common Agricultural Policy. The large price difference between cereals and imported feed is declining, which might create a comparative advantage for mixed farming by feeding on-farm produce and fertilising with self-produced manure.

Options outside the existing farming system require, in addition to the farmer being well informed, high investments in new equipment (which is not always available) and surrounding services for alternative farming systems. Building up these services is time consuming.

Conclusions

It is important that the farmer creates productive work for his own labour input on the farm with an efficient production system. This requires a certain minimum size of the farm. Arable farming with the present cropping plan requires at least 80 ha per labourer. An efficient organisation requires three labourers per unit, including the farmer. So, an efficient farm needs about 240 ha. However, most farms are smaller. Increasing farm size or part-time farming are ways of changing the land/labour ratio. Part-time farming is easier to implement than an increase in area at the present high land prices. Only large farms will be able to maintain the existing farming system in the near future.

Livestock production is another option to increase income capacity per labourer. Dairy farming is only possible for farmers who buy land and transfer their milk quota to the PRD, for which scope exists in the PRD. Exchange of land between arable farmers and dairy farmers may alleviate the existing problem of soil diseases. It may be attractive for an arable farmer to increase the production capacity by starting pork or broiler production. Changes in agricultural policy and manure legislation provide new opportunities in this area provided that regional authorities approve. Greenhouse horticulture, which is concentrated in centres, is not a solution to the land use problem and needs large investments in buildings. For the current farmer the option closest to his existing abilities is field production of vegetables. However, this is a low yielding activity with the added risk of oversupplied markets.

Arable farmers in the region have asked for a general solution to their problems. It is clear that such a solution does not exist. It is necessary to intensify production for the creation of income opportunities and reduce the input of environmentally unfriendly production factors at the same time. Given these constraints, outflow of labour from the agricultural sector will be substantial in the next decade. The remaining farming system will be much more diverse than the current system. Markets do not allow a general solution. Each farmer will select a solution that is suitable for him, the farm in general having more than one source of income. The land will largely be used as in the current system; labour will partly be spent in higher yielding activities which use less land, though still meeting environmental standards.

LAND USE OPTIONS OUTSIDE AGRICULTURE

Introduction

The provincial authorities in the northern part of the Netherlands have come to the conclusion that the problems in the rural areas cannot be solved solely within the agricultural sector. Alternatives, such as silviculture, recreation and nature development, and their role in conservation and improvement of the quality of life in the country should also be considered (Ministry of VROM, 1991). The study 'Future Perspectives of Oldambt and Peatland Reclamation District' (Van Niejenhuis et al., 1991) also investigated the land use options outside agriculture, a summary of the approach, the results and conclusions of which are given in this section. This study is also a contribution to the new Regional Town and Country Plan for the Province of Groningen, which is in preparation at present.

Integrated or landscape approach

Changes of functions in rural areas will not only affect employment or the production structure, but also the organisation of the rural area, such as the planning of the kinds

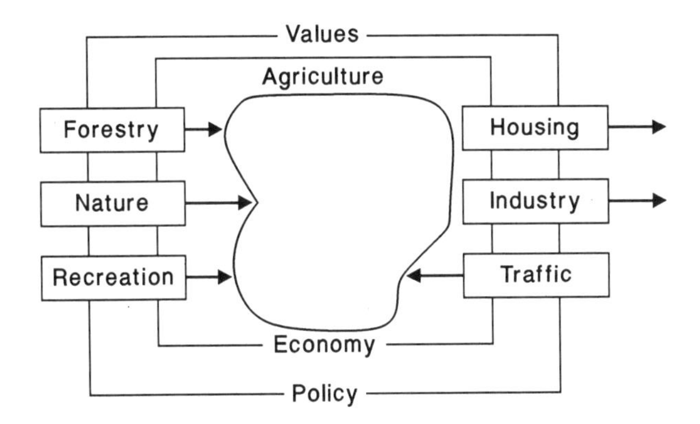

Figure 17.2 Land use dynamics

of land use and the regulation of function and purpose. Two parallel activities were initiated:

(1) Developing a *vision* on future landscape. Economic, social, transport and other problems cannot be solved in isolation, but should be integrated to reach a realistic balance. These issues and their complexities require a vision of the landscape and its uses as a whole: the rural area as an integrated multifunctional system (Figure 17.2). This brings us to the concept that proper land use planning is applied (human) ecology.
(2) Investigating developments in land use. To identify constraints for a well-balanced landscape organisation, trends and topics in each single land use were investigated from a socio-economic point of view. Landscape is not only a direct expression of land use, it is also a reflection of long-term developments in society's needs and demands. Some national indices are given in Table 17.1, giving an impression of the significance of some land use options outside agriculture.

Table 17.1 Some characteristics of the Netherlands

		1950	1990
Population (10^6)		10.0	15.0
Cultivated area (10^6 ha)		2.4	2.0
Turnover in 10^9 DF			
Agriculture		8	80
Tourism		0	35
Forestry and nature		—	4
	Prices in DF per ha	**Employment (working years per 100 ha)**	**Value added (per ha)**
Agriculture	25 000	10–30	4000
Tourism	75 000	8–10	6000
Forestry and nature	8 000	0.5–2	– 6000

Prior to specific developments in the Oldambt and the PRD, a target for the entire northern region was set. In the studies, *Bosrijk Noord-Nederland* ('the Woody North of the Netherlands': Krant et al., 1987) and *Landelijk Noord-Nederland* ('the Rural North of the Netherlands': Langbroek et al., 1990), outlines of the environmental, ecological and economic main structures for the rural areas in the three northern provinces were given.

Within these main structures a so-called regional development plan was worked out for a number of sub-regions in the area. For example, regional structure plans were developed for south-west Friesland (Krap and Thijsen, 1990), north-east Friesland and north-west Groningen (Stuurgroep NOF/NWG, 1993) and the Oldambt and the Peatland Reclamation District (Van Niejenhuis et al., 1991).

PRD Land use scenarios

In the regional plan for the Oldambt and the PRD, three different land use scenarios were formulated. The significance of the functions forestry, nature and recreation as well as the level of integration with the agricultural functions vary between scenarios.

- *Scenario 1: Agricultural diversification and intensification (yellow scenario).* In this scenario the subsectors of agriculture will show stronger development than when developing autonomously. Because of differentiation and specialisation, stronger agricultural subsectors will develop. For example, in this scenario dairy cattle farms will be moved into the area. Also greenhouses and outdoor horticulture will increase considerably.
- *Scenario 2: Multifunctional amelioration of rural areas (blue scenario).* In this scenario, in addition to agriculture, functions such as silviculture, recreation and tourism will play a role in the area. Afforestation will be taken on: multifunctional forests by the government and temporary afforestation by individuals. In addition, the digging of a lake is planned in accordance with the theme 'Netherlands Water Land' to stimulate tourism and housing opportunities.
- *Scenario 3: Nature development (green scenario).* In this scenario it is nature that determines the direction. Accelerated execution of the Nature Policy Plan is envisaged. Agriculture, too, will partly be adapted to nature, for example in buffer zones around the nature reserves ('*beheersgebieden*').

Based on agro- and socio-economic statistics on farm-size structures, demographic development and marketing opportunities within single land uses, a spatial-economic input–output model was developed. Each scenario was evaluated according to its:

- financial and socio-economic effects
- country planning effects
- social effects
- environmental effects

The scenario technique is a multiple goal planning instrument, providing indications for a range of possible developments. It indicates the consequences of certain policy

Table 17.2 Comparison of autonomous development and the three scenarios

	Autononous development	Yellow	Blue	Green
Economic effects				
Employment in working years	3500	4500	4000	3000
Regional income	0	+	0	—
Private investment level	0	+	+	–
Social effects				
Visual planning quality	0/–	0/+	++	+
Level of facilities	–	0	0	–
Quality of the environment	0/–	0/–	+	+
Means and instruments				
Level of government investments (in millions of DF)	0	0	130–400	3100
Availability and financial possibilities of deploying instruments	0	–	—	—

Explanation of symbols: 0, comparable to the present situation; +, better/higher than in the present situation; –, worse/lower than in the present situation.

decisions, the scope for policy-making and the need for new sets of instruments. The land use scenarios are planning instruments and do not pretend to predict the future. The planning horizon is approximately 10 years. The results are presented in Table 17.2.

The provincial authorities were advised to aim at a combination of scenarios 1 and 2. Besides intensification and diversification of agriculture, a landscape-oriented infrastructure of forests and lakes should be developed to enhance opportunities for tourism and housing.

Conclusions

In a strictly economic sense there are no alternatives outside agriculture, such as the creation of lakes, forests, nature reserves and recreational areas (Table 17.2); these functions do not offer a real solution to the current economic problems. The relatively high prices of land for these kinds of functions and their low production value, besides the fact that there is no compensation for labour and capital, are the major limiting factors. This does not imply, however, that the land use options outside agriculture are not of interest. Solutions for the problems in the PRD can only be found in a well-balanced combination of all land uses. Of course, the problem of how to achieve this balance in an actual physical landscape remains. The judgement of the desirability of developing new kinds of land use in the PRD requires administrators and policy-makers to think in a future-oriented manner; this appeals to their ability to consciously create the environment. Then the question to be asked is not, 'How much does it cost?' but 'What are we, as society, prepared to pay?'

In our urbanised and densely populated country, we are now moving towards an all-embracing concept of the relationship between man and environment. Recognising the need for a complex but integrated setting where we can live in balance with all the different factors constituting our lives. Although opposition and conflicts will be unavoidable, forestry, nature and recreation all belong to that overall concept.

RURAL REDEVELOPMENT IN EAST GRONINGEN AND THE PEATLAND RECLAMATION DISTRICT

A description of the reconstruction area

The reconstruction area of East Groningen and the Peatland Reclamation District of Groningen and Drenthe is situated in the north-eastern part of the Netherlands and covers an area of 130 000 ha (Figures 17.1 and 17.3). Administratively, it belongs to two provinces and 30 municipalities, and has approximately 275 000 inhabitants. The reconstruction area consists of three entirely different regions: the Oldambt, the Peatland Reclamation District and Westerwolde.

The *Oldambt*, the northernmost part, is a region reclaimed from the sea, with mainly clay soils. It is characterised by vast open spaces between the villages. Its arable and dairy farms are relatively large to very large compared to the average Dutch farm size, varying from 50 to more than 100 ha. It is generally known as a wheat-growing district.

The *PRD*, the central part, consists of reclaimed peatland. The farms here are smaller than in the Oldambt region. In the past, main canals, tributary canals and secondary drainage canals were dug for draining the peatland and for transport of peat and, later, agricultural products. Houses and also farms were built along the canals. Arable farming, in particular potato growing, began to flourish on the reclaimed peat land. The potato starch industry developed alongside the farming activities, as did the strawboard industry. The use of water for transport led to the building of shipyards with their associated activities.

Westerwolde is the eastern part alongside the German border. It is a postglacial sandy area with an attractive landscape, which makes it suitable for recreation. The type of farming resembles that in the PRD.

Historical background

In earlier centuries there was a large area of bog in the south-east of the province of Groningen and the neighbouring part of the province of Drenthe known as the *Bourtangerveen*. This almost uninhabited and inaccessible region acted as a natural boundary, hence an official, arrow-straight border between the Netherlands and Germany was not drawn until 1824. A start to its development had already been made in the Middle Ages by local monks and others by the digging of peat, which, dried, made good fuel. The exploitation of peat took off in the seventeenth century, mainly in response to the heavy demand for fuel from the rapidly growing towns in the Netherlands and Germany. A dense network of canals was constructed, which first served to drain the bogs and then to transport the peat.

The city of Groningen played an important part in this development of the 'peat colonies'. It acquired large areas of land and was also the owner of the roads and canals, from which it received a considerable income in the form of tolls. This is reflected in

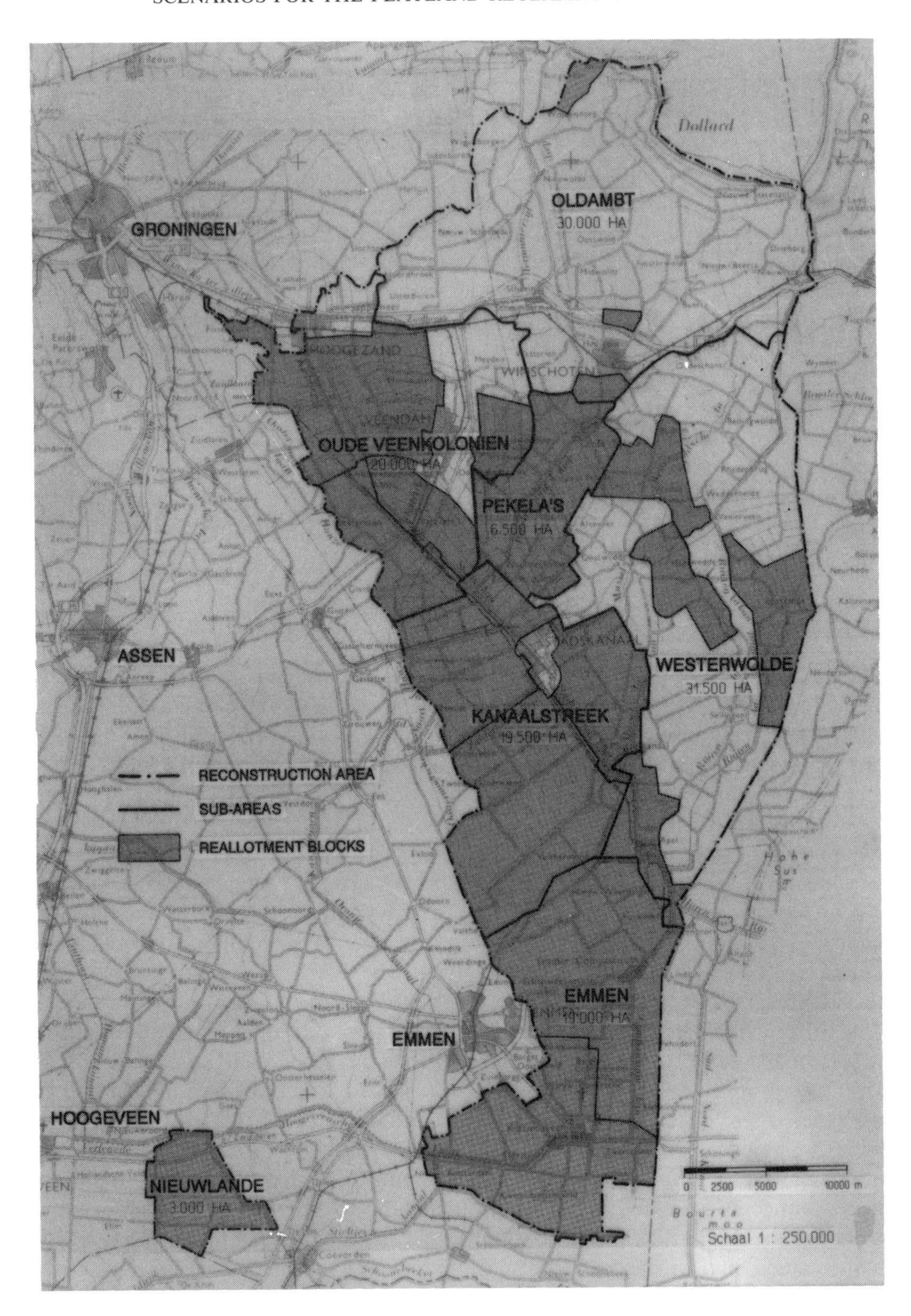

Figure 17.3 East Groningen and the Peatland Reclamation District

the name of a major town in the district, Stadskanaal, which can be translated as 'canal of the city'—the city being Groningen. Under a city regulation, peat diggers were obliged to restore the land for agricultural use after they had finished digging out the peat. This was done by saving the top layer, which largely consisted of heather turves, and subsequently mixing it with the underlying sand after removal of the peat. The soil thus created, known as 'dal' soil, was chemically infertile, but physically favourable with a good crumb structure. Its fertility was increased first by the addition of manure and urban refuse (brought in as a return cargo by the peat boats) and, from the end of the nineteenth century, by the addition of large quantities of artificial fertilisers.

This was the basis for a very specific agricultural region, with its characteristic geometrical pattern of canals and ribbon villages that developed along the main canals in a generally narrow and very strung-out form. Farm parcels were originally very narrow and nearly all transport was by water, but during the 1950s and 1960s water transport was replaced by road transport. Many canals were filled in, opening up the area for road traffic. Moreover, many of the narrow cross-canals were filled in to increase the size of the farm parcels. The most important arable crop is still the industrial potato, which is processed in local factories to produce potato flour and a large number of derivatives, including adhesives and binding agents.

Problem region

For a long time East Groningen and the PRD of Groningen and Drenthe have been facing problems which, by Dutch standards, are extensive and complex. Employment has disappeared entirely in the peat industry and decreased in agriculture. In the 1950s and 1960s, employment in the strawboard and potato starch industries also declined. New jobs in manufacturing and service industries have not sufficiently compensated the total loss of jobs, so that the area has the highest percentage of structural unemployment in the Netherlands. Although farms in the region are relatively large, they are still too small to provide a decent income when cultivated in the current farming system. Economic development in the reconstruction area is stagnating, and the average income is substantially below the national average. The composition and physical structure of the reclaimed soils, especially in the PRD, in some cases make them unsuitable for modern and economic farm management.

In 1970 the farmers in the PRD presented a request to the national government and the provincial authorities of Groningen and Drenthe for a land consolidation scheme covering an area of 70 000 ha, which was regarded as a possible solution for their problems. These problems included excessively long, narrow and often scattered lots, poor accessibility and a poorly functioning system of water management. The agricultural industry faced, in addition to the general economic recession of the 1970s, rising costs incurred in combatting the serious environmental pollution caused by the discharge of factory effluent into the canals. Boat building, once an important industry along many of the canals, had shifted to more spacious navigable water in the north-east of the province of Groningen. The residential amenities were deficient in many respects, mainly because of the ribbon development characteristic for most of the settlements and because of a lack of facilities for open-air recreation. The presence of some 9 000 ha of land encumbered with the privileges of the city of Groningen also formed an obstacle to the further development of the region.

In addition to agricultural needs, there is the need for a good infrastructure for road transport. Other problems in the reconstruction area are inadequate water management, lack of scenery and recreational facilities (except in Westerwolde), pollution of canals by industrial and household wastes, decay of old buildings, an ageing population, a relatively low level of education, migration of younger and better educated people, and a lack of skilled workers. All this creates a negative image of the area.

Preparation of the Rural Redevelopment Act

Because the problems of the area were not confined to the agricultural sector, at the request of both provinces and after consultations with the population, the Government decided on large-scale reconstruction of the PRD, the Oldambt and Westerwolde. A special act, the Rural Redevelopment Act, was drafted for this purpose, and became effective on 1 January 1979.

The objective of the Rural Redevelopment Act is to stimulate the economic and social development of the region and to create a good living and working climate. The Act provides for the coordinated and, where possible, integrated implementation of the following measures and facilities:

(1) improvement of the infrastructure and the agricultural structure;
(2) reallotment of estates situated in the reconstruction area;
(3) abolition of the historical land use rights of the city of Groningen;
(4) conveyance of ownership of roads and canals owned by the city of Groningen, but situated outside its boundaries;
(5) provision of facilities for landscape conservation and outdoor recreation;
(6) conservation and improvement of nature and landscape and of elements of cultural and historical value;
(7) drafting and implementing of a socio-cultural plan;
(8) reconstruction of town and village centres;
(9) demolition or renovation of poor housing facilities, and providing funds for this purpose;
(10) creating public facilities for the discharge and treatment of waste water.

Only part of the plan will be executed by the Reconstruction Committee itself. Other parts of the plan are the responsibility of municipalities, provinces, polder-boards, or the Ministry of Public Works. It became clear that the cooperation, integration and coordination of the project needed a special legal basis. It was also important that this special act made clear that politicians were giving maximum attention to this region, as the political situation was rather unstable.

The Redevelopment Act distinguishes two stages for its implementation. First, the redevelopment programme for the whole region, and then redevelopment plans for each of the seven sub-regions. Implicit to the programme are the starting points for reconstruction and the distribution of costs. After that, a reconstruction plan based on the programme for a particular sub-region can be made. Such a plan includes:

(1) a description of the spatial, economic and social situation in the area;
(2) a description of the agricultural situation in those parts of the area that will be realloted (the so-called reallotment 'blocks');
(3) a provisions scheme with maps to be executed by the Reconstruction Committee;
(4) other means and provisions to be executed by other authorities;
(5) an estimate of total costs and the distribution of these costs.

Execution of the rural redevelopment plan is entrusted to a reconstruction committee of 31 members and 29 advisors, representing the interested parties in the region. The members include representatives from the two provinces, the municipalities, farming organisations, polder-boards, chambers of commerce, unions, environmental organisations and socio-cultural institutions.

The advisors represent different ministries and provincial authorities. The Minister of Agriculture, Nature Management and Fisheries provides the manpower for the secretariat of the committee. This secretariat prepares the decisions to be taken by the committee and takes care of their subsequent implementation. An extensive public participation procedure gives the local population and bodies established in the area an opportunity to comment on and object to the draft plans.

Implementation of the Redevelopment Act

The redevelopment programme

In the spring of 1980 the Reconstruction Committee submitted a provisional redevelopment programme for discussion. After consultations, this provisional programme was sent for advice to the Central Land Consolidation Committee (CCC, now the Central Land Development Committee (CLC)) in the autumn of 1980. In April 1981, this committee presented the provisional programme, together with its advice, to the provincial governments of Groningen and Drenthe, who approved it in May 1983. During that period both provincial governments negotiated with the national government the financial support to be contributed by the state for the reconstruction. In December 1983 the national government decided on the reconstruction programme, after having limited its scope, especially in the financial sense. The programme contains provisions in the following fields, according to the different goals of the Redevelopment Act.

Improvement of the infrastructure

An adequate transportation infrastructure is undoubtedly a first requirement for economic development. Considerable improvements in this field in the Redevelopment Area are necessary, particularly on existing link roads (widening, duplication, etc.). Special attention is paid to the reduction of traffic congestion in a number of urban centres (Hoogezand, Winschoten, Stadskanaal) by the construction of ring roads and by the improvement of the links with Germany. Maintenance of rural roads is also very important in this predominantly agricultural region. The aim is to make all farm lots accessible by metalled road, which means that 230 km of rural roads will have to be improved and 150 km of new roads constructed. Further proposed improvements relate to the network of cycle tracks and the waterways system (widening of canals, enlargement of locks, etc.).

Land consolidation and improvement of the agricultural structure

Of the 130 000 ha constituting the Redevelopment Area, 70 000–80 000 ha are to be subject to land consolidation, involving the consolidation of the scattered lots belonging to each holding, improvement of land quality on about 20 000 ha, enlargement of holdings and relocation of farms.

Abolition of the privileges of the city of Groningen

As explained above, the development of the peat colonies in the seventeenth century took place mainly at the initiative of the city of Groningen and the city still enjoys rights of ownership over canals (about 60 km), roads (about 30 km), bridges and some 9500 ha of land. Under the Redevelopment Act these historical rights will be abolished and farmers who farm such land and have, until now, paid rent to the city, will be able to buy the land. Generally speaking, roads will be transferred to the municipalities and canals to the water control boards.

Provisions relating to landscape, recreation, nature and historical features

These provisions include conservation of certain landscape types, the planting of woodlands in the vicinity of the larger population centres, creation of open space in ribbon-strip villages and tree planting on farmsteads. A total of 4000 ha of woodlands will be planted in this treeless part of the Netherlands. In addition, the existing canals and ponds will be adapted, where possible, to the needs of water recreation. Another important provision is the conservation of the few remaining areas of raised bog, such as that near Klazienaveen. Furthermore, measures are included for facilities for both nature development and nature conservation, needing 6400 ha of agricultural land.

Socio-cultural activities

The programme stimulates activities for better functioning or use of existing facilities, which includes items like work, housing, education, public health, social welfare, and recreation. One specific issue was the desire to stimulate the participation of the population in public discussions about town reconstruction. All these items are now the responsibility of local authorities, to whom all the available funds have been delegated. Additional funds are no longer available.

Renewal of the built-up areas of towns and villages

The Redevelopment Area contains some 8500 substandard dwellings which must be improved or replaced. The urban structure and the residential amenities of many of the built-up areas along the canals are ripe for improvement. This demands costly provisions and a well-coordinated and integrated programme. Nearly all the towns and villages situated along canals wanted to participate in this part of the programme.

Environmental improvement

Work to improve the treatment and discharge of industrial effluent—which until recently was a source of considerable nuisance because of the smell—is already in progress outside the framework of the Rural Redevelopment Act. Redevelopment plans include proposals to improve the discharge of domestic sewage, by constructing sewers in thinly populated areas, and by building sewage pumping stations, etc.

Budget

The total costs of the Rural Redevelopment Programme were estimated in 1980 at over 2.5 billion guilders. The largest items are urban and village renewal, agricultural provisions and improvements to the infrastructure. Under the Programme, central government will contribute two-thirds of the cost while the remainder will have to come from the two provinces, the municipalities, the water control boards and the private sector. The programme is also eligible for a grant from the Regional Development Fund of the European Community.

Regional redevelopment plans

The redevelopment area was divided into seven sub-regions, each having a subcommittee under the supervision of the Reconstruction Committee. The task of these subcommittees was to prepare a plan for their own area. After approval of the plan they have the responsibility for its proper implementation. For each of the sub-regions a plan has been developed. After consultations with the population and after an assessment by the national government, these plans had to go to the provincial governments for final approval. The plan for Westerwolde was approved in February 1993.

It must be emphasised, that a strong relationship exists between reconstruction plans and physical planning of the provinces. Changes in the main function of any place, for example, from an agricultural function to a non-agricultural function, must be established formally by the provincial authorities, after public discussion, before the development of the new function can become part of any redevelopment plan. As has already been shown, the reconstruction of an area takes a long time.

To prevent delays due to changes in public opinion for whatever reason, it is possible, if necessary, to adapt the plan to new ideas or to elaborate the plan in agreement with actual views at the moment of execution. The Redevelopment Act contains provisions to assure flexibility.

All seven plans are now being carried out. The total estimated costs amount to 1845 million Dutch guilders. About 600 million guilders are conventional land reclamation costs. Of this, 175 million guilders will have been invested by the end of 1993. Total implementation of a reconstruction plan will take 10–20 years, depending on the size and complexity of the region concerned. Today (1993), 14 years after the Redevelopment Act became effective, it has wide public support. The last plan is expected to be completed in the year 2005.

Conclusions

Implementation of redevelopment in East Groningen and the Peatland Reclamation District covers a wide range of measures and provisions. Only if all participants are willing to cooperate can this complex programme be successful. This also holds for the financial aspects. Wide public support is necessary for preparation and implementation of the plans. So far, preparation and implementation of the Redevelopment Plans have been successful. The population has been encouraged by the newly structured situation, and after renewal of the public infrastructure people began to take care of their own properties again.

The influence of man is clearly visible in nearly all Dutch landscapes, but there are few regions where such a dominant stamp has been left and where both the appearance and the function of the region have repeatedly changed so drastically over the course of a few centuries as in this region: from peat bog to peat digging area, to agricultural area old style (narrow lots and water transport) and, thanks to the present reconstruction, to an area that can face the challenges of future decades, whether it is by the intensification of arable farming, dairy farming, nature development, landscape and recreation, or forestry.

REGIONAL LAND USE PLANNING IN THE PEATLAND RECLAMATION DISTRICT: CONSEQUENCES FOR FARMERS AND AGRIBUSINESS

Introduction

Until the beginning of the 1970s the PRD was an example of a prosperous agriculturally oriented area. The majority of the inhabitants was employed in agricultural jobs or related activities. The use of labour, capital and land for agriculture and agro-industry was undisputed. A large number of local straw cardboard and potato starch factories were very much in evidence. Increases in labour costs and the necessity of capital investment were the causes for an expulsion of labour. Farms changed from mixed farming to purely arable crop cultivation. However, compared to the adjacent Oldambt region, the PRD remained a relatively small-scale agricultural region.

From the moment returns on agricultural products started to decrease because of increased production costs, the number of farmers began to decline. Many small-scale factories shut their doors, the once prosperous straw cardboard industry closed down completely, and the potato starch industry was urged to adapt very rapidly in order to maintain its market position. Scaling up was inevitable and many small starchmills closed gates.

The lack of alternative crops left no other alternative than to intensify the use of arable land by growing more potatoes and sugar beets, the only arable crops that gave sufficient returns. Also the government encouraged farmers to intensify. The introduction of agrochemicals, such as fumigants, allowed potatoes to be grown once every two years, with sugar beets grown once every four years on the same piece of land. Today, crop rotations are still more or less the same. Since 1970 the starch potato has become the major agricultural crop on which the farmer depends.

Farmers' perspectives

Clearly, the more narrow rotation of potatoes, allowed since 1968, has been dragging the process of farm size extension. Nevertheless, farm size has increased from about 25 ha to more than 40 ha in the last 10–15 years. It is most likely that this tendency will continue. According to statistical information, less than 50% of the farmers will have a successor. So there is reason to believe that the number of farms will drop another 40% or 50% over the next 10–15 years. As a consequence of decreasing EC guaranteed prices for the main arable products since 1984, agriculture in the PRD will also remain under pressure. The minimum price for starch potatoes is linked to that for cereals. Since 1983 this price has dropped by more than 20%. As farmers depend on the returns of this crop for 50–70% of their income, it is obvious that earning capacity in agriculture has decreased dramatically. The only way out is cost reduction, scaling up and/or a second agricultural activity next to arable farming. It seems that the younger generation is more willing to leave the traditional pathways.

Arable farms in general are run using a large amount of the farmer's own capital. Acute financial problems are rare and bankruptcy seldom occurs. Successors can overcome financial problems by excellent management and skills, but not everyone is blessed with these. The new EC agricultural policy will force small arable farmers to increase their farm size beyond 100 ha. For most farmers, however, a mix with horticultural crops or pig and poultry husbandry will be the only real perspective. Apart from economic problems, strict legislation on the use of agrochemicals within the government environmental policy puts farmers under pressure as well. Both factors undoubtedly will contribute to a less intensive crop rotation of potatoes on a number of farms. This will have a negative influence on the raw material position of the starch industry.

Position of the agro-industry

Until now the PRD's image was dominated by the agro-industry. Each farmer in one way or another is a member of a producer's organisation or industrial cooperative. Most farmers depend on the agro-industry for more than 70% of their income. In the potato starch industry at the moment, some 2000 people are employed. Since 1979, after a dramatic period, only one company has remained—the cooperative AVEBE, with five factories in the region. Total linked and derived employment is estimated to be some tens of thousands, farmers included. The Dutch potato starch and starch derivatives industry is the largest in the world. Annual production accounts for about half the production in the EC. Because it is concentrated in the PRD, the starch industry is highly dependent on an optimal raw material supply from the region itself, as potatoes are a bulk product. Market growth and loss of area in the district have been compensated by raw material acquisition in the adjacent German Weser–Ems Region. It seems that these opportunities outside the PRD have now reached their limit. Operating on a commodity market for the majority of its products, also in competition with cereal and maize starches, the potato starch industry will not be able to completely compensate decreasing minimum prices for starch potatoes. Profit margins in the agro-industry in most cases are smaller than in other industrial branches. Only through continuous efficiency improvement, changes in product mix, innovation and added value can the industry maintain its position. Consolidation of its position in the Netherlands and extension both in and outside the EC, including other raw material sources, is a main priority.

Lack of other industrial activities is of great concern in this area. Many projects have been started up in search of feasible processing opportunities for new industrial crops. Also regional agro-industries and agribusinesses are involved in these activities. If a project turns out to be promising, it will certainly get the full support of the established agro-industry. A sound agricultural sector is in their interests too, and practically all agro-industries in the region have a cooperative background. However, there is great awareness that in general only large-scale processing of an industrial crop is feasible. Therefore, it is necessary to look beyond the borders of the PRD.

The results of the projects that started a few years ago are not available at this moment, but expectations are that some might achieve successful implementation in the next 10 years. The PRD offers a rich experience in agriculture and affiliated industries. It is natural that this region has a certain attractiveness for potential agro-industry, especially after completion of the Reconstruction. More industrial activity may reinforce existing industrial activities.

Land use options

The success of the Common Agricultural Policy of the EC, which has resulted in substantial overproduction, has led to extensive discussions in the Netherlands about taking more arable land out of production in certain regions. Taking into account the limited area of arable land in the Netherlands compared to the total EC potential, this discussion bears a few unrealistic elements.

Nature and forestry lobbyists (not interested private parties, but politicians, public officers and nature conservationists, mainly sponsored by government funds) have already fallen upon the supposed bankruptcy of arable farmers like vultures. In their opinion a large supply of cheap land could become available for new wildlife or nature preservation or for large-scale afforestation over a short period of time. More extreme scenarios for 'de-agriculturisation' have already been presented. These expectations may well turn out to be over-optimistic.

More and more people, including the experts mentioned already, realise that the future supply of food products in the EC and the world is not consistent with a permanent set-aside strategy. It cannot be denied that agriculture, and especially arable farming, is under intense pressure at the moment. It is evident that this also holds for the Netherlands, but this does not mean that good arable land will be put up for sale for purposes other than agriculture.

Conclusions

Arable land in the PRD is regarded as having good production potential, because of a moderate climate, high organic matter content in the soil and a good water supply. Therefore, it is unrealistic to expect that the land in this region will cease to have a prime agricultural function in the EC. To keep up with Europe's agricultural development, it is necessary to scale up farming in the region. Designation of good arable land for other purposes does not match this objective. Also, in areas where less intensive cultivation of potatoes is regarded desirable for economic or environmental reasons, it is not sensible to implement this by diverting arable land to other uses. It is very important that the regional agro-industry has access to the existing potential arable area to maintain its raw material position. It is clear that it will strongly oppose plans to divert large areas of arable land to nature development and afforestation. It is most likely that trade unions will also take a stand against these plans when they realise the consequences for employment.

This does not mean that the agribusiness opposes plans for refurbishment of the region with non-agricultural land use for nature, woods and recreation. On the contrary, a sound life environment is also in the interests of the agro-industry and its employees. The region must in a certain way appeal to its inhabitants and potential newcomers as a place to work and live and to spend their leisure time. However, farmer's organisations and the agro-industry believe that more than ample marginal land is available to satisfy the needs for nature and recreation areas as identified in the Reconstruction Plan. Incidentally, these marginal lands are situated for the greater part in areas where the landscape has a touristic value or where nature and ecological value is undoubtedly present. Sooner or later, these marginal lands will loose their agricultural value anyway, as other comparable areas in Europe.

A lot of money has already been invested in the Reconstruction Plan. Implementation of the Reconstruction Act not only results in improvements in infrastructure and social life, but also in new agricultural employment. All elements have their place under general accordance. It is the task of the regional agricultural, industrial and governmental leaders to create a consensus on future development of the region. It is not in the interests of the region that the government is influenced by lobbyists who will take actions that were not agreed upon previously. Recent plans by the national government for large-scale afforestation will be met by strong opposition in the PRD. In no way can these plans be proven to be an asset to the district's economy. On the contrary, they are more likely to have a negative influence.

It is obvious that the basis of a sound economy in the PRD lies in agricultural development. It is also clear that only diversification of land use can help farmers to realise a reasonable income and may attract more and diversified agribusiness.

REFERENCES

Ippel, B. M. and Noordam, W. P., 1990. Kwantitatieve informatie voor de akkerbouw en de groenteteelt in de volle grond: bedrijfssynthese 1990–1991. IKC-AGV & PAGV, publicatie 53. PAGV, Lelystad, The Netherlands.

Kamminga M. R., Hetsen, H., Slangen, L. H. G., Bischoff, N. T., Van Hoorn, A. S., 1993. Toekomstverkenningen ruraal grondgebruik. Wageningen Agricultural University, Wageningen.

Krant P., Postuma, J. S. and Thijsen, S., 1987. *Bosrijk Noord-Nederland*. Grontmij, Groningen, The Netherlands.

Krap, A. and Thijsen, S., 1990. *Nationaal Landschap Zuid-west Friesland, Produktontwikkelingsplan*. Grontmij, Drachten, The Netherlands.

Langbroek, E., Meester, R., Strijker, D. and Thijsen, S., 1990. *Landelijk Noord-Nederland*. LB&P, University of Groningen, and Grontmij, Groningen, Drachten, The Netherlands.

Ministry of VROM, 1991. *Vierde Nota over de Ruimtelijke Ordening*. VROM, Den Haag, The Netherlands.

Van Der Ploeg, J. D., 1993. Over de betekenis van verscheidenheid. Inaugural address, Wageningen Agricultural University, Wageningen.

Van Niejenhuis, J. H., Smit, A. B., Strijker, D., Van Dullemen, M. J., Moens, R. P., Thijsen, S. and Wisse, J., 1991. *Toekomstperspectieven Oldambt en Veenkoloniën*. Provinciale Besturen van Groningen en Drenthe, The Netherlands.

Struik P. C., Van Niejenhuis, J. H., De Hoogh, J., Veerman, C. P., Schouls, J., Van Arkel, H. and Renkema, J. A., 1991. Problematiek en vooruitzichten van de Nederlandse akkerbouw. Mededeling 99 van de Vakgroep Landbouwplanteteelt en Graslandkunde. Wageningen Agricultural University, Wageningen.

Stuurgroep NOF/NWG, 1993. *Ontwikkelingsplan NO-Friesland/NW-Groningen*. Grontmij, Groningen, The Netherlands.

Wennekes, W., 1993. *De Aartsvaders*. Atlas, Amsterdam, The Netherlands.

CHAPTER 18

Competing for Limited Resources: Options for Land Use in the Fifth Region of Mali

F. R. Veeneklass,[a] H. van Keulen,[b] S. Cissé,[c] P. Gosseye[b] and N. van Duivenbooden[d]

[a]AB-DLO. Present address: The Winand Staring Centre (SC-DLO), Wageningen, The Netherlands.
[b]Research Institute for Agrobiology and Soil Fertility (AB-DLO), Wageningen, The Netherlands.
[c]Étude sur les Systèmes de Production Rurales en 5ème Région (ESPR), Sévaré, Mali.
[d]AB-DLO. Present address: Wageningen Agricultural University, Wageningen, The Netherlands

INTRODUCTION

Regional agricultural development can be defined as the dynamics of changes in infrastructure and technologies that are pursued to improve existing agricultural systems and the welfare of the population. In general, a wide variety of technologies (i.e. well-defined ways of converting inputs into agricultural products) is available, both in terms of the commodities and in terms of the mix of inputs used to produce them, but not all of these are technically feasible and/or economically viable in the physical environment, as characterised by relevant soil properties, prevailing weather conditions, the socio-economic context characterised by the regional capital and human resources and constraints, and the prices of inputs and products, of a given region.

Analysis of options for agricultural development in a region, as a basis for land use planning, aims at selection of the 'best' development option, i.e. that resulting in the most complete realisation of the development goals. However, generally, various actors with different interests are involved in the development process, leading to at least partly conflicting objectives, thus precluding the unequivocal definition of 'best'. To arrive at an acceptable choice in this situation, we must identify, on the one hand, the technical possibilities in a region, as dictated by the natural-resource base and the available production techniques, and on the other hand, all possible development objectives or goals, while taking into account the socio-economic situation.

In this chapter, a method using interactive multiple goal linear programming (IMGLP: de Wit et al., 1988) is illustrated to explore the agro-technical possibilities and quantify the necessary technical coefficients, to identify the possible degree of goal attainment,

The Future of the Land: Mobilising and Integrating Knowledge for Land Use Options
Edited by L. O. Fresco, L. Stroosnijder, J. Bouma and H. van Keulen.

and to judge the economic viability of different development options, as an aid in land use planning for one of the provinces of Mali in West Africa (henceforth in this chapter referred to as 'the Region'). The results of the study have been extensively reported (Cissé and Gosseye, 1990; Veeneklaas, 1990a; Van Duivenbooden et al., 1991; Veeneklaas et al., 1991), so this chapter only highlights some methodology and results.

SOURCES OF DATA

The soil resource was defined on the basis of an extensive inventory of land resources (PIRT, 1983). Seven main soil groups were distinguished on the basis of soil texture, and were further subdivided into a total 18 subgroups on the basis of properties of the top soil (gravelly), soil fertility status and depth of the groundwater table. For each of these units a representative set of soil physical (referring mainly to the water transport and storage characteristics) and chemical (referring to the supply of plant nutrients from natural sources (soil fertility) and the recovery of applied fertiliser) properties was defined. The soil physical characteristics were introduced in a crop-growth simulation model, applied to calculate potential and water-limited production (Erenstein, 1990).

Weather data required for the simulation model are daily values of minimum and maximum temperature, solar radiation, atmospheric humidity, wind speed and precipitation. Except for rainfall, the spatial and temporal variability in weather characteristics within the Region is small. Hence, for those variables, long-term average monthly data, recorded at Mopti, the capital of the Region, were used. Daily precipitation values for a period of 30 years (1959–1988) were available from seven weather stations in the Region. On the basis of that information, four rainfall zones were distinguished, characterised by long-term annual averages of 300, 320, 382 and 485 mm, respectively. For each of the zones, 'normal' and 'dry' years were defined, comprising the middle 60% and the driest 20% of the rainfall years, respectively.

Combining the soil resources and the rainfall zones, 11 agro-ecological zones were distinguished which form the basis for the agro-physical characterisation of the Region (Figure 18.1).

AGRICULTURAL ACTIVITIES

Agricultural activities in the Region comprise arable farming, animal husbandry and fisheries. Each of the activities may take place anywhere in the Region, provided that the required inputs are available. For each of the activities, different production techniques have been defined, characterised by a specific combination of technical coefficients quantifying their inputs and outputs.

All agricultural activities are assumed to be sustainable, so that their production capacity in the long run is not jeopardised. For arable farming activities, sustainability is defined in terms of nutrient elements, i.e. the total store of each of the macro-elements in the soil is assumed to be constant. This implies that in the absence of the application of nutrients from external sources (manure or chemical fertiliser), export in agricultural products should not exceed the supply from natural sources, which is achieved by defining sufficiently long fallow periods. For animal husbandry systems, sustainability implies

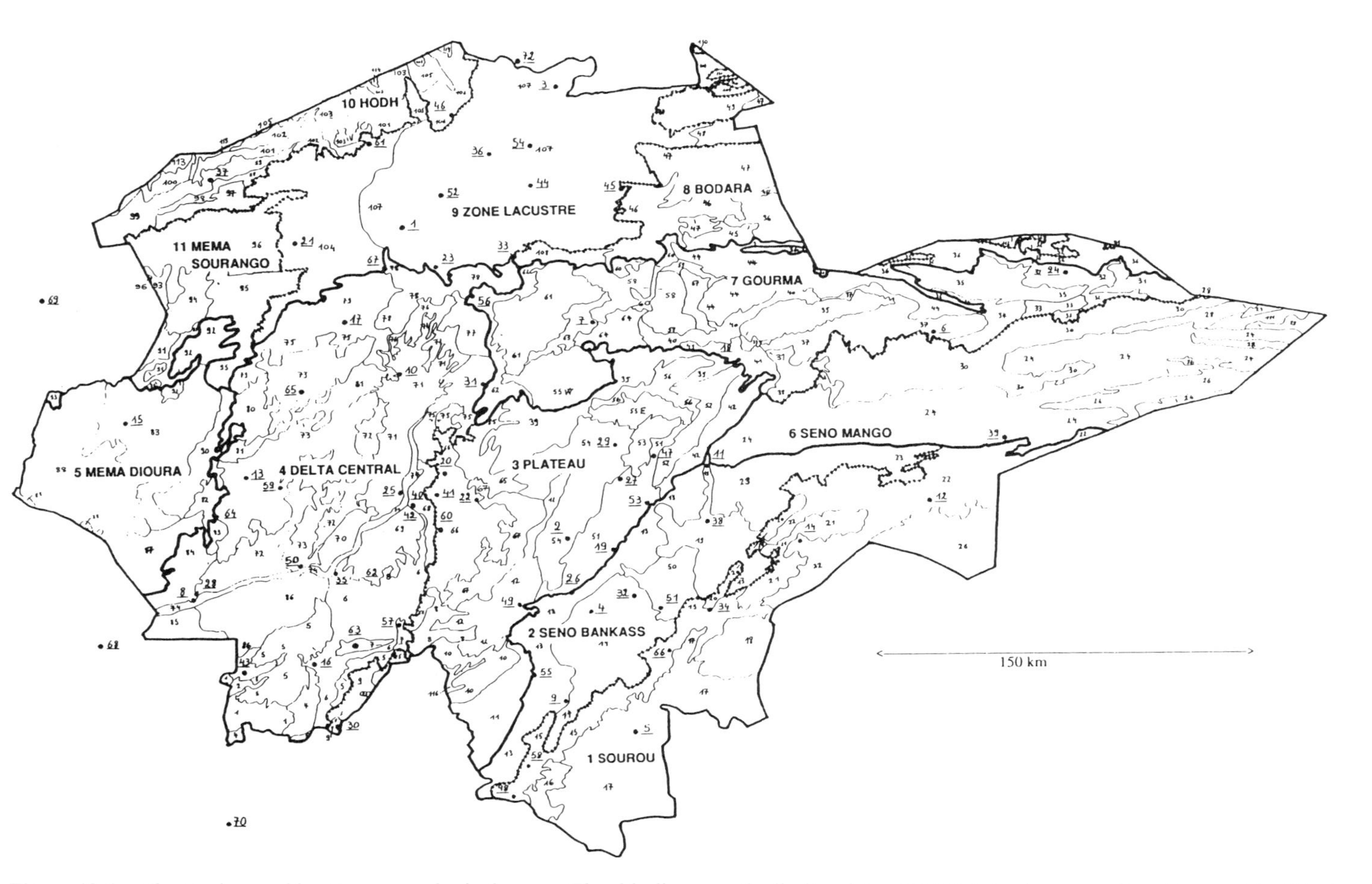

Figure 18.1 The Region and its 11 agro-ecological zones. The thin lines are the limits of the 116 PIRT basic map units which are identified by the small numbers

Table 18.1 Input–output table of millet production techniques on soil type C1

Characteristic	Extensive				Semi-intensive	Intensive
	1	2	3	4	5	6
Animal traction	–	–	+	+	+	+
Manure	–	+	–	+	+	+
Chemical fertiliser	–	–	–	–	+	+
Fallow	+	–	+	–	–	–
Inputs (per ha/yr)						
Fallow/manure/fertiliser						
Ratio fallow years/year cultivated	5	—	6	—	—	—
Manure (kg DM)	0	1930	0	2290	2530	1930
Fertiliser N (kg)	0	0	0	0	12	96
Fertiliser P (kg)	0	0	0	0	0	12
Fertiliser K (kg)	0	0	0	0	0	56
Labour (man-days)[a]						
6 Cleaning the field	5	1	5	1	1	1
6 Transport and application of manure	—	17.5	—	21	15.5	12
1 Basic dressing	—	—	—	—	1	1
1 Land preparation	3	3	4.0+2 At	4.0+2 At	4.0+2 At	12.0+6 At
1 Sowing	5	5	5	5	5	2.0+1 At
2 Weeding 1	15	15	10.0+2 At	10.0+2 At	10.0+2 At	10.0+2 At
2 Top dressing 1	—	—	—	—	4	4
2 Pesticide spraying 1	—	—	—	—	—	0.5
3 Weeding 2	12	12	12	12	12	12
3 Top dressing 2	—	—	—	—	—	4
3 Pesticide spraying 2	—	—	—	—	—	0.5
4 Harvesting	5	5	6	6	5	12
6 Transport, threshing and winnowing	16.5	16.5	13.5	13.5	19.5	46
Total	61.5	75	55.5+4 At	72.5+4 At	77.0+4 At	117.0+9 At

Monetary inputs (FCFA)						
Capital charges						
Small equipment	700	700	700	700	1000	1500
Plough	—	—	1670	1670	1670	5620
Sowing machine	—	—	—	—	—	1600
Sprayer	—	—	—	—	—	1200
subtotal	700	700	2370	2370	2670	9560
Operating costs						
Seeds	60	60	60	60	60	60
Pesticides	100	100	100	100	250	6500
subtotal	160	160	160	160	310	6560
Total	860	860	2530	2530	2980	16120
Oxen (ox)	—	—	0.33	0.33	0.33	0.75
Outputs (per ha/yr)[b]						
Grain (kg DM)	500	500	600	600	960	2390
Straw (kg DM)	1750[c]	1750[c]	1980[c]	1980[c]	2800[c]	4570[d]

[a]Numbers in front of operations refer to the period of the year (see text); At refers to oxen team days.
[b]In a normal year in rainfall zone I (average precipitation in May–October: 530 mm).
[c]Average N content is 3.9 g/kg.
[d]Average N content is 5.1 g/kg.

stable animal numbers for each of the species distinguished, based on sustainable forage production, which in addition to nutrient limitations, is subject to persistency limitations, so that only a fraction of total annual production can be exploited. For fisheries, sustainability refers to a maximum quota of fish that can be caught annually.

In arable farming three cultivation types are considered: rainfed, flood retreat and irrigated or inundated. Each of these is further subdivided by crop species: millet, rice, sorghum, fonio, cowpea, groundnut, forage crops, shallots and 'other vegetables' (comprising tomatoes, tobacco, cassava, cabbage, etc.), although not every combination is relevant. For each crop, different technologies are defined on the basis of the following criteria: (i) application of fallow periods, (ii) oxen traction, (iii) application of farmyard manure, and (iv) application of chemical fertiliser. This resulted in the formulation of techniques at three levels of intensity: extensive, without any application of external inputs; intensive, with high levels of such inputs; and semi-intensive, with intermediate levels. Moreover, in intensive techniques a high degree of innovative practices is included. The degree of differentiation considered for each crop depends on its relative importance: for millet, a major crop in the Region, six different production techniques have been formulated, whereas for a minor crop like fonio (*Digitaria exilis*), only one production technique is included. An example of an input–output table for millet production techniques (applying to a specific soil type in a specific agro-ecological zone) is given in Table 18.1.

Animal husbandry systems in the Region comprise cattle, sheep, goats, camels, donkeys, horses, pigs, poultry and wild game at varying degrees of importance. Different production techniques have been formulated on the basis of four criteria:

- animal species (cattle, sheep, goats, donkeys and camels)
- main production objective (meat, milk, traction or transport)
- mobility of the animals (migrant, semi-nomadic or sedentary)
- production level (low, intermediate or high)

The inputs in these systems consist of required feed, specified according to season, spatial distribution and quality, labour and monetary costs. As an example, the input–output table for cattle activities is given in Table 18.2.

The fishing activities are distinguished on the basis of the main occupation and the mobility of the households engaged in fishing:

(1) households practicing fishing as the main occupation, migrant (MFF);
(2) households practicing fishing as the main occupation, sedentary (MSF);
(3) households practicing fishing as a side activity, sedentary (SSF).

The three household types vary in capital endowment and in productivity. In contrast to the arable farming and animal husbandry activities, the available data did not allow calculation of target yields, hence, the amount of fish captured and the required labour have been used as the basis for the input–output coefficients.

Table 18.2 Input–output table for cattle activities (TLU/yr); required quality of diet, intake of forage and concentrates, total labour in the wet and dry season and money, meat, milk, number of animals, and manure

Code	Main production	Mobility	Intake (kg dry matter)			Total labour (man-days)		Money (1000 FCFA)	Meat (kg liveweight)	Milk (kg)	No. of animals	Manure (kg dry matter)
			Diet	Forage	Concentrates	Wet season	Dry season					
B1	Oxen	sedentary	I	2000	—	2	8	2.3	22	0	0.55	442
B2	Meat	semi-mobile	I	2000	—	3	8	2.3	37	0	—	298
B3	Meat	semi-mobile	II	2000	—	3	9	3.5	56	92	—	285
B4	Meat	migrant	I	2000	—	3	8	2.3	37	0	—	230
B5	Meat	migrant	III	2100	—	3	9	3.5	71	219	—	222
B7	Milk	sedentary	II	2100	—	3	9	2.3	54	165	—	444
B8	Milk	sedentary	III	2200	—	4	10	3.5	62	376	—	445
B9	Milk	migrant	II	2100	—	3	9	2.3	54	165	—	232
B10	Milk	migrant	III	2200	—	4	10	3.5	62	376	—	232
B11	Milk	sedentary	IV/c	1820	380	5	14	22.0	61	520	—	415
B12	Milk	sedentary	IV	2200	—	5	14	22.0	61	520	—	415

REGIONAL CONSTRAINTS AND RELATIONS

Agricultural exploitation of the Region is subject to various constraints, of which the known physical and technical ones have been incorporated in the model, but the institutional and/or social constraints—which can be equally important—cannot be taken into account directly in a formal quantitative model because of lack of reliable quantitative insight. Hence, these must be taken into account in a post-model analysis.

The basic restriction for land is that it can be used for one purpose at a time only, thus introducing the competition for land. Moreover, not all land is suitable for agricultural exploitation due to, for instance, degradation, hence, each soil type in each agro-ecological zone has been assigned a so-called 'utility index', ranging from 1 (if all land can be used) to 0 (if none is suitable). Specific soil types may in addition only be suitable for certain crops, such as land that is temporarily flooded. Finally, all land outside a radius of 6 km from a permanent water point is considered suitable for pasture only. If techniques requiring a fallow period are selected, the specified area per unit cultivated land is used for fallowing.

With respect to labour, the year has been subdivided in six periods, on the basis of the cropping calendar. The labour requirements for arable farming activities are specified in terms of these periods. For animal husbandry, only the wet and the dry season are considered separately. Where fishing is practised as the main occupation, labour is required throughout the year, whereas for the third activity, labour is required during off-peak periods only. In the model the specified condition is that in each agro-ecological zone, in each of the periods labour demand should not exceed labour availability.

Animal traction is assumed to be provided solely by oxen, under the condition that in each agro-ecological zone the number of oxen pairs required should not exceed the number available, which is an output of animal husbandry activity (B1, Table 18.2). Manure is also specified as output of animal production activities (Table 18.2) and used in arable cropping activities and in some agro-ecological zones as a substitute for firewood. In the model the condition is specified that per agro-ecological zone, demand should not exceed the supply.

Forage availability from natural pastures is specified per agro-ecological zone, taking into account its temporal distribution (wet season and dry season) and its quality (Table 18.2). In addition, availability and quality of crop residues and concentrates (both in the dry season only) are taken into account to estimate total feed availability. The total forage demand (differentiated in time and quality) per agro-ecological zone should not exceed the supply.

Total catch of fish in the Region is subject to an upper limit, which depends on flood level and is different, therefore, for 'normal' years and 'dry' years.

Transport animals—donkeys and camels—are indispensable for daily life in the Region. However, it is difficult to assign them directly to specific agricultural activities. Therefore, the required number of transport animals is related to population density; for camels the required number is defined for the Region as a whole, for donkeys per agro-ecological zone.

Crop yields, comprising economic product and crop residues, are specified for each agro-ecological zone, per soil type, whereby harvest and post-harvest losses have been taken into account. A (crop-specific) part of the crop residues is available for animal consumption and has been specified in terms of quality (Table 18.2). Inputs into agricultural activities consist of four groups—labour, monetary inputs, traction and nutrients—specified per activity, and (for nutrients and labour) as a function of yield. Subsistence needs of the local population, specified per sub-region, have been defined in terms of energy intake (equivalent to 626 g millet/person/day), animal protein intake (25 g meat/person/day) and variation in the diet (specified in terms of minimum intake of the various products produced in the Region).

GOALS

In the optimisation model, 20 variables have been defined that, in principle, can be optimised or can be assigned limiting values. In the actual study, only a few of these variables ('goals') are actually optimised (maximised or minimised); for most variables only minimum or maximum levels are set.

The set of goal variables can be classified in various categories.

Physical production targets in normal years (metric tons = 1000 kg/year)

For crop production, the physical production targets include:

- total millet, sorghum and fonio production
- total rice production
- total marketable crop production, i.e. total crop production (except fodder crops) minus the subsistence needs for crop products

For animal husbandry, the physical production targets (liveweight for meat), include:

- total meat production, comprising beef, mutton and goat meat
- total beef production
- total milk production, comprising cow, sheep and goat milk
- total herd size, expressed in tropical livestock units (TLU)

Monetary targets (10^6 FCFA/year)

A pivotal goal variable in the optimisations is total monetary revenue in normal years, i.e. the balance, in monetary terms, of all marketable outputs of crop, livestock and fisheries activities plus incoming money from emigrants, and all inputs that have to be paid for. As labour inputs are not priced, they are not included in this accounting scheme, neither are organic manure and crop by-products produced within the Region (which are only included in the input–output framework in physical terms, taking care that the balance is correct) and land (except in cases of depreciation costs of polders or irrigation works). Therefore, monetary revenue can—when the incoming money from emigrants is excluded—be interpreted as the remuneration of labour in the Region.

The term 'emigration' is here reserved for those members of the base population that leave the Region (or the agricultural sector) and do not return to work during peak labour periods. Emigrated labour does not demand locally grown food, thus reducing subsistence needs. Moreover, one can expect that emigrated labour brings in a certain amount of money. In other words, the Region, in addition to exporting agricultural products, can also export labour at a certain price.

Emigrants, by definition, cannot be employed in the agricultural activities in the Region. While maximising monetary revenue, the model weighs the gains in terms of lower subsistence needs and more outside income against the loss in terms of lower labour availability for agricultural activities.

In addition to total monetary revenue, three other monetary variables have been defined. They refer to monetary inputs and serve mainly to restrict their value:

- total monetary inputs in crop activities (seed, fertiliser, other operating costs, depreciation of equipment)
- total monetary inputs in livestock activities (veterinary care, concentrates, etc.)
- total monetary inputs in crop, livestock and fishery activities (including depreciation and maintenance of fishing equipment and fuel for motor boats)

Risks in dry years

Goal variables in this category are primarily used to restrict their values to desired limits. They refer to physical crop production and the number of animals at risk.

- total production of millet, sorghum and fonio in dry years
- total rice production in dry years
- total food crop production in dry years

Another goal related to risk avoidance is minimising grain deficits in dry years, i.e. total grain production (millet, sorghum, fonio, rice, peanut and cowpea) minus the subsistence needs. The unit of measurement is millet-equivalents, expressed in energy content. Grain deficits can be defined in two ways: (i) the difference between the regional subsistence needs and total production or (ii) the sum of sub-regional grain deficits ignoring possible sub-regional surpluses.

The total number of animals at risk in dry years, i.e. the number of animals (TLU) for which insufficient feed—quantitatively or qualitatively—is available from pastures, fodder crops and crop by-products produced in the Region, is defined as a goal variable also. Its value may not be equated to mortality, as animal migration or imported supplementary feed may offer solace.

Employment (man-years) and emigration (number of persons)

Restricting the number of people leaving the Region might be a national objective. This can be formulated in two ways: indirectly, by maximising gainful employment in the agricultural sector (for instance, by setting a lower limit to monetary revenue) or, directly, by limiting emigration. Total employment is expressed in man-years: the labour input in any of the activities is multiplied by the length of the period for which labour is required. Summation over all periods, all activities and all agro-ecological sub-regions results in total labour input over the year or total employment.

Nature reserve (ha)

The possibility of reserving an area in the delta for wildlife protection was introduced by creating nature reserves. When a positive lower limit is set to this goal variable, part of the land is not available for crop cultivation or grazing. Moreover, the upper limit of fish catch will be lower, reflecting the impact of the smaller area of surface water that can be fished.

CONSTRUCTION OF THE SCENARIOS

The two base scenarios

On the basis of the possible objectives for development of the Region and the constraints and relations included in the model, technically feasible scenarios for agricultural land use with their associated production and input levels can be generated. Here, only results referring to the Region as a whole are discussed.

Each scenario is characterised by the goal optimised and the set of restrictions imposed on the other objectives. In other words, the results represent a situation where one goal variable is optimised subject to a particular set of restrictions on the other goal variables and, of course, subject to all model restrictions.

We focus on optimisation of one goal, maximisation of total monetary revenue in normal years, under two sets of goal restrictions. One set represents a more risk-taking attitude; the other emphasises risk-avoidance under unfavourable weather conditions. Moreover, in the latter scenario a higher premium is placed on restricted emigration. Satisfying these additional requirements implies a reduction in the optimum value of monetary revenue. In technical terms, the feasible area is more restricted and, hence, the optimum value of the goal will be lower. To what extent this happens, i.e. what price one has to pay for lower risks, is illustrated below.

First, the two base scenarios, or main development strategies, for the agricultural sector of the Region, are introduced.

The *R-scenario* is a more risky, high-revenue scenario, characterised by:

- high production surpluses (in monetary terms) in normal rainfall years;
- permitted emigration of at most 250 000 persons (almost one-fifth of the present population of the Region);
- no strong demands on the degree of food self-sufficiency in either normal or dry years;
- acceptance of relatively large grain deficits and relatively large numbers of animals at risk in dry years.

The *S-scenario* is a self-sufficiency, safety-first scenario, characterised by:

- a high level of self-sufficiency in basic food, also in dry years;
- a low level of grain deficit and number of animals at risk in dry years;

- homogeneous distribution of production over the agro-ecological sub-regions;
- some degree of diversification among the main crops;
- emigration restricted to 50 000 persons;
- high employment.

In the R-scenario, total monetary revenue is maximised under relatively loose restrictions on the other objectives. The S-scenario is constructed, starting from the

Table 18.3 Stepwise reduction in revenue going from the R-scenario towards the S-scenario (billion FCFA)

		MM[a]
R-scenario		66.7
Step 1	Emigration ⩽50 000 persons (was ⩽250 000 in R-scenario)	45.7
Step 2	Total regional grain deficit <110 000 t millet equivalents (t m.e.) (was <150 000) and sum of sub-regional grain deficits <130 000 t m.e. (was <150 000)	43.1
Step 3	Total number of animals at risk in dry years <100 000 TLU (was <400 000 TLU)	36.0
Step 4	Rice production in a normal year >42 000 t (was >20 000 t)	35.2
Step 5	Inputs in crop activities <15 billion. FCFA (was <20 billion)	33.7
Step 6 (S-scenario)	Employment >336 000 man-year (was >300 000 man-year)	32.5

[a]Maximum attainable monetary revenue.

Table 18.4 Values of (selected) goal variables in the two base scenarios

	R-scenario		S-scenario	
	Restriction	Goal value	Restriction	Goal value
Production, normal year (1000 ton)				
Marketable crop	—	45	—	101
Meat	—	125	—	87
Animals (1000 TLU)	—	1762	—	1491
Monetary targets (billion FCFA)				
Monetary revenue[a]	—	66.7	—	32.5
Money inputs, all	—	15.2	—	23.6
Risks in a dry year				
Rice production (1000 t)	>10	10[b]	>10	12
Regional grain deficit (1000 t m.e.)	<150	141	<130	130[b]
Sum sub-regional grain deficit (1000 t m.e.)	<150	150[b]	<110	110[b]
Animals at risk (1000 TLU)	<400	400[b]	<100	100[b]
Other objectives				
Employment (1000 man-year)	>300	336	>336	336[b]
Emigration (1000 persons)	<250	250[b]	<50	50[b]
Nature reserves (km^2)	>1	1[b]	>1	1[b]

[a]Goal optimised.
[b]Means binding restriction (shadow price >0).

R-scenario (Table 18.3), by successively, in six steps, tightening the restrictions on selected objectives (see Veeneklaas, 1990b).

RESULTS OF THE BASE SCENARIOS

Table 18.4 gives an impression of the key results of the two base scenarios with respect to the attainment of the objectives.

Total monetary revenue

Total monetary revenue of the agricultural sector in the Region ranges from 66.7 billion FCFA (US$ 222 million, at the exchange rate at the time of the study) in the R-scenario to 32.5 billion FCFA (US$ 108 million) in the S-scenario, equivalent to a per capita monetary income of 64 000 FCFA (US$ 212) per year in the R-scenario, in which emigration of a quarter of a million people is allowed, and of 26 000 FCFA (US$ 87) per year in the S-scenario with 50 000 emigrants. In addition to monetary income there is income in kind (subsistence needs).

The difference in revenue between the two scenarios can largely be explained by the restrictions on emigration and on the number of animals at risk in dry years. Tightening the emigration restriction from 250 000 to 50 000 reduces revenue by 21 billion FCFA. Adding the restriction of only 100 000 instead of 400 000 TLU at risk in dry years, reduces revenue by a further 7.1 billion FCFA. The results, both with respect to land use and income levels, strongly depend on the assumptions with respect to prices of inputs and outputs. The effects of different price regimes have also been analysed in the study (see the section on Variants, p. 244).

The low revenue in both scenarios is to a large extent due to the low profitability of arable farming, which is, in addition to the unfavourable price ratios, due to satisfaction of subsistence needs for grain and the requirement of sustainable exploitation in terms of nutrients. The former implies that only a limited part of the crop products are marketed and, thus, contribute to income. The latter implies that soil exhaustion is not permitted; application of (expensive) fertiliser is often necessary to attain target yields because fallowing and organic manure are insufficient to satisfy the nutrient requirements as dictated by export from the field and unavoidable losses (Table 18.5).

Shadow prices

In both scenarios, a number of goal restrictions is binding, albeit more frequently in the S-scenario (characterised by tighter constraints on the goal variables) than in the R-scenario. A binding restriction indicates that a more favourable value for the goal variable would have been obtained in the absence of that restriction. To what extent the restriction limits the value of the goal optimised, is numerically expressed by its shadow price, defined as the change in the value of the goal variable at a relaxation of the restriction by one unit. The dimension of a shadow price is, therefore, the unit of the goal variable; in this case million FCFA per year/unit of the restriction.

Table 18.5 Composition of total monetary revenue (billion FCFA per year)

Source	Value marketable output	Monetary inputs	Monetary revenue
R-scenario			
Livestock	37	2	35
Fisheries	22	7	15
Crops	3	6	−3
Emigration			19
Total			66
S-scenario			
Livestock	24	2	22
Fisheries	21	7	14
Crops	7	15	−8
Emigration			4
Total			32

For example, the shadow price of the restriction 'total rice production >10 000 ton in a dry year' in the R-scenario is 0.458 million FCFA per ton (US$ 1525). Hence, the 'price' of safeguarding one ton of rice production in a dry year is 458 000 FCFA. Because this refers to a hypothetical 'if . . . then . . .' situation, it is referred to as the 'shadow price' of a restriction.

All model restrictions can, in principle, show non-zero shadow prices, but here we only discuss those of the goal restrictions. High shadow prices are exhibited by the restriction 'number of animals at risk in dry years', as expected from the sharp decline in attainable gross revenue when this goal restriction is tightened. In the R-scenario the shadow price is 18 000 FCFA per TLU; in the S-scenario, 54 000 FCFA per TLU (US$ 60 and 180, respectively).

The shadow price for the upper limit to emigration is 96 000 FCFA (US$ 320) per person in the R-scenario and 236 000 FCFA (US$ 768) in the S-scenario. The direct effect of restricted emigration on monetary revenue is the smaller total amount of money generated by the emigrants (75 000 FCFA/person). The higher shadow price implies that an additional effect exists, originating from the higher subsistence needs in the absence of emigration, which is not fully compensated by the higher labour availability.

The additional binding goal restrictions in the R-scenario are rice production and the limit to the sum of sub-regional grain deficits both in dry years. However, the shadow price for the latter restriction is low: 2 FCFA per kg millet-equivalent. This is different in the S-scenario, where the restrictions on grain deficits in dry years are tighter. Especially, the requirement that total regional grain deficit should not exceed 110 000 ton millet-equivalents is a major constraint for realising a higher value of monetary revenue. Its shadow price is 502 FCFA (US$ 1.67) per kg millet-equivalent, which far exceeds the actual producer price of 55 FCFA (US$ 0.18) per kg millet.

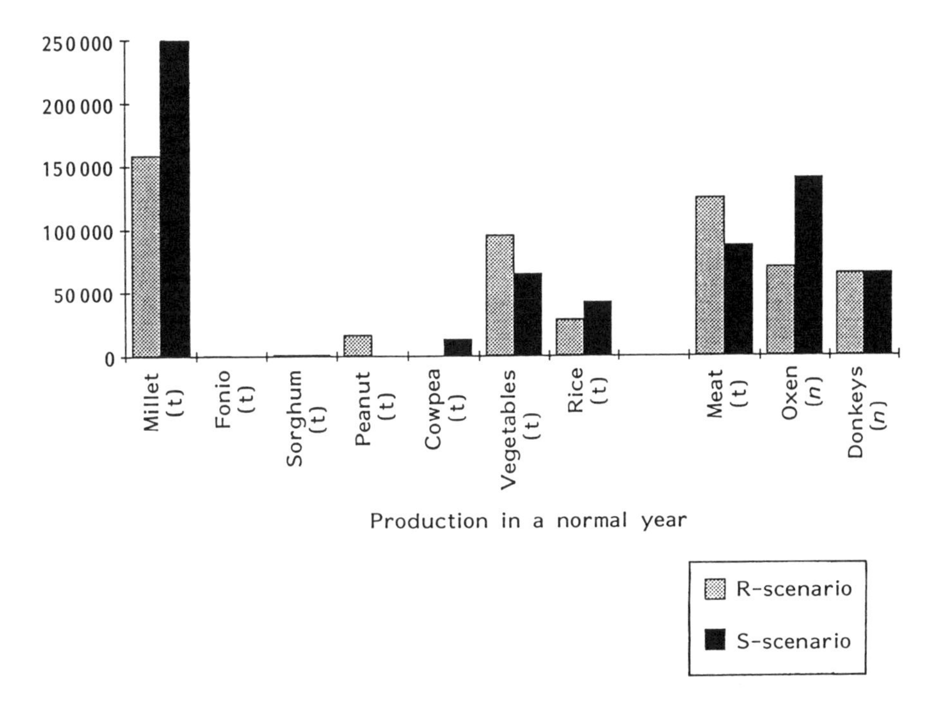

Figure 18.2 Total production of various commodities in a normal year, given in ton dry matter except for vegetables (ton fresh weight); meat (ton liveweight); oxen and donkeys (number of animals)

Self-sufficiency in basic food

Can the Region provide the minimum basic food needs of its rural population, presently numbering about 1.3 million? For animal protein, the answer is 'yes'; for grains, however, the answer is 'hardly'.

Subsistence needs for animal protein, set at 175 g of meat (carcass weight) or 600 g of fish (fresh weight) per person per week, can be easily satisfied, even under unfavourable weather conditions. Moreover, in both base scenarios an average of 3 kg of milk per person per week is available.

For grains, however, in the R-scenario, even in years with normal rainfall and flood, an overall deficit of 23 000 ton millet-equivalents exists, at a total regional grain production of 215 000 ton. In dry years the deficit is 141 000 ton. In this scenario the combined demand of sustainability and maximum total monetary revenue results in: (i) a relatively small area under cultivation, (ii) a rather low level of intensification, and (iii) selection of the most profitable, but not necessarily the most energy-rich crops.

In the S-scenario, the upper limit to total grain deficit in a dry year is set at 110 000 ton of millet-equivalents. At current prices such a deficit would be equivalent to grain imports worth at least 6 billion FCFA (US$ 20 million). In normal years a surplus of 65 000 ton of millet-equivalents is produced, at an overall grain production of 349 000 ton. But even in this scenario the Region is not a major grain exporter.

Total production levels, including subsistence needs, for the various commodities in normal years, are presented in Figure 18.2.

VARIANTS

In formulating the two base scenarios, choices had to be made on the numerical values of technical coefficients and parameters. These choices have been based as much as possible on observation, simulation results and theoretical considerations, but for various reasons they are, and always will be, to some extent disputable.

Moreover, those coefficients that can be affected by policy measures, such as taxes, subsidies and prices, are always disputable. Assuming those to remain constant, the general procedure in the base scenarios is not always fully satisfactory in a policy-oriented study. One might be interested in the potential effects of instruments in this field, e.g. with respect to intervention prices of outputs or prices of crucial inputs such as fertiliser. Some of these effects have been examined in the study.

Finally, there may be discussion about normative choices such as the desirability of reserving part of the delta for wildlife protection. The model and the analysis cannot, of course, be used to judge that desirability; the trade-offs with other objectives, however, can be made explicit. Modifications of the base scenarios are called 'variants' in this study. We discuss one of them here; refer to Veeneklaas et al. (1991) for further information.

Creation of nature reserves in the delta

In its Sahel Studies, the World Conservation Union formulated the following recommendation for priority action on protected areas (IUCN, 1989, p.102): 'Establish a network of protected areas in the Niger flood plain at Lac Debo, Lac Horo and Seri. This is the largest flood plain in West Africa and an important habitat for manatees, wart-hogs and a wide range of migratory birds.' Lac Debo and Seri are located within the Region; the areas involved comprise the 'site de Walado' in the north of the Central Delta, which includes Lac Debo, and is 1031 km^2 in area, and the 'site de Seri', 400 km^2 in area, in the mid-western part of the Central Delta.

The effect of reserving these areas for nature protection, on agricultural and fisheries production and income of the Region has been examined, under the assumption that protection of wildlife implies exclusion of all agricultural (including fisheries) activities. The results are presented in Table 18.6, where the designations R1 and S1 refer to this variant of the two base scenarios.

The effect on the values of the goal variables is of course different in the two development scenarios, i.e. the impact is much greater in the S-scenario, where more claims are put forward, than in the R-scenario. In the S-scenario, creation of a nature reserve with a total area of 1431 km^2 results in a decrease in annual monetary revenue in a normal year of 5.5 billion FCFA (US$ 18 million), whereas in the R-scenario that amounts to 2.1 billion FCFA (US$ 7 million). In other words, setting aside land for nature will cost annually between US$ 50 (R-scenario) and US$ 130 per ha (S-scenario) in terms of foregone net agricultural and fisheries production.

Table 18.6 Effect of the creation of nature reserves in the Delta on goal achievement, differences with base scenarios (R1-R and S1-S)

	R1-scenario		S1-scenario	
	Goal value	Difference with base scenario	Goal value	Difference with base scenario
Production, normal year (1000 ton)				
Millet, sorghum and fonio	160	—	280	−2.1
Marketable crop pr.	45	−0.0	85	−15.3
Meat	123	−1.5	75	−11.6
Animals (1000 TLU)	1717	−45	1320	−171
Monetary targets, normal year (billion FCFA)				
Monetary revenue[a]	64.6	−2.1	26.9	−5.5
Money input, all act.	14.5	−0.7	22.6	−1.0
Risks in a dry year				
Crop products (1000 ton)	190	+0.3	222	−13.4
Region grain deficit (1000 t mil.eq.)	140	−0.3	110	—
Number of animals at risk (1000 TLU)	400	—	100	—
Other objectives				
Employment (1000 man-year)	334	−2.2	336	—
Emigration (1000 persons)	250	—	50	—
Nature reserves (km^2)	1431	+1430	1431	+1430

[a] Goal variable optimised.

For a fair assessment of these results, however, one must bear in mind the limitations of this analysis. First, in this study only the impact on the agricultural sector, which by definition is negative, has been examined. The creation of nature reserves will have positive effects outside that sector, both in terms of monetary income (tourism) and employment (management). Secondly, the final impact is sensitive to assumptions with regard to the effects on fisheries. In this analysis it has been assumed that the reduction in total catch is proportional to the reduction in available area due to the creation of nature reserves (9%). This may be an overestimate due to, for example, mobility of fish in reality, but that is difficult to quantify.

By the creation of nature reserves the fish catch in normal years is estimated to be 8300–8500 ton lower, representing a value of about 2.3 billion FCFA. Monetary inputs in fisheries, however, will be reduced also, by about 670 million FCFA, so that the loss in income from fisheries is between 1.6 and 1.7 billion FCFA, which in the R1-scenario represents the larger part of the total reduction in revenue. In the S1-scenario, on the other hand, the loss in income from animal husbandry is more important. The reduced area of dry season pastures in the Delta Central results in a reduction in animal population from 698 000 to 539 000 TLU.

In the R1-scenario, the northern part of the Region (Zone Lacustre) serves to a limited extent as an alternative dry season home-base for migrant cattle. In the S1-scenario that is not possible, due to additional restrictions. Total annual meat production in that scenario is consequently considerably lower (12 000 ton liveweight), than in the S-scenario, implying a reduction in the value of marketable meat of 3.4 billion FCFA. The effect of the reduction in total meat output is slightly mitigated by the larger proportion of small ruminants in the total population, whose meat makes a better price than beef.

Crop production is hardly affected by the creation of a nature reserve, with the exception of a shift in vegetable cultivation in the Zone Lacustre from shallot to 'other vegetables'. Their high quality crop residues that can be used as fodder outweigh, in the final analysis of conflicting claims in this scenario, the higher yields of shallots.

Other variants

In addition to the variant described above, the impact of the following variants has been analysed in the study (Veeneklaas et al., 1991):

- reducing the price of fertiliser by 50%
- a 50% increase in the producer price of crop products
- alternative coefficients for livestock activities
- reduced production of inundated pastures following a series of dry years

Additional variants suggested in the main study are:

- pasture production: mowing of inundated pastures
- pasture production: fire control on rainfed pastures
- pasture production: intensified management (fertiliser application, introduction or re-introduction of leguminous species or perennial grasses, abatement of wind and/or water erosion)
- expansion of the irrigated area
- introduction of herbicides

CONCLUDING REMARKS: PROSPECTS FOR APPLICATION OF THE MODEL RESULTS

As indicated in the introduction, one of the major arguments for the initiation of the present study is the increasing competition for the limited natural resources among the various agricultural activities. Especially, the competition between arable farming and animal husbandry for the limited land resources, in both the rainfed and flooded areas, has led to acute problems.

As explained elsewhere (Van Keulen, 1990), results of the multiple goal optimisation model cannot be used directly to guide regional development planning. Translation is necessary, in a post-model analysis, in which especially those aspects that cannot be translated in 'hard' relations have to be taken into account, to arrive at explicit recommendations for policies that will result in the desired developments.

Although a unique blueprint for the development of the Region (i.e. an overall land use plan) cannot be presented, the results of the study indicate the scope for development under the condition of sustainability and different development goals. Two such scenarios, characterised by different boundary conditions with respect to goal achievement, have been constructed. The solutions presented are optimal with respect to regional monetary revenue, under the formulated boundary conditions and the constraints imposed.

Both situations, referred to as the R- and the S-scenarios, differ from the present situation. One of the major reasons is that optimum conditions have been assumed, aimed at maximum goal achievement, contrary to real-life situations. Moreover, mainly physical and technical constraints and relations have been taken into account. In other words, the results refer to the maximum potentials from a technical point of view. Apart from the fact that quantification of the applied relations may be subject to debate, there is a tendency in this type of analyses to overestimate the potentials *vis-à-vis* the actual situation where other constraints, such as institutional and/or socio-economic constraints, may also play a role. It may be argued, on the other hand, that technical innovation has not been taken into account, i.e. the production techniques defined and their technical coefficients are based on current knowledge, but the scope for improvements in this respect is fairly well known.

Another important reason for deviation of the results from the current situation is that the defined production techniques are based on sustainable exploitation of the natural resources, which at present is clearly not the case. Hence, exhaustive exploitation of the natural resources under the current conditions provides the opportunity to achieve temporarily higher yields and income than realised under the conditions assumed in the model (Van der Pol, 1992).

Comparison of model results—the scenarios and the variants—with the present situation is, therefore, only relevant in relation to the question of which way a transition can be achieved from the current exhaustive mode of exploitation to one of the selected modes of sustainable exploitation. The differences between the present situation and the prospective one should provide indications for the necessary efforts. It is evident that in such an analysis, in addition to the technical constraints, socio-economic considerations will have to be taken into account more explicitly.

The main objectives of the present study were to gain insight into and formulate proposals for reduced competition for limited natural resources among the various activities on the one hand, and to promote the introduction of sustainable production systems that would satisfy the development needs of the rural population, on the other.

From a scientific and technical point of view, answers have been given to questions pertaining to the possibilities for exploitation of the natural resources in the Region; from a technical point of view, it has been shown that it is possible to reduce the problems of competition and develop sustainable activities both in normal and dry seasons. However, there is still a long way to go before scientific demonstrations can be replaced by implementation of the results of the optimisation model:

(1) The results of the optimisation model cannot be used as such by regional development planners, because the post-model analyses dealing with socio-economic aspects have not been incorporated and also the computer model has never been implemented in Mali, so the planners have not had a chance to actually use it. Moreover, the gap between the technical results of the optimisation model and the day-to-day reality of the rural production systems has not as yet been bridged.

(2) In addition to the technical and socio-economic aspects, development, sustainable or not, also has to take into account political aspects at the national level or on the part of other stake-holders in the development process who may have objectives different to those of the producers. The model deals with an administrative region of a highly centralised political system. The development plan is in fact not conceived at regional level, but at the national level, where the development objectives do not always correspond to the views at regional or local level. Moreover, even if a regional plan has been established by regional authorities, its implementation brings to the fore families and individuals that have not participated in the development of the model, nor in the discussions about the development plan, and whose main concern is not avoidance of the conflicts associated with the exploitation of the common resources, but maximisation of the gain from exploitation of those resources.
(3) One of the opportunities for greater impact of the model results on the actual situation might have been further elaboration of post-model analyses. However, the donors, the research institute involved and the development projects (Organisation de Développement d'Élevage au Mopti, ODEM, and Organisation Riz Mopti, ORM) are currently not interested in further elaboration of the post-model analyses or in implementation of the results of the study, despite their support for the recommendations and resolutions of the regional and national seminars in this vein.

Despite the disappointing positions taken by the partners (donors, research institute, and national counterparts), the results of the study are used in a dual fashion:

(1) Some international (FAO, ISNAR) and non-governmental organisations have expressed an interest in use of the database that has resulted, for application in analyses relating to their own specific interests. The documentation of ESPR and AB-DLO, as well as the teamleader of ESPR, are consulted regularly.
(2) Other non-governmental organisations, especially the Canadian (SUCCO), request assistance of the teamleader in analysis and research on solutions for certain sectorial development programmes and rough outlines of local management schemes. This is an important step in the implementation of the results for the Region—though two swallows (one teamleader and one foreign non-governmental organisation) are unable to make spring, even under the illuminated sky of the 5th region.

REFERENCES

Cissé, S. and Gosseye, P. A. (eds), 1990. Competition pour des ressources limitées: Le cas de la cinquième Région du Mali. Rapport 1: Ressources naturelles et populations. Étude sur les Systèmes de Production Rurales (ESPR), Mopti, Mali. DLO-Centre for Agrobiological Research (CABO-DLO), Wageningen.

De Wit, C. T., Van Keulen, H., Seligman, N. G. and Spharim, I., 1988. Application of interactive multiple goal programming techniques for analysis and planning of regional agricultural development. *Agricultural Systems* **26**: 211-230.

Erenstein, O., 1990. Simulation of water-limited yields of sorghum, millet and cowpea for the 5th Region of Mali in the framework of quantitative land evaluation. Report Dept. Theoretical Production Ecology, Wageningen Agricultural University, Wageningen.

IUCN, 1989. *The IUCN Sahel Studies 1989*. IUCN/NORAD, Gland, Switzerland.

PIRT (Projet Inventaires des Ressources Terrestres au Mali), 1983. *Les Ressources Terrestres au Mali*, Vol. I: Atlas, Vol. II: Rapport technique; Vol. III: Annexes. Mali/USAID, TAMS, New York.

Van der Pol, F., 1992. *Soil Mining: An Unseen Contributor to Farm Income in Southern Mali*. Bulletin 325, Royal Tropical Institute, Amsterdam.

Van Duivenbooden, N., Gosseye, P. A. and Van Keulen, H. (eds), 1991. Competing for limited resources: The case of the fifth Region of Mali. Report 2: Plant, livestock and fish production. Étude sur les Systèmes de Production Rurales (ESPR), Mopti, Mali. DLO-Centre for Agrobiological Research (CABO-DLO), Wageningen.

Van Keulen, H., 1990. A multiple goal programming base for analysis of agricultural research and development. In: Rabbinge, R., Goudriaan, J., Van Keulen, H., Van Laar, H. H. and Penning de Vries, F. W. T. (eds), *Theoretical Production Ecology: Reflections and Prospects*, pp. 265–276. Simulation Monographs, Pudoc, Wageningen.

Veeneklaas F. R., 1990a. Competing for limited resources: The case of the fifth Region of Mali. Report 3. Formal description of the optimization model MALI5. DLO-Centre for Agrobiological Research (CABO-DLO), Wageningen, Netherlands/Etude sur les Systemes de Production Rurales (ESPR), Mopti, Mali.

Veeneklaas, F. R., 1990b. Dovetailing technical and economic analysis. Ph.D. Thesis Erasmus University Rotterdam, The Netherlands.

Veeneklaas, F. R., Cissé, S., Gosseye, P. A., Van Duivenbooden, N. and Van Keulen, H., 1991. Competing for limited resources: the case of the Fifth Region of Mali. Report 4: Development scenarios. CABO/ESPR, CABO-DLO, Wageningen.

CHAPTER 19

Extrapolation of Maize Fertiliser Trial Results by using Crop-Growth Simulation: Results for Murang'a District, Kenya

R. P. Roetter[a] and C. Dreiser[b]

[a]University of Trier, Trier, Germany. [b]Mozart Str. 7, Sinsheim, Germany

INTRODUCTION

In view of scarce arable land and a rapidly increasing food demand in Kenya, maintenance of the soil nutrient balance is an important requirement for agro-ecological sustainability (Smaling, 1993; Von Urff, 1993). Fertiliser research can only contribute to increased productivity and maintenance of production capacity over a period of time, if results are efficiently transferred to target groups and areas. Extrapolation of site-specific experimental results to representative areas is, therefore, one of the high priority tasks and, at the same time, one of the most challenging tasks for applied agricultural research.

Evident progress in this field has for a long time been prevented by applying inadequate methods, based on the concepts of 'transfer by analogy' (Nix, 1985) or multivariate statistical analysis (Monteith and Scott, 1982).

Though new techniques (dynamic crop-growth simulation models and geographic information systems) have been developed that could overcome earlier methodological shortcomings, the problem of a lack of suitable data sets to apply these techniques still exists (Nix, 1985), especially in the tropics (Driessen, 1986; Van Diepen et al., 1989; Liu et al., 1989).

A comprehensive soil fertility project, the Fertilizer Use Recommendation Project (FURP), was launched in Kenya during 1985–1986 to develop crop- and area-specific fertiliser recommendations for the major food crops in areas with relatively favourable agro-climatic conditions.

Trial site selection was preceded by examination and synthesis of information on earlier fertiliser trials, climate, soils, population pressure, etc. (FURP, 1987/88; Smaling and

The Future of the Land: Mobilising and Integrating Knowledge for Land Use Options
Edited by L. O. Fresco, L. Stroosnijder, J. Bouma and H. van Keulen.

Van De Weg, 1990). During 1987–1992, field trials were conducted at 70 sites, representing widely varying climate–soil environments, to generate an empirical basis on yield responses to organic and mineral fertilisers (FURP, 1987/88, 1990; Smaling et al., 1992). For long-term recommendations, extrapolation of the trial results should not be restricted to space, but should also include the temporal component, i.e. the extrapolation of five-year trial results to expected long-term weather conditions.

The objectives of this study are, first, development of a method of data synthesis that makes full use of the readily available data with a rather inexpensive infrastructure and, secondly, to apply that method in a regional case study to demonstrate what can immediately be achieved by applying appropriate methods to the available data. This is to supplement the rather descriptive agro-ecological zonations as introduced by FAO (1978), adapted and applied to 31 districts of Kenya by Jaetzold and Schmidt (1982/83) and FURP (1987/88), with quantitative information on production risk, mean yield and gross margin expectation under present and alternative management practices and economic situations. The study crop is maize, which is the main food crop and the reference crop in most of the fertiliser experiments conducted in Kenya (Allan et al., 1987; KARI, 1991).

By linking results from crop-growth simulation model WOFOST (van Diepen et al., 1989; Roetter, 1993) to the geographic information system (GIS) IDRISI (Eastman, 1992), an attempt has been made to establish a framework for the use of a simulation model coupled to geo-coded agro-ecological and agro-economic databases in Kenya, which is applicable at the regional as well as at the farm level.

The case study will contribute to the discussion on how to mobilise and integrate experimental data as decision-making support for extension officers and farmers. A sound validation of the results has not been attempted, but it is hoped that this case study may form the basis of target-oriented research activities.

MATERIALS AND METHODS

The methods used have been developed and described in detail by Roetter (1993). The step from point to spatial analysis was tailored to the type and quality of readily available data for the Murang'a District: historical rainfall records, climatic norms, soil and rainfall maps, and fertiliser trial results for maize from soil–climate–cultivar combinations relevant to the district.

Murang'a District in Central Kenya was selected, because maize is cultivated here under widely varying agro-ecological conditions. Farmers use various cultivars, and data from two farm surveys allow comparison of calculated results with some real-world data. Various maize cultivars were considered: Hybrids 625, 613c, 512, 511 and Katumani composite B (Acland, 1980; Kiarie, 1985; Roetter, 1993).

The crop-growth simulation model

For simulating maize growth and yield in Kenya, the standard version of model WOFOST (Van Diepen et al., 1988) has been modified and validated for the main maize cultivars grown in the country (Roetter, 1993). In the model, crop yields are calculated for three well-defined production situations:

1. Production situation 1 (PS1): potential production (growth is determined by crop characteristics, light and temperature only), i.e. the production ceiling for intensive irrigated arable farming.
2. Production situation 2 (PS2): water-limited production (additional to the factors in PS1, moisture availability is taken into account as a yield-limiting factor), i.e. the production ceiling for rainfed arable farming.
3. Production situation 3 (PS3): nutrient-limited production (additional to the factors in PS1 and PS2, availability of macro-nutrients N, P and K is considered as a yield-limiting factor), i.e. the production ceiling for a given soil fertility status that is characterised by crop uptake of N, P and K in the absence of water stress.

Detailed descriptions of the complete program (Version 4.1), principles underlying the model and possible applications are given by Van Keulen and Wolf (1986), Van Diepen et al. (1988, 1989) and Roetter (1993).

Input data to the simulation model

As daily rainfall is the driving variable with the strongest impact on model results (Roetter, 1993), rainfall records were exploited as far as possible. Sufficiently long (≥20 years), complete (daily) records were readily available for 11 stations in the Murang'a District (Roetter, 1993).

Except for rainfall, climatic norms were used as input data: long-term monthly means of minimum and maximum air temperature, global radiation, dew point temperature and wind speed for meteorological stations Kimakia Forest (2435 m a.s.l.), Sagana State Lodge (1850 m a.s.l.), Ruiru Jacaranda Coffee Research Station (1610 m a.s.l.) and Thika Horticultural Research Station (1550 m a.s.l.) were obtained from KMD (1984).

Each data set served as input for those points with similar agro-climatic characteristics. However, to adequately account for the variability in temperature-dependent crop phenology, the altitude dependence of minimum and maximum air temperature as described by Braun (1986) was taken into account to adjust the temperature data for running the model at 11 altitudes, i.e. the long-term rainfall stations. Other climatic elements were not adjusted for the specific situation of those stations.

Data on soil moisture retention, soil depth and particle size distribution from eight soil profiles (FURP, 1987/88), representing the main soil types in the district, were analysed to define a 'standard soil type' with regard to soil hydraulic characteristics (Table 19.1). Moreover, the individual data of the eight soil profiles have been used to determine the influence of different soil moisture characteristics on simulated yields under the rainfall regimes recorded in the Murang'a District.

For other inputs required for the model, default values have been used, assuming that crop husbandry is practised as in the FURP experiments (FURP, 1990) and that soil and water conservation measures are carried out at a moderate level of intensity (Mati, 1986; Ngugi and Kabutha, 1986). As only potential and water-limited production situations were simulated, inputs required for production situation 3 were not relevant.

Table 19.1 Soil hydraulic characteristics of 'standard soil type' for Murang'a, derived from eight soil profiles

Transmission zone permeability (cm/day)		Volume fraction of moisture at suctions (pF)				
Topsoil	Subsoil	0.0	2.0	2.4	3.4	4.2
5.25	3.15	0.550	0.444	0.402	0.310	0.275

Source: FURP (1987/88).

Digitising and transforming mapped information into geo-coded databases

In this study the UTM (Universal Transverse Mercator) projection was applied, characterised by equidistant x and y units and perpendicular coordinate axes, making digitising convenient. The unit of analysis of the map was 1 ha/pixel (pixel = picture element), resulting in an image of 623×448 pixels with corner UTM coordinates 98 7700, 24200 and 99 4100, 33100.

Topographical point, line, and polygon data were transformed into digital format and stored in vector format. 'Location' boundaries (lines) derived from the administrative map of the Murang'a district (scale 1:100 000; CBS, 1989), rainfall isolines (lines) from rainfall maps (scale 1:360 000) of first and second rains, respectively, and soil mapping units (polygons) from the soil map (scale 1:360 000; FURP, 1987, Vol. 21), were digitised.

Basic calculation procedure

The core of the method is the relationship between the arithmetic mean of simulated relative grain yields (PS2/PS1, = s-ratio) and seasonal rainfall at 66% probability (Roetter, 1993). The spatial distribution of the 11 rainfall stations used for modelling, covering practically the entire value range displayed on the rainfall maps, was such that model results could be easily related to rainfall values. Required steps for linking simulated relative yields to spatially differentiated rainfall data were:

(1) running the model for determining potential (PS1) and water-limited (PS2) grain yield using 'standard settings' of input parameters for each rainfall station, except for starting dates (emergence) and the corresponding soil moisture contents (Roetter, 1993);
(2) calculation of relative grain yield (PS2/PS1 or s-ratio) for each simulated season and the arithmetic means of s-ratios, for first and second rains, respectively, for each rainfall station;
(3) extracting, for each station, rainfall data, for first and second rains, from existing maps (seasonal rainfall at 66% probability);
(4) analysing the data pairs 'seasonal rainfall at 66% probability' and 'average s-ratio' statistically, to establish a relationship that can be incorporated in a GIS to generate s-ratios from digitised rainfall information.

The relationship between seasonal rainfall at 66% probability (X) and simulated average s-ratio for first (Savg1) and second (Savg2) rains for the selected rainfall stations in the Murang'a district is described by nonlinear regression equations:

$$\text{Savg1} = 0.003\,392\,98\ X - 0.000\,002\,06\ X^2 - 0.55 \qquad (19.1)$$

$(n = 11;\ R^2 = 0.9876)$

$$\text{Savg2} = 0.002\,395\,86\ X - 0.000\,001\,41\ X^2 - 0.250\,47 \qquad (19.2)$$

$(n = 11;\ R^2 = 0.9489)$

Map output-oriented calculation procedures

Four map types were produced, each showing synthesised information for first and second rains.

Risk of crop failure

Calculated s-ratios are fitted to a theoretical distribution function, the beta distribution, defined over the interval (0,1) which can represent different distribution types (Johnson and Kotz, 1970). For each station, the probabilities of a relative yield less than or equal to 0.05 are calculated for first rains. The value 0.05 for the s-ratio was chosen rather arbitrarily to define the threshold for crop failure. Since simulated potential grain yields of maize range between 6 and 12 t/ha in the Murang'a District, the absolute yield level, corresponding to an s-ratio of 0.05 is low, i.e. 300–600 kg/ha. Thus, error in the absolute values representing crop failure is negligible.

Subsequently, a relationship between risk of crop failure in first rains (Rcf1) and seasonal rainfall at 66% probability (X) was established, similar to steps (2) and (4) of the basic procedure, yielding the regression equation:

$$\text{Rcf1} = -0.0023\ X + 0.000\,001\,647\,09\ X^2 + 0.802\,248 \qquad (19.3)$$

$(n = 11,\ \mathrm{R}^2 = 0.8817)$

Average water-limited yield (as a percentage of non-water-limited yield)

This data surface (soil-specific Savg) is calculated by multiplying the rainfall-determined average s-ratio (Savg) by a soil unit-specific modifier (Mswr). This modifier expresses the difference between values of Savg resulting from (a) simulation with 'standard moisture retention data' (Table 19.1), and (b) simulation with retention data of the soil types representing the various soil units (Roetter, 1993), as a fraction of the 'standard'. For example, in first rains, Savg calculated for a seasonal rainfall of 500 mm (66% probability) and 'standard soil type' is 0.63 (using equation (19.1), model results for soil type-specific retention data do not deviate from the standard run under the given rainfall regime, then: Mswr is 1.0 and soil-specific Savg is 63% (Savg $\times$ 100 $\times$ Mswr = soil-specific Savg (%)).

Average yield increase resulting from fertiliser application (here: N50P22), under soil and rainfall conditions (as a percentage of maximum yield at N0P0)

The data surface for this map is created by integrating experimental data that have been assigned to the different soil units. Observed soil type-specific yield differences of treatments N0P0 and N50P22 in seasons without water stress at various FURP trial sites were used to calculate average yield increase (Ayincr) resulting from fertiliser application (N50P22) under the given soil and rainfall conditions (as a percentage of maximum soil-type-specific yield at N0P0). From the trial results given by Roetter (1993), yield (t/ha) and yield increase at N50P22 (MaxgyinN50P22 in %) were derived and assigned to soil units represented by one or more trial sites. Only the predominant soil type within each unit was considered. Subsequently, the following equation was applied:

$$\text{AyincrN50P22 } (\%) = \text{Savg} \times \text{Mswr} \times \text{MaxgyinN50P22}$$

Mean gross margin expectation (Ksh/ha) resulting from fertiliser application (N50P22), under given soil and rainfall conditions

For calculating the 'mean gross margin data surface', the equation

$$\text{MGM} = (\text{Savg} \times \text{Mswr} \times \text{MaxgyN50P22} \times \text{Mp}) - \text{Vcost}$$

was applied, where MGM is mean gross margin (Ksh/ha), MaxgyN50P22 is the soil-type-specific grain yield (t/ha) attainable at N50P22, Mp is the maize price (Ksh/t grain) and Vcost is the cost of fertiliser, labour and pesticides (Ksh/ha). For this map the constant values used were: a maize grain price of 3900 Ksh/t, costs of 2700 Ksh/ha for fertiliser N50P22 (36 Ksh/kg N, in calcium ammonium nitrate and 41 Ksh/kg P, applied as triple superphosphate) reported for Nairobi in spring 1992; and costs for labour and insecticides of 1000 Ksh/ha.

Techniques employed for constructing geo-coded data surfaces and the procedure of combining simulated and experimental geo-coded data are described by Roetter (1993).

RESULTS AND DISCUSSION

Map output has been produced to fit A4 format, resulting in a scale of approximately 1:360 000 (Roetter, 1993). In first rains, high risk occurs in the Highlands in the west due to excess of water and in the south-east due to water deficits. A 15–20 km wide band in the centre of the district shows only slight or no risk of crop failure, whereas east and west of this band, risk increases (Figure 19.1). In second rains, high risk only occurs in the eastern part of the district, in areas with less than 270 mm rainfall at 66% probability. Areas with a high risk of crop failure during first rains are those with seasonal rainfall at 66% probability of less than 310 and more than 1000 mm. Map type (a) only reflects 'climatic risk', i.e. differences in soil hydraulic properties have not been taken into account. Nevertheless, the results are not likely to deviate much from reality, since in low rainfall years the positive effect of favourable soil physical properties is usually

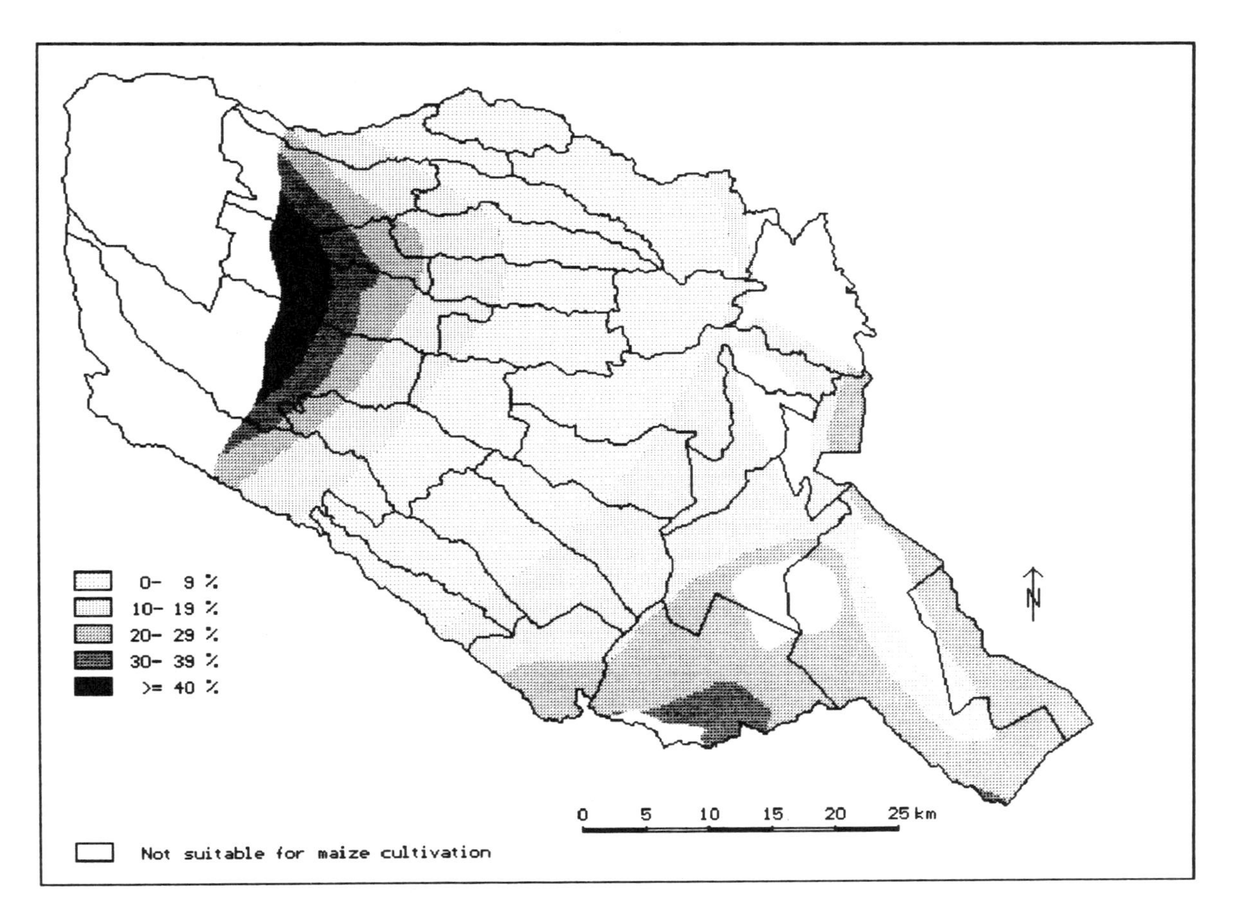

Figure 19.1 Maize, first rains: risk of crop failure (% of all years) as determined by rainfall conditions (map type (a)). Crop failure is determined as grain yield <0.5 t/ha

negligible (Cooper et al., 1987). With respect to high risk due to high rainfall, the picture may be too pessimistic: only one station with rainfall exceeding 1000 mm (at 66% probability) was used in modelling and these results have a strong influence at the upper end of the non-linear regression so that the presented output is rather uncertain.

Comparison of the results of the present study to existing land use zonations shows, for instance, that the areas with high risk due to water deficits coincide with Jaetzold's subzones characterised by agrohumid periods of 85–104 days or less (Jaetzold and Schmidt, 1982/83). Comparison to the agricultural zoning presented by Ngugi and Kabutha (1986), reflecting the actual situation in the district in 1985, shows that high risk areas were mainly used for estate farming (focus irrigated coffee, pineapple), (focus sisal) and low-density dryland farming/ natural vegetation.

The areas *a priori* classified as 'not suitable for maize cultivation' comprise the tropical alpine (TA), the forest (UH0) and the sheep–dairy zones (UH1) in the west of the district as presented by Jaetzold and Schmidt (1982/83), and additionally, all soil mapping units that mainly consist of either very poorly drained heavy clay soils (with vertic properties) or shallow soils (FURP, 1987/88; Roetter, 1993). Highest simulated average yields (first rains: 81–90% of the non-water-limited yield (map type (b)) cover only a narrow, 3–7 km wide band in the centre of the district, whereas the next class (first rains: 71–80%) stretches 10–15 km east and west of this central band. In second rains, the highest simulated average yields reach 61–70% of the non-water-limited yield. Lowest values (20–30% in the first and 11–20% in second rains) appear in the south and east of the district.

Highest values occur in the area that is also considered most favourable for coffee, according to both agro-ecological zonation (UM2) and actual land use, i.e. zone 2 'smallholder coffee zone' (Ngugi and Kabutha, 1986). The same applies to average yield increase at treatment N50P22 (map type (c), Figure 19.2), illustrating the problem that food self-sufficiency is seriously constrained as long as the most promising environments are occupied by cash crops that are, for the time being, economically more attractive to the farmer than food crops.

The average yield increase (%) resulting from fertiliser application (N50P22), in relation to soil and rainfall conditions, shows interesting spatial patterns. While there are rather marked boundaries of yield increase classes towards the west of the central band of highest values (61–70% as compared to unfertilised maize), the picture towards the east partly resembles the geometry characteristic for nonlinear systems. The most distinct boundary in the west coincides with the boundary between two soil mapping units, illustrating that here soil characteristics have a more pronounced effect than rainfall conditions. In second rains, spatial variability in average yield increase at treatment N50P22 is rather small, ranging between 11 and 40%.

An important assumption in calculating average yield increase (map type (c)) and gross margins (map type (d)) was that yield levels as recorded under non-water-limited conditions in FURP experiments on the soils representative for each mapping unit are constant in time. Of course, this is unrealistic for continuous cropping of maize without fallow periods or use of external inputs, as presently practised by the majority of farmers. However, it is likely that yield levels at both N0P0 and at N50P22 will decline proportionally over time, if nutrient depletion is not balanced by proper management

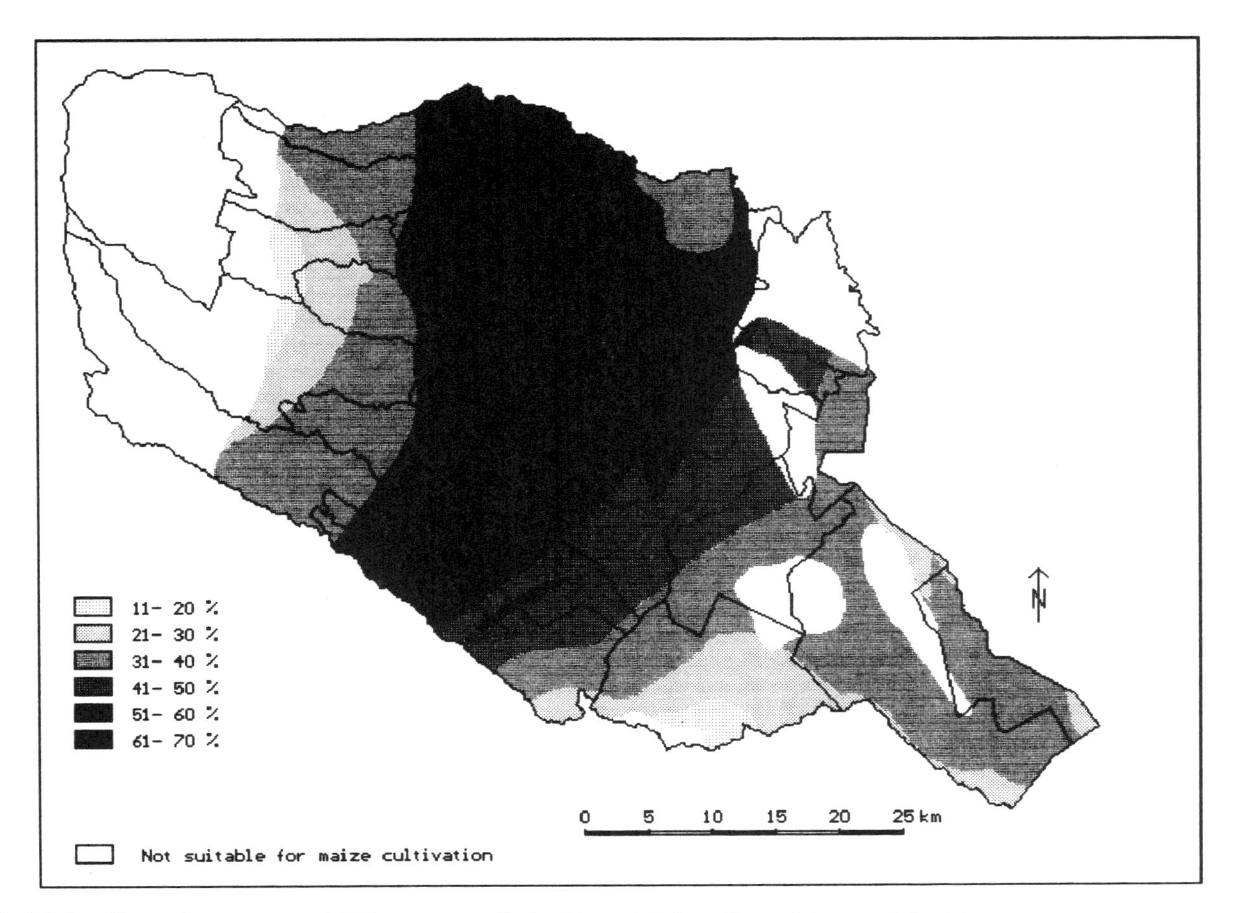

Figure 19.2 Maize, first rains: average yield increase, as limited by rainfall, soil conditions and fertiliser rate, at treatment N50P22 as a percentage of maximum soil-specific yield at treatment N0P0 (map type (c))

(including fertiliser application, both inorganic and organic, incorporation of crop residues and soil and water conservation measures). Moreover, it should be kept in mind that nutrient depletion in areas characterised by low rainfall is usually a minor problem.

In first rains, eight classes of mean gross margin (Ksh/ha) have been distinguished with increments of 2000 Ksh/ha, ranging from −1000–999 to 13 000–14 999 Ksh/ha. The highest class includes, to a large extent, one soil mapping unit, except for small areas receiving high rainfall in the west and the south-east of this unit, appearing in two bands parallel to the isohyets. Mean gross margins below 1000 Ksh/ha are only found in the south of the district—where there is a 66% probability of less than 350 mm rainfall. In second rains, mean gross margins below 1000 Ksh/ha occur in areas with less than 300 mm of rain. Highest values (7000–8999 Ksh/ha) are restricted to the centre of the district.

Highest mean gross margin expectations for maize are found in zones 2 (smallholder coffee), 3 (subsistence agriculture; focus maize) and 4 (subsistence agriculture; focus maize + beans) as identified by Ngugi and Kabutha (1986). Agricultural zone 7 (low-density dryland farming/natural vegetation) coincides with areas showing mean gross margin expectations of less than 3000 Ksh/ha in first rains and less than 1000 Ksh/ha in second rains.

CONCLUSIONS

The output of this integrated approach, comprising information on risk, average yield and gross margin expectations for maize (monoculture) is, in principle, suitable for decision-making support at various levels, though here, in the first instance, aiming at regional level.

The choice of map scale 1:360 000 was based on the resolution of available soil and rainfall maps. As results of more detailed soil surveys or rainfall analyses become available, that information can be incorporated and the scale of the analysis and hence of map output can be adjusted accordingly. The same applies to the results of additional fertiliser trials with relevance to the district which may become available in the course of time. The contents of maps (b), (c) and (d) are directed at the target group 'profit-oriented farmers', assuming that financial losses in some years can be compensated in others. Results should, first of all, be considered as an indication of what might be achieved by using a well-defined production technique. In any real decision situation, there are alternatives. Considering, for instance, constraints to the proposed production technique for increasing food production and for decision-making support, far more than mere technical information is required. Among the socio-economic constraints relevant to many farmers, lack of money to buy the necessary means of production, and lack of economic incentives for their application, deserve special attention.

REFERENCES

Acland, J. D., 1980. *East African Crops*. 5th edn. Longman, London.

Allan, A. Y., Duerr, G. and Stroebel, H., 1987. Compilation of results from former fertilizer trials. Ministry of Agriculture/NAL/FURP, Final Report, Annex I. Ministry of Agriculture, Nairobi, Kenya.

Braun, H. M. H., 1986. Some characteristics and altitude relationships of temperatures in Kenya. Kenya Soil Survey, M18, Ministry of Agriculture, Nairobi, Kenya.

CBS (Central Bureau of Statistics), 1989. *Administrative Map of Murang'a District*. Ministry of Economic Planning and Development, Nairobi, Kenya.

Cooper, P. J. M., Gregory, P. J., Tully, D. and Harris, H. C., 1987. Improving water use efficiency of annual crops in the rainfed farming systems of West Asia and North Africa. *Experimental Agriculture* **23**: 113–158.

Driessen, P. M., 1986. Soil data. In: Van Keulen, H. and Wolf, J. (eds), 1986. *Modelling of Agricultural Production: Weather, Soils and Crops*, pp. 212–234. Simulation monographs, Pudoc, Wageningen.

Eastman, J. R., 1992. *IDRISI Version 4.0 User's Guide*. Clark University, Graduate School of Geography, Worcester, MA.

FAO, 1978. *Report on the Agro-Ecological Zones Project. Methodology and Results for Africa*. FAO World Soil Resources Report 48/1, FAO, Rome.

FURP (The Fertilizer Use Recommendation Project), 1987/88. Final Report. Annex III: Description of First Priority Trial Sites in the Various Districts. Ministry of Agriculture/NAL/FURP. National Agricultural Laboratories, Nairobi, Kenya.

FURP (The Fertilizer Use Recommendation Project), 1990. Methodology and inventory of existing information. Ministry of Agriculture/NAL/FURP, Final Report—Main Report. National Agricultural Laboratories, Nairobi, Kenya.

Jaetzold, R. and Schmidt, H. (eds), 1982/83. *Farm Management Handbook of Kenya*, Vol. II/A-C. Ministry of Agriculture/GAT Nairobi, Kenya and GTZ Eschborn, Germany.

Johnson, N. L. and Kotz, S., 1970. *Distributions in Statistics. Continuous Univariate Distributions—2*. Wiley, New York.

KARI (Kenya Agricultural Research Institute), 1991. *Annual Report 1990*. KARI/HQ/Pub/1/91, Kenya Agricultural Research Institute, Nairobi, Kenya.

Kiarie, N., 1985. Development of early maturing maize cultivars. *East African Agriculture and Forestry Journal* **48**: 40–50.

KMD (Kenya Meteorological Department), 1984. *Climatological Statistics for Kenya*. Kenya Meteorological Department, Nairobi, Kenya.

Liu, W. T. H., Botner, D. M. and Sakamoto, C. M., 1989. Application of CERES-maize model to yield prediction of a Brazilian maize hybrid. *Agricultural and Forest Meteorology* **45**: 299–312.

Mati, B. M., 1986. Soil conservation—cultural and structural aspects: an evaluation report. Paper presented on the 3rd National Workshop on Soil and Water Conservation in Kenya, No. 16. University of Nairobi, Department of Agricultural Engineering, Nairobi, Kenya.

Monteith, J. L. and Scott, R. K., 1982. Weather and yield variation of crops. In: Blaxter, K. and Fowden, L. (eds), *Food Nutrition and Climate*, pp. 127–153. Applied Science Publishers, London.

Ngugi, A. W. and Kabutha, C. N., 1986. Agroforestry and soil conservation on small-scale farms in Murang'a District. Paper presented on the 3rd National Workshop on Soil and Water Conservation in Kenya, No. 17. University of Nairobi, Departement of Agricultural Engineering, Nairobi, Kenya.

Nix, H. A., 1985. Biophysical impacts: agriculture. In: Kates, R. W., Ausubel, J. H. and Berberian, M. (eds), *Climate Impact Assessment: Studies of the Interaction of Climate and Society*, pp. 105–130. Scientific Committee on Problems of the Environment (SCOPE), Chichester, UK.

Roetter, R., 1993. Simulation of the biophysical limitations to maize production under rainfed conditions in Kenya. Evaluation and application of the model WOFOST. PhD Thesis, University of Trier (Materialien zur Ostafrika-Forschung, Heft 12), Trier, Germany.

Smaling, E. M. A., 1993. An agro-ecological framework for integrated nutrient management with special reference to Kenya. PhD Thesis, Wageningen Agricultural University, Wageningen, The Netherlands.

Smaling, E. M. A. and Van De Weg, R. F., 1990. Using soil and climate maps and associated data sets to select sites for fertilizer trials in Kenya. *Agriculture, Ecosystems and Environment* **31**: 263–274.

Smaling, E. M. A., Nandwa, S. M., Prestele, H., Roetter, R. and Muchena, F. N., 1992. Yield response of maize to fertilizers and manure under different agro-ecological conditions in Kenya. *Agriculture, Ecosystems and Environment* **41**: 241–252.

Van Diepen, C. A., Rappoldt, C., Wolf, F. and Van Keulen, H., 1988. Crop growth simulation model WOFOST documentation (version 4.1). Staff working paper SOW-88-01, Centre for World Food Studies, Wageningen, The Netherlands.

Van Diepen, C. A., Wolf, J., Van Keulen, H. and Rappoldt, C., 1989. WOFOST: A simulation model of crop production. *Soil Use and Management* **5**: 16–24.

Van Keulen, H. and Wolf, J. (eds), 1986. Modelling of agricultural production: weather, soils and crops. Simulation monographs, Pudoc, Wageningen, The Netherlands.

Von Urff, W., 1993. Nachhaltige Nahrungsmittelproduktion—Auch ein konzeptionelles Problem. *Entwicklung + ländlicher Raum* **27** (1): 2.

CHAPTER 20

A Support System for Planning Sustainable Agricultural Land Use in Flanders

M. van der Velden, J. van Valckenborgh, J. van Orshoven, K. Smets, A. Grillet, L. Hubrechts, J. A. Deckers, D. van den Broucke and J. Feyen

Institute for Land and Water Management (K. U. Leuven), Leuven, Belgium

INTRODUCTION

The increasing, often conflicting, demand for land by various sectors of society is forcing the Flemish authorities to formulate a new and consistent regional land use plan. With respect to this plan, it is of prime interest to the farming community and the related agricultural sector to propose an Agricultural Main Structure (AMS) or master plan, expressing its own specific land claims.

Such an AMS is expected to be valuable not only in defending the land claims of the agricultural sector as a whole against the claims of other sectors, but also in improving the spatial organisation within the sector. Questions which have to be addressed in this respect are, for example, the future use of idle farmland and the consequences of the conversion of extensive into intensive or zero-land production systems.

Geographical Information System (GIS) technology is potentially suitable to develop an AMS because of its capability to manage and model geographic and attribute data in a quantitative and dynamic way. Hence, to assist the Flemish agricultural community in:

(1) defining its land requirements,
(2) confronting the land requirements with the physical, socio-economical and legal constraints, and
(3) bringing about a land use reallocation,

an interactive GIS-based support system, called ILIAS, is being elaborated. This support system is developed using pcARC/Info and dBase software.

The Future of the Land: Mobilising and Integrating Knowledge for Land Use Options
Edited by L. O. Fresco, L. Stroosnijder, J. Bouma and H. van Keulen.

METHODOLOGY

The support system is based on three key elements:

- mobilisation of data and integration of knowledge
- formulation of decision rules
- geographical analysis and presentation of the results

Mobilisation of data and integration of knowledge

Geographic data

The geographic data consist of three cartographic layers, i.e. a municipality map, a soil association map and a river basin map. The spatial analysis and operations are performed on so-called 'elementary spatial planning units' (ESPU), resulting from the geographic overlay of the soil association map and the municipality map (Figure 20.1). At a later stage in the land use allocation process, an additional overlay with the map of hydrological units can be made.

Attribute data

Attribute data of the geographic layers which are taken into account for the development of the AMS are related to:

- socio-economic aspects
- physical aspects
- legal aspects

Socio-economic aspects

Eighteen relevant agricultural subsectors were identified which can be grouped into seven Agri-Business-Complexes (ABCs). Each ABC integrates the economic activities for producing inputs to an agricultural sector, the agricultural production process itself and the activities for the distribution and processing of the agricultural outputs. The farming activity is considered the core activity in this vertical chain and is, therefore used to identify the Agri-Business-Complexes. Table 20.1 lists the subsectors and ABCs that are important for Flanders. All socio-economic data are aggregated at the level of the municipalities (Table 20.2).

To describe the future development of the agricultural sector, internal and external factors have to be taken into account. Internal factors are related to structural changes which take place on the level of an individual farm. These micro-changes happen in a macro-environment determined by external factors such as the reform in the EC Common Agricultural Policy, the GATT agreements, expected foreign trade with Eastern Europe, and the EFTA countries' concern for environmental problems and changing consumption patterns. The specific impact of these external factors for the year 2000

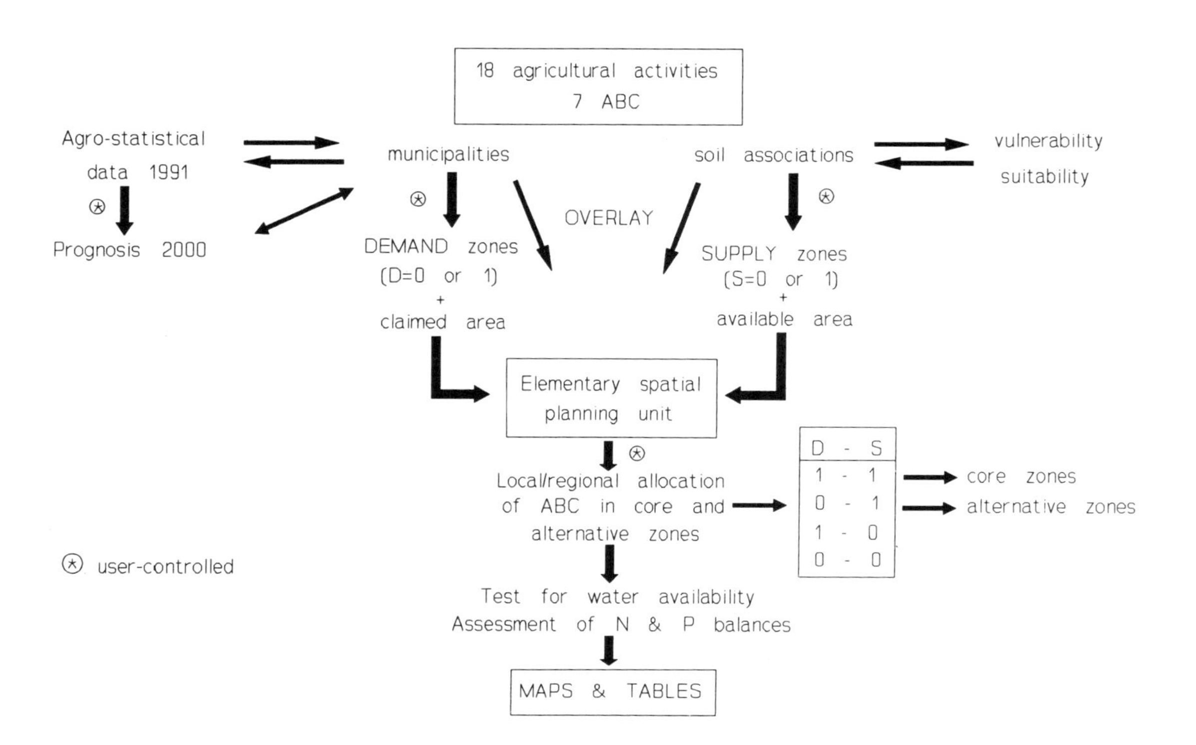

Figure 20.1 Schematic outline of the ILIAS tool

Table 20.1 Agricultural subsectors and Agri-Business-Complexes for Flanders

Agricultural subsector	Agri-Business-Complex
1 Cattle, meat 2 Cattle, milk 3 Other animals	Land-linked animal husbandry
4 Fattening calves 5 Pigs 6 Poultry, eggs 7 Poultry, meat	Land-less animal husbandry
8 Grain crops 9 Sugar beets 10 Potatoes 11 Other crops	Field crops
12 Intensive vegetable growing 13 Extensive vegetable growing 14 Strawberries, open air	Outdoor horticulture: edible products
15 Vegetables, under glass 16 Strawberries, under glass	Indoor horticulture
17 Fruit crops	Fruit crops
18 Non-edible horticulture products	Non-edible horticulture products

Source: Viaene et al. (1993).

is evaluated and expressed as a percentage increase or decrease per subsector (Viaene et al., 1993) for the whole of Flanders. For each municipality, an overall internal agricultural stability-index has been determined, derived from a cluster analysis of six variables related to continuity in agriculture (Deklerck and Lhermytthe, 1992).

Physical aspects

Data on crop-specific soil suitability, vulnerability for soil degradation and breakthrough of chemicals (Van Orshoven et al., 1992) and requirements for supplementary irrigation water (Hubrechts and Feyen, 1992) are available for the 45 soil associations in Flanders. At the level of the hydrological basins, water availability has been calculated based on the magnitude of the high-water wells (Voet, 1992).

Legal aspects

The 'Manure Redistribution Regulation', known as the '*Mestdecreet*', specifies the maximum admissible quantities of nitrogen and phosphate that can be applied on agricultural land. It is defined as a maximum of 400 kg N per hectare per year; 200 kg P_2O_5 per hectare per year for fodder maize or pasture land and 150 kg P_2O_5 per hectare per year for other crops. These regulations apply to Flanders as a whole.

Table 20.2 Attribute data for the elementary spatial planning units (ESPs)

1. *Attributes of municipalities*
   Current area occupied by subsector and current number of livestock
   GSR[a] per subsector (BEF/ha)
   Internal agricultural stability-index
   External subsectorial stability-index
   Agro-typological characteristics
2. *Attributes of soil associations*
   Crop-specific soil suitability
   Crop-specific soil vulnerability
   Crop-specific irrigation requirements
   ABC-specific water consumption
3. *Attributes of hydrological units*
   Total volume of water available for irrigation and other use in animal husbandry or agro-industry

[a] Gross Standard Return, defined as an average value of the 'gross return' for a given region (Belgium is taken as one region). The 'gross return' of an agricultural product is defined as the monetary value of the gross production, of which specific costs have been subtracted.

Table 20.2 presents the attribute data of the elementary spatial planning unit, related to their socio-economic and physical aspects.

Formulation of decision rules

The basic idea underlying the formulation of the AMS is the search for a compromise between (see Figure 20.1):

(1) the need for land by the agricultural subsectors in the year 2000, a 'demand' that can be translated into demand zones, and
(2) the supply of land (supply zones) to the agricultural community by society.

Decisions have to be taken in order to:

(1) delineate the demand zones,
(2) assess the claimed areas within the demand zones,
(3) delineate the supply zones and assess their area, and
(4) allocate land to the different ABCs in a balancing exercise of the geo-referenced demand and supply.

One set of decisions represents one land use allocation scenario.

Demand 2000

The subsectorial demands for land in the year 2000 are defined at the municipality level using the following attributes:

(1) present area for farming activities or present number of livestock

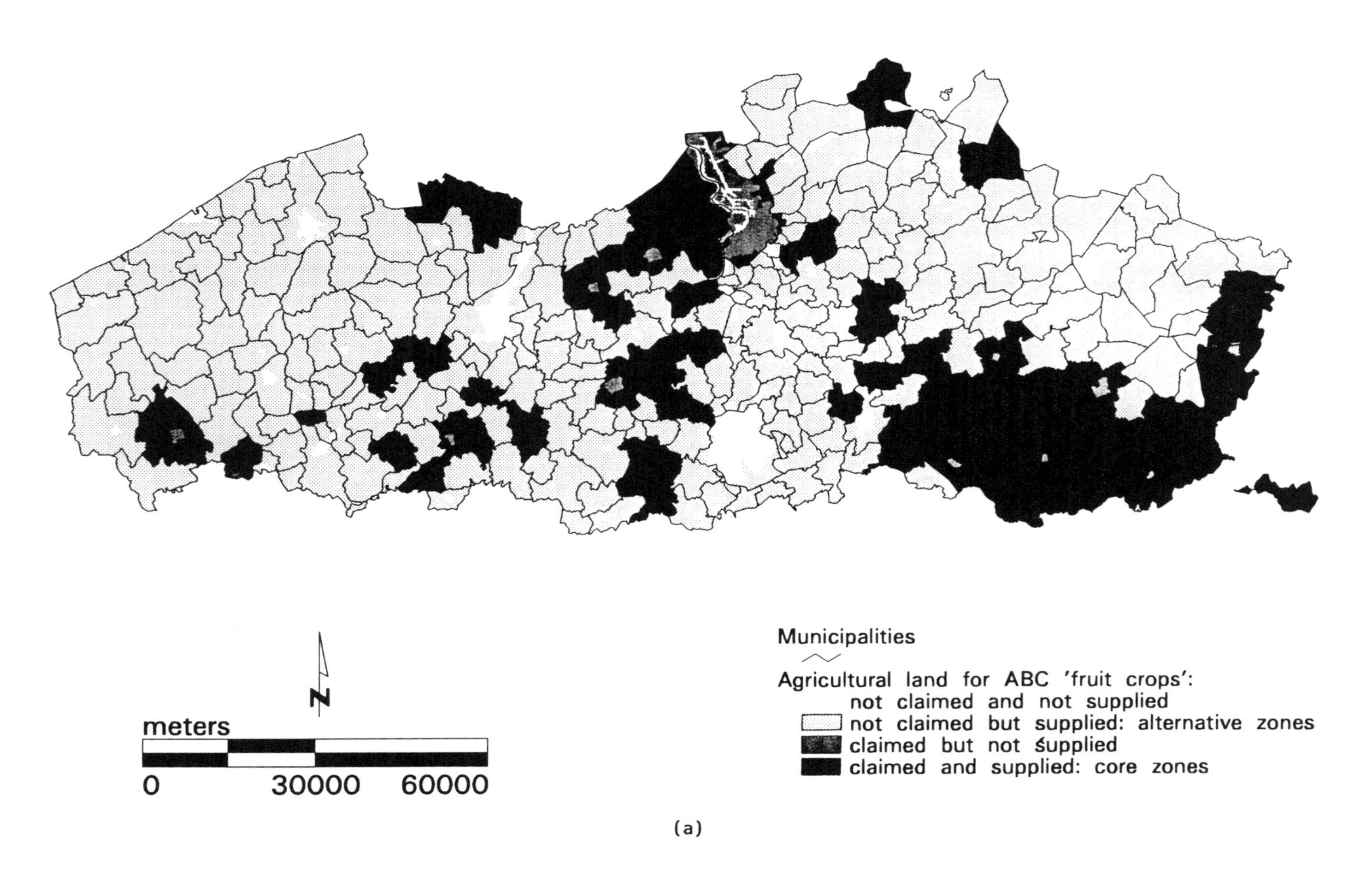
Municipalities
Agricultural land for ABC 'fruit crops':
not claimed and not supplied
not claimed but supplied: alternative zones
claimed but not supplied
claimed and supplied: core zones
N
meters
0
30000
60000

(a)

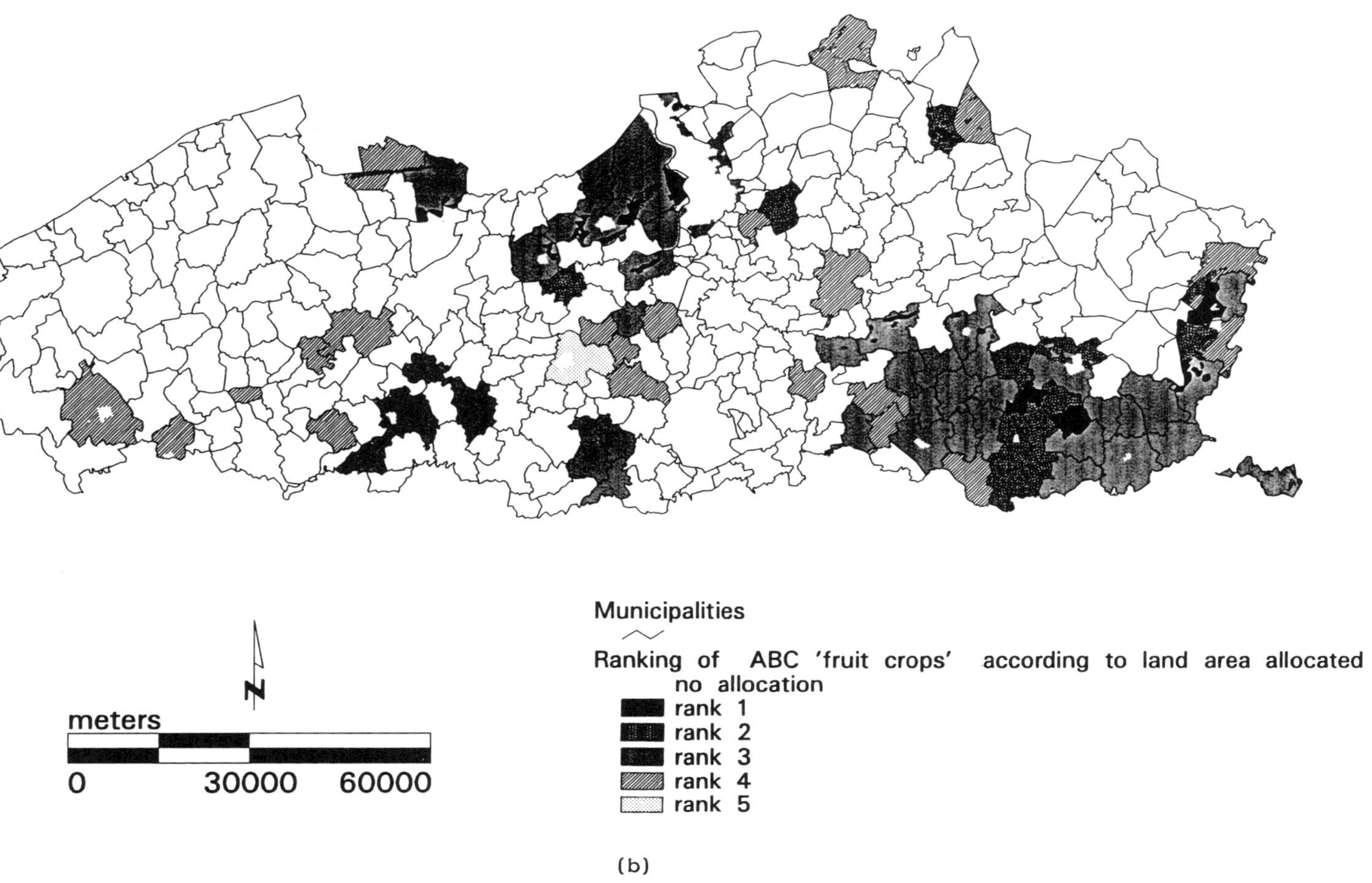

Figure 20.2 (a)Supply and demand zones for the ABC 'fruit crops'. (b) Land allocation to the ABC 'fruit crops' (ranking from very suitable to less suitable)

(2) external stability of the subsector
(3) internal stability of the agricultural sector

Supply 2000

The supply zones for each of the subsectors are derived from the soil association map. Ordinal data on crop-specific soil suitability and vulnerability per association are used as threshold values to define those zones.

Allocation of land

Claimed areas of land for a given ABC must be allocated in the appropriate supply zones, taking into account the simultaneous claims of different ABCs on the same land. This land use allocation process can be limited to the level of the ESPU, i.e. to the core zones (see Figure 20.1) or extended to the alternative zones at the level of municipalities, provinces or the complete Flemish region. The current ILIAS-version works with user-defined priority settings for the six land-linked ABCs.

Geographical analysis and presentation of the results

The output of a scenario-analysis consists of maps and tables displaying for each ESPU, municipality, province, soil association or the entire Flemish region:

(1) the input data and the projected agro-statistical data towards the year 2000,
(2) demand and supply zones,
(3) intermediate and final results of the land use allocation, such as the magnitude of the allocated area per ABC and the magnitude of the non-allocated (land surplus) or the non-available area (land shortage), and
(4) the reasons for existence of surplus land (land that is not fully claimed or supplied due to its too low suitability or too high vulnerability).

Figure 20.2(a) shows the core, alternative, 'demanded but not supplied' and 'neither demanded nor supplied' zones for the ABC 'fruit crops' in a hypothetical scenario, before reallocation. Figure 20.2(b) shows the resulting land allocation to this ABC in relative terms.

OPPORTUNITIES

The ILIAS tool is designed for interactive land use planning. It enables the elaboration and analysis of alternative scenarios resulting from alternative choices made, as described in the previous section. The application programs, integrated in the GIS, allow the user easy updating of information, knowledge and objectives, and hence, generation of renewed versions of the AMS for discussion and amendment. The final AMS should be considered as a 'desirability' plan which determines broad areas in which specific agricultural developments can be encouraged. The main objective of the AMS is not to prohibit, but rather to stimulate given farming activities in given regions.

To improve the performance of the ILIAS tool for land use planning, more options can be made available to compute the claimed areas per municipality and per subsector, i.e. to be more precise in defining the demand zones. Criteria other than soil-related can be introduced to define the supply zones, e.g. landscape and agro-typological features. Land use allocation can be improved by the introduction of concepts of multiple-goal planning and raster-based search algorithms. The requirements for water by the allocated land use combination can be assessed against water availability. Finally, the calculation of the nitrogen and phosphorus balances, according to the legal regulations, should be fully integrated in the system.

The presented procedure and analysis leads to AMSs at a scale of 1:500 000: 5415 polygons are delineated for 13 512 km^2. This scale level is certainly suitable to characterise the agricultural land use requirements at regional level. It is, however, impossible to draw any conclusions about the exact allocation within the different spatial units.

ACKNOWLEDGEMENTS

All studies have been executed at the request of the Ministerie van de Vlaamse Gemeenschap, Departement Leefmilieu en Infrastructuur, Administratie Milieu, Natuur- en Landinrichting (AMINAL), Bestuur Landinrichting en -beheer.

REFERENCES

Deklerck, D. and Lhermytthe, K., 1992. Typologie landelijke gebieden, Rapport 1991. Stichting Plattelandsbeleid v.z.w., Leuven, Belgium.

Hubrechts, L. and Feyen, J., 1992. Raming van de waterbehoefte voor de irrigatie van vollegrondsgroenten met toepassing op de provincie West-Vlaanderen. Rapport nr 5. Instituut voor Land- en Waterbeheer, Katholieke Universiteit Leuven, Leuven, Belgium.

Van Orshoven, J., Verbeke, P., Meylemans, B., Faes, L., Deckers, J. A. and Feyen, J., 1992. Beoordeling Agrarisch Bodemgebruik. Een geografisch informatiesysteem betreffende agrarische bodemgeschiktheid in Vlaanderen. Rapport Instituut voor Land- en Waterbeheer, Katholieke Universiteit Leuven, Leuven, Belgium.

Viaene, J., De Craene, A. and Devolder, V., 1993. Landbouw en ruimte in Vlaanderen. Rapport Afdeling Agro-Marketing. Universiteit Gent, Gent, Belgium.

Voet, M., 1992. Beschikbaar oppervlaktewater uit hoogwatergolven. Werkgroep voor Wetenschappelijk onderzoek inzake landinrichting. Centrum voor landbouwkundig onderzoek, Merelbeke, Belgium.

CHAPTER 21

Mobilising Knowledge for Environmentally and Socially Adapted Land Use Planning in a Semi-arid Region

T. Vetter

GTZ, QRDP, Marsa Matruh, Egypt

INTRODUCTION

In 1988, the governments of the Arab Republic of Egypt and the Federal Republic of Germany agreed on a joint cooperative project on agricultural and rural development on the Mediterranean coast of Egypt. Project representative is the Egyptian Environmental Affairs Agency (EEAA). The German contribution is given by the Deutsche Gesellschaft für Technische Zusammenarbeit (GTZ) on behalf of the German Ministry for Economic Cooperation (BMZ). The Egyptian implementing agency is the Matruh branch of the Egyptian Ministry for Development, Housing, and New Settlements, which has been promoting land reclamation and agricultural development on the Mediterranean coast for decades. The name of the project is the Qasr Rural Development Project (QRDP), El Quasr being the biggest village in the project area. This chapter describes the activities of the project in the field of land use planning.

THE AREA

Between the Mediterranean coast of Egypt and the Western Desert, there is a small fringe of land which receives scattered rain showers in the course of the winter months. This is sufficient to create a limited potential for agriculture and hence, the region has been inhabited for centuries. The region is called the North-West Coastal Zone of Egypt (NWCZ) and extends for approximately 500 km from Alexandria to Salloum, the border town between Egypt and Libya. Its southern extension is some 30 km from the coast (Figure 21.1). The project area of QRDP was selected as a representative part of the

The Future of the Land: Mobilising and Integrating Knowledge for Land Use Options
Edited by L. O. Fresco, L. Stroosnijder, J. Bouma and H. van Keulen.

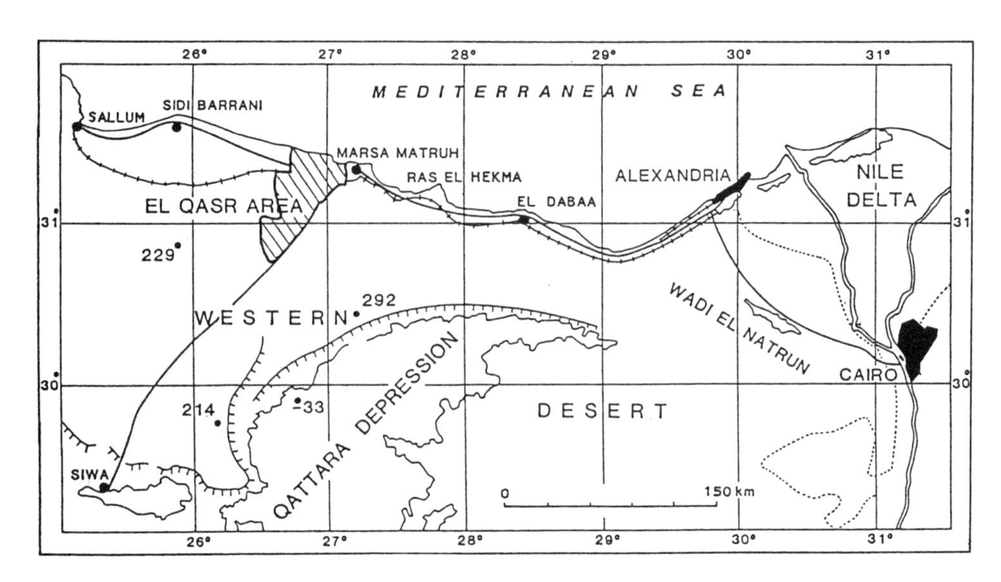

Figure 21.1 Location of the El Qasr area

North-West Coastal Zone covering about 1750 km^2. The boundaries were defined according to tribal boundaries.

THE LOCAL POPULATION

The local population of about 20 000 consists of Bedouins of the Awlad 'Ali tribe. The population density is highest near the coast where the largest favourable agricultural areas are located. It gradually declines towards the south up to approximately 30 km from the coast, where aridity increases sharply. Some widely scattered settlements are found even further south, however. The Bedouins changed from a nomadic way of living to the sedentary one during the last decades. Their main source of agricultural income was and is animal husbandry, accounting for about two-thirds of their agricultural income. Accompanying the process of sedentarisation, crop production has become increasingly important. Mainly figs, olives and barley are cultivated.

Although sedentarised, Bedouin society is still organised according to tribal hierarchy. The tribe of Awlad 'Ali is subdivided into subtribes (*qabila*, pl. *qaba'il*), clans (*'aila*, pl. *'ailat*) and extended families (*bait*, pl. *buyut*). The boundaries between the 16 subtribe areas in the project area are well established. However, claims of individual landowners frequently create problems since individual landownership boundaries are not defined unequivocally.

THE SETTING

Exclusive winter rainfall brings an annual average of 140 mm at Marsa Matruh with a steep gradient towards the south. As typical for semi-arid climates, rainfall is erratic in its temporal and spatial distribution. The situation can be characterised as follows: for most of the year there is not enough water and for short periods there is too much.

Rain water is the only source of fresh water. Drinking water for Marsa Matruh is supplied by a water pipeline and trains from the Nile Delta.

Geologically, the area consists of different types of tertiary limestones and marls. The layers are concordant, striking E–W and dipping ~1° N. There are no major dislocations or faults. Karst weathering and sea-level changes were the dominant geomorphological processes in the Pleistocene.

A typical topographic cross-section from the coast to the south comprises low elevations in the coastal areas, interspersed with flat and partly saline depressions and dune ridges. The slope to the south increases up to a scarp with an elevation of about 100 m. Beyond this, the topography is extremely flat, generally rising very gently towards the south. Slopes of less than 1% prevail. The tableland is dissected and drained by wadis that are rarely longer than 20 km and run prevailingly in a N–S direction.

At irregular intervals old marine terraces occur, forming smooth steps in the topography. At approximately 20 km distance from the coast the linear pattern of drainage ceases. The transition zone to full arid conditions is characterised by a pattern of non-contiguous depressions. Despite the aridic moisture regime, typical processes of karst morphodynamics are still active there. A typical character of the terrain is that the depressions are some 0.1–1 m below their vicinity and if not for different vegetation would hardly be recognisable. In general, the calcareous underground and the low rainfall resulted in the development of shallow and stoney soils. Loamy sands and sandy loams are the prevailing textures. Consequently, salts of aeolian origin are leached and salinity is generally moderate. Calcium and magnesium are above optimum, other nutrients are medium to low, and organic matter is very low.

Soils are differentiated mainly by depth, water-holding capacity and relief position. Yermosols are the prevailing soil type, being calcic on hillcrests and haplic and luvic on steep slopes. In wadis, calcaric Fluviosols or luvic Yermosols occur; calcaric Regosols are especially common in sand sheet areas. Since infiltration capacities are low, the winter rainstorms generate overland flow and wadi runoff.

ACTUAL LAND USE

In accordance with the climate conditions, major parts of the region are utilised as a natural pasture. The Bedouins traditionally are breeders of sheep and goats. The taste of the meat produced is highly appreciated in the Arab world. Overstocking of the herds is a threat to sustainability of the rangeland.

Arable farming is possible under favourable conditions such as an additional supply of surface runoff water and a soil depth of at least 50 cm. The occurrence of such conditions is exceptional, hence, only approximately 10% of the total land can be considered suitable for sustainable arable farming. Although marginal, the existing potential is not yet fully utilised. Figure 21.2 shows the principle of runoff-irrigated agriculture in a wadi.

Many ancient terraces, dikes and even channels are remains of highly sophisticated water management schemes. They were built and maintained by ancient inhabitants, most probably Romans, and provide evidence that the precious water and sparse soil resources were utilised in a highly efficient way. But they also show clearly that there

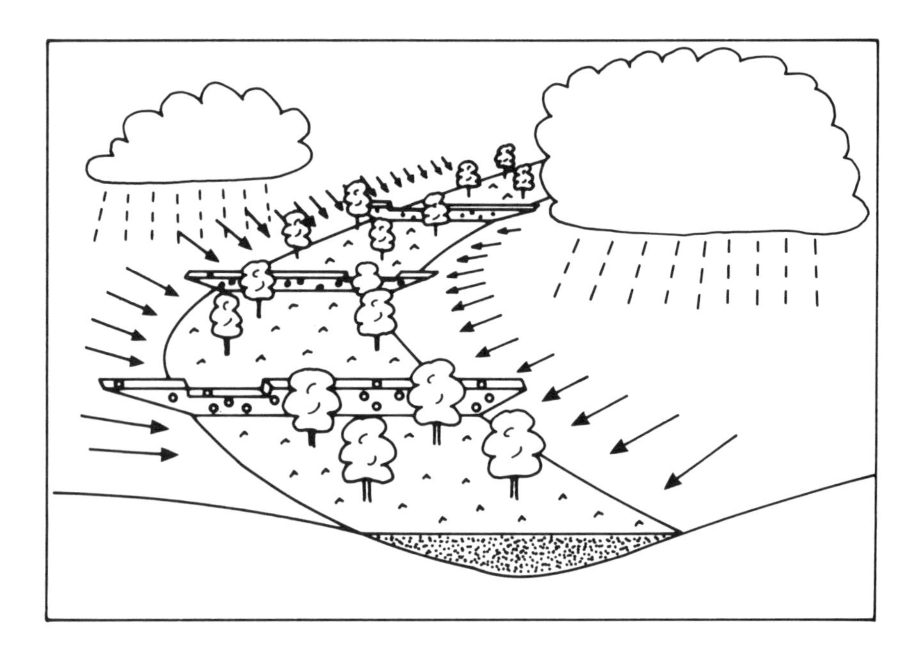

Figure 21.2 The principle of runoff irrigated farming in a wadi (arrows indicate sheet runoff)

must have been a well-organised society with the administrative authority and instruments to plan, implement and maintain water management schemes and to distribute the water. Last but not least, it proves that there actually is enough water to distribute.

The pattern of actually cultivated areas is patchy, restricted to coastal plains, alluvial fans, wadis, terraces, ancient gardens on the tableland, and the many depressions in the south. Agricultural mechanisation and subsidies have led to extensive ploughing of shallow soils on the tableland for barley production. Consequently, soils are exposed to erosion, and the natural vegetation of annual herbs—an important natural fodder for the herds—is impoverished.

LAND USE PLANNING AND ENVIRONMENTAL MONITORING

To monitor natural and man-induced processes and to formulate alternatives in land utilisation, a working group was established within the QRDP, named according to its activities: Land Use Planning and Environmental Monitoring (LUPEM).

As a first step, an information basis was set up. Satellite images and aerial photographs, weather records of a meteorological network, wadi runoff measurements, and soil data, in addition to existing maps, field surveys and special scientific reports are being evaluated. Among these are studies on the rural income situation, ground- and surface-water resources, marketing and rangeland. Spatial information has been integrated in a Geographical Information System (GIS).

A major activity was the formulation of a land use plan. It consists of a set of five maps at a scale of 1:75 000 and a report, and is based on a subdivision of the project

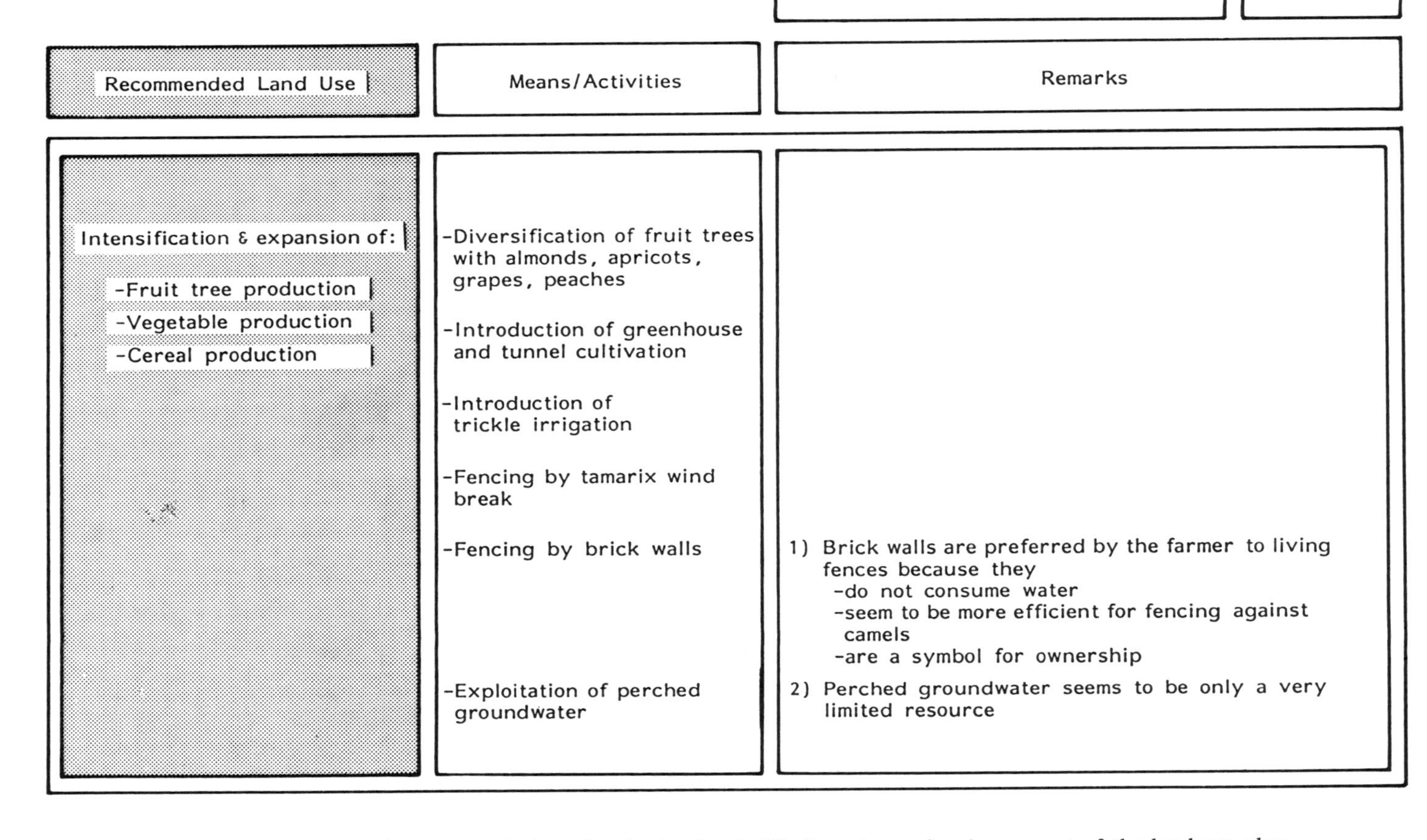

Recommended Land Use	Means/Activities	Remarks
		Shallow depressions — **SPd**
Intensification & expansion of: -Fruit tree production -Vegetable production -Cereal production	-Diversification of fruit trees with almonds, apricots, grapes, peaches	
	-Introduction of greenhouse and tunnel cultivation	
	-Introduction of trickle irrigation	
	-Fencing by tamarix wind break	
	-Fencing by brick walls	1) Brick walls are preferred by the farmer to living fences because they -do not consume water -seem to be more efficient for fencing against camels -are a symbol for ownership
	-Exploitation of perched groundwater	2) Perched groundwater seems to be only a very limited resource

Figure 21.3 Table of recommendations for the land unit 'Shallow depressions' as a part of the land use plan

area in land units. The land units are of similar or equal geomorphological and ecological origin and are subject to comparable dynamics of geomorphology and ecology. They have similar agro-ecological characteristics. Several types of dune formations, depressions, plains, slopes, wadis, terraces, tablelands and alluvial fans were mapped.

The southern shallow depressions mentioned above are a good example of one of the land units. They collect surface runoff from the surrounding barren areas. The resulting soil moisture favours plant growth. Both water and plant growth stimulate chemical weathering of the calcareous mother rock and thus induce soil development. The plant cover not only protects the soils from wind erosion, but also favours trapping of sand during sand storms. The resulting sand sheet has a mulching effect, which in turn reduces soil surface evaporation and encourages leaching of wind-deposited salts.

From an ecological point of view such a depression is a stable and self-sustaining system with a considerable potential for agriculture; they can be mapped with the help of satellite imagery. Other systems of land units are much less stable and have other limitations for development and requirements for protection.

Physical characteristics, present land use, problems and limitations for both present and possible future land use, and also recommended land use and reclamation measures for each of a total of 23 different land units are described in the tabular part of the land use plan (Figure 21.3).

The tribal boundaries have been mapped, which allows allocation of potentials, input requirements and investments to tribal areas. Equal treatment of the tribes is imperative for the Egyptian development agencies.

IMPLEMENTATION OF ALTERNATIVE LAND USE OPTIONS

How can the land use planning proposals of QRDP be implemented in the field? How is a land use plan negotiated at a semidetailed scale? The land units are characteristic agro-ecological units for which recommendations are given. The aim was neither to carry out detailed surveys of individual areas nor to arrive at an agreement with the stakeholders. The semidetailed scale is too general for such an objective. To focus on the local level would mean to change to a more detailed scale and to conduct systematic negotiations with groups of stakeholders. This would result in efforts that cannot be justified if done for large areas and area covering at the same time. Concentrating on parts of the area will conflict with the aim of equal treatment of the tribes.

At local scale, planning concentrates on key problems such as water management. This has been done in the past and has been successful. However, particularities should be taken into account.

Decisions on land use are made by individual landowners. The latter are addressed by neither the project nor the governmental development agencies within a top-down approach. Rather, these landowners address the project and governmental agencies and apply for reclamation and development measures (bottom-up). Individual landownership is a new phenomenon for a nomadic society and the existing legal frame for landownership is insufficient to guide the process of privatisation of land without problems. This has had severe implications for our field work in the past. Unclarified claims on land have led to a complete stop of activities already started in some cases.

A technical difficulty for changing land utilisation practices is the specific method of rainwater harvesting without which arable farming would not be possible. The respective catchment area might be an uphill slope, a part of the tableland or the entire catchment area of a wadi. Downstream farmers consider wadi runoff as 'their' water. If an upstream farmer starts to develop his land, downstream farmers might complain that 'their' water is taken away. On the other hand, particularly larger catchment areas may concentrate runoff water to devastating flash floods after heavy rainstorms, regularly causing heavy erosion and damage to water management structures. Thus, the struggle for more water may lead to less. In a climatic region regularly experiencing drought, farmers are extremely difficult to convince that their fields will always receive sufficient amounts of water under carefully planned water-harvesting conditions. However, the aims achieved so far are encouraging.

In addition to water management but still at local scale, implementation of improved range utilisation practices face difficulties as well. Rangeland is common property. Thus, the main fodder source is free, although farmers use supplements for fattening and in periods of poor pasture. An individual animal breeder would only consider investments for the range if returns are guaranteed. So far no measures are known to increase fodder value of the natural pasture, except for protection against overgrazing. But even this is difficult, if not impossible, since from the individual point of view it is profitable to exploit the resource and not to invest.

How do we inform or possibly convince breeders and herdsmen of the benefits of land use planning? The answer to that question is yet to be found. Bedouins as a group are unlikely to address or to organise themselves. In their society, oral communication is the usual way to spread local news, and it would be pointless to send an invitation to all members of a tribe to discuss proposals together (e.g. at an extension meeting or a discussion about land use). On the other hand, the tribal organisation is not tight enough to guarantee spreading of the news. Communication processes are not transparent, particularly not for outsiders from western industrialised countries, and appear unreliable.

At regional scale, implementation of land use proposals has to be initiated at the political level. Instruments to guide changes in land use have to be devised by policy- and decision-makers. This is certainly supposed to happen in close cooperation with land use planners.

Monitoring of actual land use practices, thorough and systematic assessment of the natural resources, formulation of alternative land use proposals and practical experience at the grass-roots level as described above is a first, valuable step. But most of the way still lies ahead.

The quest for knowledge for a better future of the land has begun. Even if the alternatives and options derived from it can be realised only to a certain degree, they will provide the guidelines to be followed for development and protection of environmental resources.

CHAPTER 22

Selected Case Studies at Regional Level

AN EDUCATIONAL EXAMPLE OF INTEGRATION OF AGRICULTURAL, ENVIRONMENTAL AND STATISTICAL KNOWLEDGE WITH A GEOGRAPHICAL INFORMATION SYSTEM

M. de Bakker
Prof. H. C. van Hall Institute, Groningen, The Netherlands

Authorities like to have the most recent information about their region. In order to incorporate recent technical developments like the use of a Geographical Information System (GIS) in higher education, a case was developed for graduate students to simulate a procedure to answer questions for regional authorities. It integrates soil and crop factors with statistical data, combined with remote sensing data for recent land use, to evaluate opportunities (location and size) for new crops. Environmental aspects such as distance to open water and vulnerability of soils are included. The case is situated in the northern part of Friesland. Bulb-growing could be an alternative land use. The flow diagram for the procedure is shown in Figure 22.1.

A soil map (1 : 50 000) is combined (3) with a classified remote sensing image. Statistical information (2) at municipal level is combined with the land use. Factors (1) determining the capabilities of the soil for bulb-growing are evaluated to present a map (5) with possible suitable locations. Alternatives (4) (e.g. distances to open water) are studied and compared. The procedure is done with the raster GIS IDRISI.

The Future of the Land: Mobilising and Integrating Knowledge for Land Use Options
Edited by L. O. Fresco, L. Stroosnijder, J. Bouma and H. van Keulen.

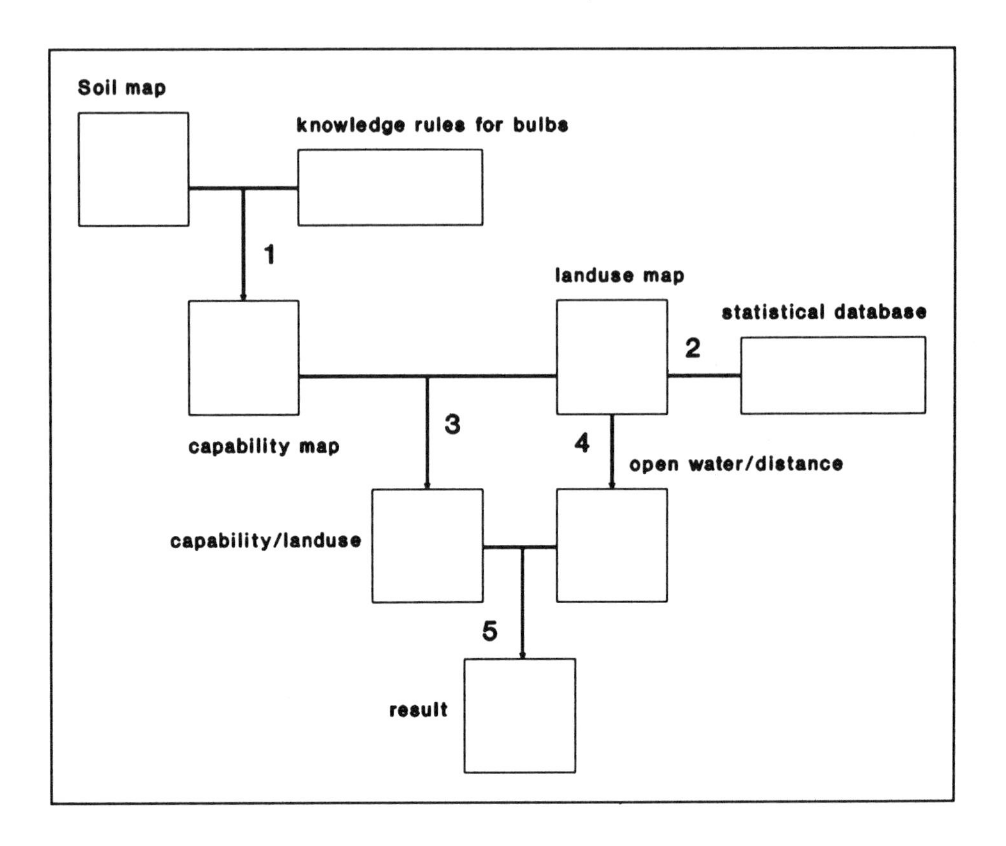

Figure 22.1 A flow diagram of the process. Numbers refer to the process described in the text

THE FUTURE OF THE AMAZON REGION: MOBILISING AND INTEGRATING KNOWLEDGE FOR LAND USE OPTIONS IN THE BOLIVIAN AMAZON

G. W. van Barneveld and P. Spijkers
DHV Consultants BV, Amersfoort, The Netherlands

Until only a few years ago, land use in the Amazon region of Bolivia (100 000 km^2) favoured environmental protection. While in other Amazon countries deforestation by small colonists has reached 0.3–0.4% per annum, in Bolivia it is 0.15%. In fact, from a regional perspective, Bolivia's tropical Amazon rainforest is still reasonably intact.

The main reason for this is the attractive economy of Bolivia's Amazon region. For more than a century, tapping of natural stands of rubber (*Hevea brasiliensis*) and the collection of Brazil nuts (*Bertholetia excelsa*) have been the basis of the regional economy. However, since rubber tapping became economically unattractive in the mid-1980s, thousands of former rubber tappers have settled in areas of primary rainforest. Here they are felling and burning thousands of hectares of primary forest annually in their efforts to produce food under a system of shifting cultivation. It has been estimated that deforestation in the Bolivian Amazon at present is just as serious as in other parts of the Amazon region.

A strategy has been developed for sustainable development and land use in the region, based on mobilisation and integration of data on physical, biological and human resources of the region. Originally there were no data on these subjects: they had to be collected. To do this an integrated, systematic and comprehensive inventory and study were carried out.

Using remote sensing techniques and field work, a 1:500 000 scale inventory was made of the natural resources of the region. This was followed by land evaluation and the formulation of alternative land use options. A forest-ecological survey resulted in the mapping of 18 different forest types, which were categorised on the basis of differences in biodiversity and the presence of non-wood forest resources. Similar inventories were made on current land use, deforestation and trends in forest degradation, and on forest concessions and logging activities. A population census (colonists, extractivists, and the indigenous population) was also carried out and the dynamics of the region's population was investigated. In-depth system studies of farming and extraction activities were carried out, as well as studies on migration, land tenancy, the indigenous population, and the role of women in the farming and extraction systems.

All the data collected have been integrated into a Geographic Information System. This allowed the development of a comprehensive master plan for the sustained development of the region. Basic to this plan is a Regional Land Use Zoning Plan, which describes alternative options for the 'best use of land', given the accepted objectives of environmental and societal opportunities and constraints. It is meant to indicate the possibilities for land use, forest exploitation, forest extraction and rural development, and requirements to ensure that these activities take place in a sustainable way. For each of the preferred land use options, specific development and/or conservation projects have been worked out, including projects on sustainable agro-foresty development, the use of non-wood forest resources, forest management, the development of nature reserves, and projects aimed at the improvement of the socio-economic situation of the most marginal groups in society. The Land Use Zoning Plan and the accompanying projects constitute a strategy for the long-term development and conservation of the Bolivian Amazon.

AEROSPACE SURVEYS AND ON-FARM RESEARCH TO MANAGE GROUNDWATER RESOURCES IN LORRAINE, FRANCE

M. Benoit,[a] F. le Ber[b] and J. Bachacou[b]
[a]INRA Station S.A.D., Mirecourt, France. [b]INRA C.R.F. Champenoux, Seichamps, France

Agriculture and forestry are the main activities in the rural area of Lorraine, France. The rural area supplies the region with mineral and drinking water. With the introduction of new EEC, measures, new preventive methods are needed to protect these water resources. This research project investigates which land use changes may significantly affect the groundwater resources in Lorraine.

LAND USE EFFECTS ON GROUNDWATER QUALITY

Since 1989, the water quality on 66 fields has been measured: 14 fields with ceramic porous cups, and 54 with nitrogen profiles. The fields represent the main cropping and grassland systems of Lorraine. All measurements are done in farmers' fields without special recommendations to the farmers: 'You do what you want and when you want'. We can classify the quality of the groundwater for the different land uses. The average values represent the integrating measurements of soil properties and farmer practices. The standard deviations represent effects of variations in the degree of variation. Results have been calculated for 1989–1992.

A CONCEPTUAL MODEL OF WATER QUALITY AT THE WATERSHED

Our method comprises three steps:

(1) determine the watershed delimitation,
(2) identify the land use of the watersheds by LANDSAT or SPOT data, and
(3) calculate the simulated value of water quality (i.e. surface of each land use × nitrate value under each land use).

At the same time, we measure the quality of water in 19 sources in the Lorraine region. The size of the watersheds varies from 16 to 984 ha. The difference between the simulated value and the measured one is less than 19 mg NO_3 per litre for these 19 sources of superficial groundwater.

CONCLUSION: TOP-DOWN OR BOTTOM-UP SCALE CHANGES

To test our water quality model, we use the field-scale observations and the simulated water quality value: we take into account the surface of all types of land use and their water quality value. Then, we can compare this simulated water quality with the measured water quality of the source of each watershed. This approach is bottom-up.

However, if we have no agronomic measurement sites in a region, how can we establish the contribution of the variants of land uses to the water quality? If we know the water quality at the watershed level in several sources and the surface of all land uses in the watersheds, we can formulate a set of n equations with n unknowns. When we solve these equations we find the value of water quality at the field level for each type of land use: this is our top-down approach.

FARMERS, LAND USE AND GROUNDWATER QUALITY: AN INTERDISCIPLINARY APPROACH

M. Benoit and M. C. Muhar
INRA Station S.A.D., Mirecourt, France

The objective of this project is to provide local actors involved in water quality with management tools and methods to improve this quality with local solutions in a preventive way. For us, these actors, especially the farmers, are 'research objects' because we want to understand their practices and their effects on groundwater quality, but they are also 'research actors' because they are users of our tools and methods to improve their management practices.

RESEARCH DIMENSIONS OF THE PROBLEM

We use geological, geographic and agronomic knowledge to determine the temporal and spatial dimensions of the groundwater improvement process. We distinguish three delays during the groundwater improvement process: the hydrogeological delay between the source and the subsoil; the agronomic delay between farmer practices on the soil; and the farmer practices delay between problem finding and changing practices.

Since the beginning of our work in 1988, in our research region, we have produced the following results:

- hydrogeological delay: from some months to one year in 28 watersheds case studies;
- agronomic delay: from some months to two years in 44 field plot experiments;
- farmer practices delay: from three to more than five years in six agricultural system case studies.

On the land, we have to manage a 'three-dimensional space' if we want to understand groundwater quality: the watershed where groundwater is generated, the land users who influence the water fluxes and qualities, and the climate that effects the quantity of water entering the watersheds.

LAND USE MANAGEMENT IN A WATERSHED

To understand land use in a watershed, we have to know the users and their logic. For the agricultural users, especially farmers, we take into account the proportion of each farm type in the watershed. After that, we describe the logics of land use for each farm type. To model farmers' logic on land use management we work as an interdiciplinary team comprising a soil scientist, an agronomist, an economist, and also a sociologist because if we try to modify land use in this watershed, we have to understand 'farmer reasoning'. By changing the conditions of farmer decisions we can change land use. For example, we can propose the composting of farmyard manure if we want to apply this to grassland, because farmers prefer to apply this manure without composting it in winter time before maize on grass ingestion by cattle.

If we want to model future land use, we have to link the future of the functioning farm in question to the part of each land use. The steps in this model are:

(1) farm typology
(2) farm land use logic: for each type of farm
(3) description of the proportion of land uses in each farm
(4) modelling of territorial growth for each farm
(5) modelling of land use location in the farm territory

CONCLUSION: COLLECTIVE MANAGEMENT OF THE FUTURE OF THE LAND IN THE WATERSHEDS

We observe that new collective land use management practices are created in watersheds to improve groundwater quality when the water users and the land users are the same human community: a village or a group of villages. If water consumers have no relation with the land users in the watershed, the delay in land use changes increases.

INORGANIC FERTILISER AVAILABILITY AS A DETERMINANT FOR LAND USE DEVELOPMENT IN THE FIFTH REGION OF MALI

N. van Duivenbooden
AB-DLO. Present address: Wageningen Agricultural University, Wageningen, The Netherlands

Inorganic fertiliser application is one of the measures to increase crop production and to prevent soil nutrient depletion. An interactive multiple goal linear programming model has been used to explore the possible impact of increased inorganic fertiliser availability on land use, and crop and livestock production in the Fifth Region of Mali. In the model, cropping, animal husbandry, and fisheries are defined in a target-oriented way. Sustainability is quantified through an equilibrium of N, P and K balances. Other aspects included in the model comprise gross revenue, labour, monetary inputs, emigration of the local population, and food subsistence needs for each of the 11 agro-ecological zones distinguished. Marketable crop production has been maximised under three scenarios: restricted (more or less the current situation), intermediate and unrestricted inorganic fertiliser availability.

Results indicate that increased fertiliser availability leads to (1) an increase in total crop production with higher diversity in crops, both in commodities and level of intensification; (2) an increase in animal production and a higher degree of diversification: the number of cattle decreases while that of sheep and goats increases, because of more efficient manure collection and the earlier time at which they can be sold; (3) increased dependence on animal husbandry and fishery activities to attain the imposed minimum gross revenue; and (4) self-sufficiency in food in seven of the 11 agro-ecological zones in normal years with respect to rainfall and flooding. In dry years (lowest 20%) this can only be achieved if in addition to emigration of about 20% of the population and unrestricted inorganic fertiliser availability, some sacrifices (e.g. lower regional gross revenue) are made.

This case study clearly demonstrates that it is necessary to evaluate options for regional land use development in a complete model that reflects the complexity of the region. It also shows the need for development of non-agricultural activities to generate additional regional gross revenue to allow sustainable agriculture.

PEDOLOGICAL ASPECTS OF AGRICULTURAL LAND USE IN SLOVAKIA

M. Dzatko
Soil Fertility Research Institute, Bratislava, Slovakia

Agricultural land use patterns in Slovakia are the result of improper goal setting aimed at self-sufficiency in food production at any cost. Often, arable lands are situated on poorly suitable and sloping sites.

The starting point for new concepts of inevitable land use changes is the system of Pedo-ecological Units of Slovakia, representing the basic system of agricultural land classification and evaluation. In general, the Pedo-ecological Units (PEU) are defined as relatively homogeneous land units on the basis of interactions of environmental components, mainly soil, climate and relief, each with its own specific production potential. The characteristics of the land units are defined in a hierarchical framework starting from the basic units (of which there are more than 10 000), through Pedo-ecological Regions (80), to Pedo-ecological Areas.

Quantitative land-capability assessment is based on target-oriented modifications of factorial and regression analyses of the relationships among crop performance and PEU properties, resulting in maps for sustainable land use.

Practical realisation of new projects is highly dependent on both economic and ecological considerations. This combination illustrates the inevitability of changes in agricultural land use under the new economic and social conditions in Slovakia.

AN OBJECT-ORIENTED APPROACH TO LAND USE MAPPING AT SUB-REGIONAL SCALE

E. J. Huising,[a] L. O. Fresco[b] and M. Molenaar[b]

[a]PO Box 339, 6708 CR Wageningen, The Netherlands. [b]Wageningen Agricultural University, Wageningen, The Netherlands

Land use planning requires information on actual land use and processes of change. GISs can be used as a tool to analyse land use if an adequate model describing the variation of land use in space and time is available. Such a model was developed based on a formal data structure and data from Costa Rica.

The definition of objects should be adjusted to the regional setting and to the role of these objects in the analysis of land use. Inventories therefore serve two aims: to help to understand the regional setting and provide information for the identification and description of the objects. Object definition, therefore, becomes an integral part of the inventory process. A generic definition of objects and their characteristics can be given on the basis of the different hierarchical levels that we can identify in agricultural systems. The spatial objects, defined in relation to the hierarchical levels, are:

- the agricultural field or pasture, associated with the crop or herd system;
- the farm unit, associated with the farm system;
- the land use zones, associated with the sub-region as agricultural system.

For the regional level the associated spatial unit has not yet been defined. The hierarchical levels in agricultural systems correspond to object aggregation levels. For Costa Rica we focused on the sub-regional level, because that has hardly been elaborated. Land use zones (LUZs) are defined as the relevant objects at this aggregation level. An inductive approach was adopted for the definition and classification of land use zones, i.e. the LUZs are not defined *a priori* but result from the inventory process. Pattern recognition techniques are used to identify and classify the LUZs. Their spatial patterns could be recognised and quantitatively described using aerial photographs. Inventories of land cover patterns were made using satellite imagery. It was possible to identify relationships between the spatial pattern and land cover pattern characteristics of each LUZ and the farming systems (and associated farm sizes) and land utilisation types. Land use patterns were defined to classify the combinations of land utilisation types per LUZ.

LUZs can be considered as aggregation types for agricultural fields, i.e. a LUZ is a composite of these fields. But LUZ can also be considered as an association type for farms, i.e. several farms may belong to one LUZ, but also several LUZs may belong to one farm. The life cycles of LUZs may last many decades. Their origin is related to the colonisation of new areas, the process still being reflected in the spatial structure of the LUZ.

For an agricultural LUZ, the geometric characteristics tend to remain rather constant during its life cycle, while the land use characteristics may change within a few years. Change in land use could be inferred from change in land cover pattern. For example, trends like a decrease in the area for maize cultivation was observed for most of the relevant LUZs. The role of the LUZ in the analysis of land use lies in the identification of possible problem areas or areas with possibilities for improvement of the land use within the framework of land use planning.

THE EFFECTS OF BIOPHYSICAL AND SOCIO-ECONOMIC PROCESSES ON LAND USE CHANGES OF THE HORTOBÁGY REGION (HUNGARY)

Z. Karacsonyi
Environmental Management Workgroup, Debrecen Agricultural University, Debrecen, Hungary

Hortobágy is the most level plain unit in the central region of the River Tisza, covering an area of 2300 km^2. Its landscape was formed by the River Tisza during the Holocene since the whole area once belonged to the flood area of this river. Its ancient appearance was characterised by vast swamps with woods, meadows, pastures and limited areas of cultivated land. As a result of the effects of biophysical and socio-economic processes on land use, Hortobágy became the largest sodic steppe (‘*puszta*’) in Central Europe. To protect the special natural value of this region, the first Hungarian national park was established here in 1973.

In terms of land use, over the centuries, numerous important and characteristic changes have taken place in the Hortobágy region. During the Tatar and Turkish wars small villages depopulated and were abandoned. These areas later became the property of

Debrecen town, and there animal husbandry developed with an important export of live animals to Western Europe. Regulation of the River Tisza, started in 1846, induced significant changes like drying up and sodification of the *puszta* (alkali soil). The Hortobágy region turned into a real *puszta* after regulation of the River Tisza and related drainage works (permanent and temporary swamps were dramatically reduced, pastures became dry and of low quality, etc.).

At the end of the 1940s, state farms were formed, associated with breaking up of pastures and the introduction of intensive arable farming. As a result of political decisions, in the 1950s, a forced rice-production programme started, which resulted in useless, deserted, destroyed areas. It is in the nation's interest to preserve the natural values of this area. Hence, in 1973 the Hortobágy National Park was established. The area of the national park is a biosphere nature reserve and most of it falls under the Ramsar Convention.

During the recent changes in land relationships in Hungary it has been ensured that nature reserves and nature values are an integral part of the Hortobágy National Park, showing that the conflict between market-oriented production at the Hortobágy State Farm and the protection of nature values can be solved. Investments started with the aim to develop tourism, first with national and foreign private capital (e.g. Epona rider-village). One of the most important ecological rehabilitation programmes in Hortobágy National Park is the rehabilitation of the original water regime.

This rehabilitation programme comprises:

(1) temporary water supply addition (through canals and reservoirs), since natural water supplies are insufficient (region 1);
(2) full-scale reconstruction: construction of water supply systems for dried up regions (region 2);
(3) reconstruction of watershed areas: reconstruction of regions now interspersed by access roads, ditches and canals, to eliminate fast drainage (region 3); and
(4) storing and holding back precipitation to restore the original water regime (region 4).

EFFECTS OF LAND USE ON PROPERTIES OF SALT-AFFECTED SOILS

F. B. Labib and I. S. Rahim
National Research Centre Dokki, Cairo, Egypt

Properties of four profiles selected from virgin salt-affected soils in the northern part of the Nile Delta and Fayoum have been compared to those of four profiles reclaimed 20–25 years ago. Texture, total and type of salts, organic matter content, $CaCO_3$, gypsum, major elements (NPK) and status of trace elements (Fe, Mn, Zn, Cu and B) have been determined. Field observations and analytical results showed that distinct changes have occurred following reclamation. The Aquollic Salorthids or Mollic Fluvaquents have developed to Mollic Torrerts or Ustic Torrifluvents. Total Fe and B are higher in the virgin soils than in the reclaimed sites, while total Zn and Cu show a reverse trend. Total Mn and DCB-Mn are somewhat higher in the cultivated soils, while the free oxides of Fe (DCB extracts) are higher in the virgin soils. Zn and Cu, which represent the adsorbed forms and not the co-precipitate with Fe and Mn oxides, are higher in the cultivated soils, particularly in the surface layers. The DTPA-extractable trace elements in the virgin soils were significantly higher than in the reclaimed soils, except Zn. Surface horizons include more DTPA-extractable trace elements due to the enrichment with organic residues.

AGRICULTURAL WATER BALANCE IN THE SAHELIAN REGIONS: A MULTIDISCIPLINARY APPROACH TO REDUCE DROUGHT RISKS

F. N. Reyniers
Water Management Research Unit, CIRAD-CA, Montpellier, France

In spite of the importance of drought damage to crops in the Sahelian regions of West Africa, quantitative assessment of the effects of water balance terms from regional to farmer level is still rudimentary. For many years this deficiency has led to incorrect diagnoses and misdirected efforts to reduce drought stress. An agricultural water-balance characterisation has been implemented in a network of national agricultural research centres and at CIRAD. The first step in the methodology is the zonation of potential cereal yields (maize, millet and sorghum) in relation to rainfall. This potential is evaluated, first by identifying an indicator of crop water supply called IRESP, a product of actual evapotranspiration during the growing period and the ratio of actual to potential evapotranspiration during the productive stage. Potential crop yield can be calculated from IRESP using a series of regressions established by agronomic multilocation trials in each country. The frequencies of IRESP have been calculated for meteorological stations in Senegal, Mali, Burkina Faso, Ivory Coast, Togo and Tchad where at least 20 years' data are available. The IRESP values are calculated for different lengths of growing period, sowing dates, and soil water reserves. The second step in the analysis is the regression between IRESP and farmer's yield. Then the diagnosis is established by identifying land and technical factors explaining the yield gap due to inefficient use of the rainfall resources. Technical measures are proposed to reduce the yield gap by improving water consumption, and reducing runoff and drainage.

In conclusion, the conditions for reducing drought risks by land use planning are investigated. The need for a better understanding of the interactions between water balances, fertility, and economic factors at low input farm levels is emphasised. This simple approach to land use planning seems to be appropriate for regions where water is a major limiting factor for crop yield even with relatively high rainfall.

MODELLING LANDSCAPE ECOLOGICAL KNOWLEDGE FOR LAND USE PLANNING: THE CENTRAL OPEN SPACE PROJECT

J. Roos-Klein Lankhorst, W. B. Harms and J. P. Knaapen
The Winand Staring Centre for Integrated Land, Soil and Water Research, Wageningen, The Netherlands

Four scenarios representing possible strategies for nature restoration in the central part of the Netherlands (the Central Open Space) have been evaluated on their impact on vegetation, fauna, land use and scenery. To support this evaluation, a computer model (COSMO) has been developed that links current landscape ecological knowledge to a geographical information system. The interface between the scenarios and the model consists of a set of target vegetations suitable for planning purposes.

The evaluation consists of four steps:

(1) The model tests the suitability of the abiotic conditions for the planned target vegetations. If a target vegetation does not match the abiotic conditions, the model proposes measures to adapt these conditions (e.g. raising the water table or removing the topsoil). It also proposes alternative target vegetations. Thus, it helps the planner to update the target vegetation map and/or the measures map. The proposed target vegetations are then translated into management types.
(2) The next step in the model simulates the temporal and spatial developments of the vegetation, according to the proposed management. This results in maps of the expected vegetations after 10, 30 and 100 years.
(3) The vegetation types are then translated into their suitabilities as habitats for animal species. The resulting maps of potential habitats (after 10, 30 and 100 years) also provide indications as to the sizes of the populations that can be attained. This is achieved by clustering continuous areas of potential habitats.
(4) By comparing the generated maps, the impact of the proposed management (required to obtain the target vegetations) on vegetation and fauna can be assessed. These results are finally evaluated using criteria that are relevant to policy-makers, such as diversity and international value and costs.

The present model is based on current knowledge of vegetation succession and animal requirements. Unpredictable circumstances, such as major climate changes, are not taken into account; nor are consequences of pollution. Therefore, it is a valuable tool for planners in assessing and comparing nature restoration plans, but it is not suited for accurately predicting future vegetation and animal developments.

PEDOGEOMORPHOLOGICAL STRATIFICATION OF THE ENVIRONMENT IN NORTH-EASTERN RORAIMA (NORTH AMAZONIA) AND ITS RELATION TO PRESENT LAND USE

C. E. Schaefer
University of Reading, Reading, UK

The pedological, geomorphological and vegetational characteristics of the north-eastern Roraima State area in North Amazonia were studied and the interactions between these factors and the present land use and fire-use were analysed, using Landsat TM-5, SLAR images and field campaigns. These data were matched with the official information on land use and new settlements in Roraima. The studied area represents the most diverse and complex environment in the whole of North Amazonia.

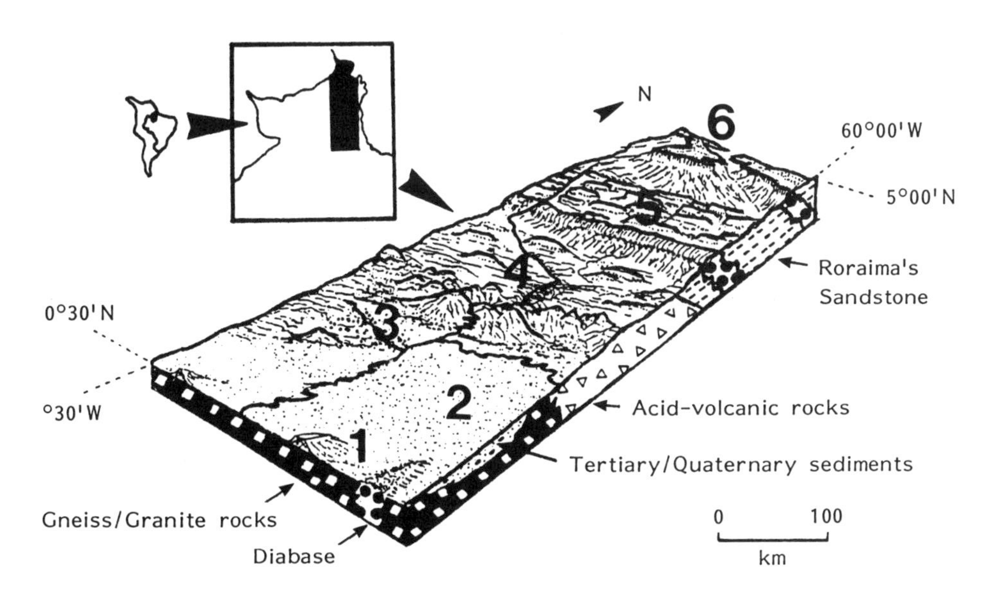

Figure 22.2 Block diagram of the north-eastern Roraima transect and the six 'pedoenvironmental' units

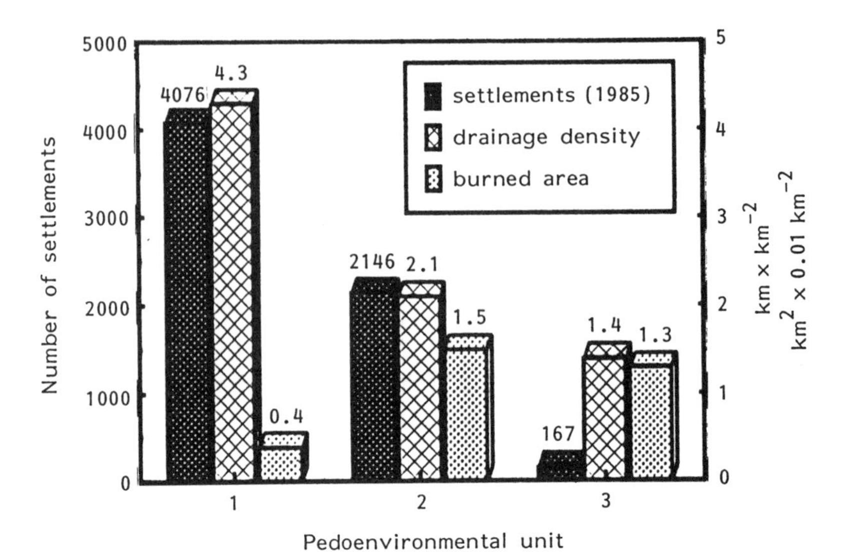

Figure 22.3 Number of nex settlements (1970–1985), drainage density and burned area in units 1–3, based on Landsat images (1991)

The study permitted stratification of the environment into six 'pedoenvironmental units', both under '*cerrado*' (savanna) and under tropical forest, at reconaissance scale (Figure 22.2). The differences in land use between the units were analysed. In the last two decades (1970–1985), there is an increasing trend in occupation towards the rainforested area (unit 1), but the great impact of land use and fire-use related to the savanna areas (units 2 and 3) is widely underestimated (Figure 22.3).

The natural conditions in the savanna areas, i.e. extremely low nutrient status (unit 2) or high sodium content (unit 3), sandy or stony soils, scarcity of vegetation and extensive flattened surface associated with the oldest colonisation and fire-use in North Amazonia, provide the conditions for a large-scale environmental degradation process.

In the broad scenario, the present ecological trend is the generalised forest expansion over the savanna area, although human interference may be changing this process, favouring the maintenance of open vegetation by the progressive exhaustion of the soils.

APPLICATION OF A QUANTITATIVE LAND USE PLANNING SYSTEM IN THE SOUTH-EASTERN ANATOLIAN PROJECT

S. Senol
Cukurova University, Adana, Turkey

The South-Eastern Anatolian Project is the largest irrigation scheme in Turkey, covering 1.8 million ha. The aim of the study was to determine suitable land utilisation types

(LUTs) and to give land use options to planners, policy-makers, etc., before the implementation of this project. A quantitative land evaluation system, developed in the light of the principles of the FAO framework, has been used. Land use is assessed with respect to specific LUTs that already exist in the area, along with others that can be applied after irrigation, and non-agricultural LUTs. Land characteristics (LCRs) of land units derived from the detailed soil map have been stored on computer. Productivity rating indices (PRIs) have been determined for each LUT and each level of LCRs on the basis of their land requirement. The PRI is set to 1.0 when the level of land characteristics is optimal for the LUT under consideration, and 0.0 for its totally inhibiting limitation level. Suitability of the land units for each LUT are calculated by multiplying LUT-specific PRIs of land units. Later, the LUTs are grouped in five land use groups (LUGs) such as: perennial horticultural crops, field crops, vegetables, rainfed crops and non-agricultural LUTs. Subgroups are formed according to the type of suitable LUTs within each LUG, to be shown on the potential land use map. The potential land use maps (1:5000 scale) produced in this study, are useful tools for planners in determining cropping patterns, in locating new settlements and industrial estates, and for the extension services in giving information to farmers on suitable crops for their land.

INTENSIVE LIVESTOCK FARMING IN THE DUTCH PROVINCE OF GRONINGEN

R. van der Veeren and W. J. M. Heijman
Wageningen Agricultural University, Wageningen, The Netherlands

INTRODUCTION

Over the past 20 years intensive livestock farming in the Netherlands has increased considerably, due to the low cost of imported animal feed. This has led to severance of the link between production capacity of cultivated land and manure production, resulting in large manure surpluses.

Intensive livestock farming in the Province of Groningen is expanding because (i) its capacity in the south and the east of the Netherlands is fully utilised, and (ii) the prices of a number of arable products, like cereals, which until recently constituted the main output of the agricultural sector in Groningen, are low, as a result of recent budget-related changes in the Common Agricultural Policy. Hence, intensive livestock farming may be an alternative. This summary reveals to what extent intensive livestock farming could alleviate the problems facing agriculture in east Groningen taking into account several standards for manure application.

Table 22.1 Manure standards

	Manure standard (kg/ha/year)
Phosphate (P_2O_5)	
Current standard	200.00[a]
Stricter	125.00
Strictest	93.75
Nitrate (NO_3)	
EC guideline	170.00
Strictest standard	65.79

[a]For arable land 125.00.

Table 22.2 Production of phosphate and nitrate per animal (kg/year)

	Current diet		Modified diet	
	Phosphate (P_2O_5)	Ntrate (NO_3)	Phosphate (P_2O_5)	Nitrate (NO_3)
Dairy cow	44.000	124.000	40.480	93.000
Pig	7.400	11.000	4.810	6.380
Broiler	0.240	0.230	0.132	0.173

ASSUMPTIONS

The main assumption underlying this study is that manure cannot be transported between regions. Hence, regional manure production cannot exceed the permitted quantity for application on the land. A labour unit is defined as a farm that can be managed by one person. In the current study this is equivalent to 50 dairy cows, 1725 pigs, or six times 62 500 broilers (six 'rounds'). Manure standards are given in Table 22.1.

Animal manure supplies 75% of the phosphate needed, hence the strictest standard is 75% of the stricter standard. For nitrate, animal manure contributes only 38.7% of the requirements. Calculations have also been performed for lower mineral concentrations in animal feed. Table 22.2 shows phosphate and nitrate production per animal.

RESULTS

Table 22.3 Number of possible additional labour units (farms) in intensive livestock farming in east Groningen, based on phosphate standards, for current (c) and modified (m) diets

	Dairy cows		Pigs		Broilers	
	c	m	c	m	c	m
Present standard	126	238	111	210	60	119
Stricter standard	116	222	102	194	58	110
Strictest standard	79	158	70	139	40	79

Table 22.4 Number of possible additional labour units (farms) in intensive livestock farming in east Groningen, based on nitrate standards, for current (c) and modified (m) diets

	Dairy cows		Pigs		Broilers	
	c	m	c	m	c	m
EC guideline	122	120	108	194	32	110
Strictest standard	16	68	14	60	8	34

CONCLUSIONS

(1) Modifying diets greatly influences the scope for expansion.
(2) Scope for expansion is strongly affected by the relative dependence on animal manure for crop nutrient requirements (stricter standard and EC guideline versus strictest standards).
(3) The large difference between scope for expansion based on phosphate standards and that based on nitrate standards is striking.

The full report on the whole province will be published as:
Van Der Veeren, R. J. H. M. and Heijman, W. J. M., 1993. Environmental constraints for intensive livestock farming: the case of the Dutch province of Groningen. Wageningen Economic Papers, Wageningen Agricultural University, Wageningen.

ASSESSING THE LAND USE POTENTIAL OF VENDA

J. P. Watson
University of Venda, Louis Trichardt, South Africa

The Republic of Venda is situated in South Africa south of the Limpopo River, where it occupies about 644 000 ha between latitudes 22°15′S and 23°30′S and between longitudes 29°30′E and 31°15′E. The topography shows a striking contrast between the rugged Soutpansberg mountains lying approximately E–W across the centre of the country and the levelled plains to the north and south. The contrasted topography is associated with differences in mean annual rainfall, which ranges from over 1000 mm in the mountains to under 500 mm on the plains. The purpose of the study was to assess the potential of using Landsat images in compiling a synoptic view of the land resources and land use potential of Venda.

The method was based on visual interpretation of a Landsat colour composite mosaic at a scale of 1:250 000 together with the use of existing maps and reports on the natural resources. Field work was confined to a few journeys along main roads to verify Landsat and map interpretations. In the interpretation of the Landsat mosaic, emphasis was placed on colour, with particular emphasis on false colour red, because it integrates the effects of vegetation and rainfall. Relatively homogenous areas of Landsat colour were classified as follows: mainly red (1037 mm), moderately red (823 mm), slightly red (631 mm), and non-red (442 mm). The figures in brackets refer to the mean of the mean annual rainfall associated with each of these relatively homogenous areas of Landsat colour.

Sketch maps of Venda were prepared showing relatively homogenous areas of Landsat colour, geomorphology, terrain types, mean annual rainfall, climatic zones and surface drainage. Terrain types were distinguished by a combination of geology, topography and Landsat colour. Landsat colour classes also served as humidity zones in the sketch map of climatic zones. Land use suitability was estimated by grouping terrain types according to topography and rainfall and then estimating the suitability of each group of terrain types for broad land use classes, i.e. grazing, forest plantation, orchard and arable. Land use potential was classified according to climatic zones and broad land use classes. For example, forestry plantations were recommended for the humid-cool zone, and sub-tropical orchards for the humid-warm zone.

In conclusion, visual interpretation, with emphasis on colour, of a Landsat colour composite mosaic provided a basis for classification of the terrain types, climatic zones and land use potential of Venda. Acknowledgement is made to the Research and Publications Committee of the University of Venda for funding the project.

PART V

FARM LEVEL LAND USE PLANNING

CHAPTER 23

QFSA: A New Method for Farm Level Planning

L. Stroosnijder,[a] S. Efdé,[a] T. van Rheenen,[a] and Liliek Agustina[b]

[a]Wageningen Agricultural University, Wageningen, The Netherlands.
[b]INRES Research Institute, University of Brawijaya, Malang, Indonesia

FARM LEVEL PLANNING IN LOW INCOME REGIONS: UTOPIA?

Introduction

In developed countries there is often an excess of cheap food due to the presence of highly efficient agricultural production systems and accompanying policies (subsidies, protection, etc.). However, residues of high inputs of fertilisers and biocides are contaminating soil and water resources. Consequently, environmental issues have become a major reason for land use planning.

In low income regions (LIRs) food security is a main reason for land use planning. Low food production in LIRs is often caused, besides unfavourable climatic conditions, by the low productivity of agricultural production systems. This is the result of little or inappropriate use of modern technology due to slow overall economic development. The production means needed to boost the agricultural sector will only be generated if other sectors of the economy develop simultaneously.

Traditional farming systems, developed by gradual evolution, are unable to cope with sudden changes in conditions such as surpassing ecological threshold values or a sudden increase in population. In traditional systems, low-resource farmers practise low-resource agriculture, a form of agriculture primarily based on the use of local resources with modest use of external inputs. Local resources include soil, water and vegetation, as well as local knowledge, labour, agricultural practices and management, and local institutions. External resources, however, refer to those technological inputs such as commercial fertilisers, pesticides, hybrid seeds, tractors, irrigation systems, and information that originates outside the local area.

The Future of the Land: Mobilising and Integrating Knowledge for Land Use Options
Edited by L. O. Fresco, L. Stroosnijder, J. Bouma and H. van Keulen.

In many LIRs natural resources are over-exploited, leading to degradation of these resources. Improving sustainability of land use requires an increase in productivity by the increased use of external resources. External resources, in order to be successful, must match the specific constraints identified in the analysis of local, social and environmental conditions. This chapter will look into land use planning, particularly in relation to external inputs/resources, in LIRs from the following three perspectives: food security, farmers' welfare and environmental concerns.

Reasons for farm level planning

A major policy reason for planning at the farm level from the *food security or technical* perspective is that the agrarian sector, as a provider of food security and/or tax revenues, will be endangered if soil and crop productivity decline as a result of inappropriate use of technology. Increased production may be achieved through an increase of the cultivated area or by increasing the yield per hectare. The first, at a constant technological level, has limited scope and contributes little to economic development and sustainable land use. The latter will undoubtedly involve increased inputs of water, capital and technology. Therefore, for a sustained improvement of a secure supply of food, changes in the production function towards higher productivity are necessary. In order to succeed, developmental policies have to activate small-scale producers, mobilise all existing resources, and create and introduce technical progress. The ultimate aim is to reduce the gap between the yield from traditional forms of production and from more productive ones.

A major policy reason for planning at the farm level from the *farmer's welfare* perspective is to avoid marginalisation of rural areas. Farmers, numerous and diverse, are inventive small entrepreneurs and difficult to catch in economic whole-farm projections unless studied and addressed in detail. Indigenous knowledge (IK) is the key to local development. Rural society has physical and human resources, a stock of past investment and skills, and institutional structure. Local knowledge is, therefore, the most important available source and includes technical knowledge held by local people, specialised knowledge of skilled resource persons and social knowledge held by the group. It is important to understand and consider the decision-making process of small-scale farmers. Traditional forms of agriculture are the result of adaptation by people to production conditions in their environment. But in the light of profound change, such as unprecedented population growth rates and national and global economic and ecological change, traditional options are often unsatisfactory. Understanding decision-making processes will enable a more rapid introduction of new technologies. Farmers are often sceptical about technologies developed and tested elsewhere because they question the compatibility of these technologies to local circumstances. Testing under realistic on-farm conditions is essential for a new technology to be accepted. In order to decrease the inequality between First and Third World, technology must be transferred. If left to market forces alone, such transfer would accentuate rather than alleviate the inequality.

A major policy reason for planning at the farm level from the *environmental concern* perspective is that farmers are the society's caretakers of some major non-renewable resources such as the soil base and, partly, the water cycle. Society has an increasing stake in what farmers do with these resources. Given the present and future food demand situation, increased use of external inputs will be needed but such inputs should be used selectively to complement, where applicable, and not to totally replace useful local resources. An astute understanding of small farmers and their environment has enabled these producers to develop complex cropping and farming systems capable of providing a continuous supply of food while maintaining soil fertility. Elements of these complex systems may be useful in developing sustainable production systems elsewhere.

Advantages and problems

Advantages

Optimal use of indigenous knowledge (IK), which is an element of farming systems analysis, is important to planning because it can improve the overall efficiency as appropriate technology is subject to socio-cultural and environmental judgement. Local farming practices were developed for a certain aim (usually subsistence), by a certain group of people (with socio-cultural values) in a specific ecological setting. In the last 20 years many aid projects have had only moderate success. One of the reasons for this is that many of the proposed new technologies do not make good use of IK and techniques already present among the local people, or simply do not fit in with these techniques (Stroosnijder, 1992). The lack of an objective and quantitative analytical framework is a drawback when studying IK (Stroosnijder, 1993).

Planning of land use in LIRs occurs at different scales, from the national level to the regional, the village and finally the farm level. The last of these is important since it interacts directly with the final decision-maker; the farmer. Addressing the farmer directly, as in participatory planning, will also result in support for planning at higher levels. However, policy-makers often raise questions of whether planning at the farm level is really needed, what the pros and cons are, and whether it is feasible to do such planning at reasonable costs.

Problems

There is definitely a space-scale problem. An analysis at the farm level will take into account the constraints that exist at farm level, and not the constraints that exist at higher levels. This may hamper the identification of realistic development potentials, and the utmost care must be taken to overcome this drawback. However, the advantages of addressing the final decision-maker may outweigh the drawbacks. Some believe that it must be possible to aggregate data of a lower scale into a higher scale. Others seriously doubt it. The reverse, i.e. appropriately addressing farmers by disaggregation of information at a higher scale, has not been achieved till now.

There is also a time-scale problem. In the case of (traditional) technology that was introduced long ago (under strongly deviating boundary conditions), it is necessary to find out the changes in the farming system over time. This can be done by studying historical sources (secondary data); but a dynamic simulation of the farming situation can also shed light on how the technical components of the system have developed over time (Stroosnijder and Van Rheenen, 1993).

Finally, there is a problem with the detailed data gathering and the related cost of this data gathering. The lower the planning scale, the more detailed the data need to be. In the study area where our new planning method was developed there are numerous small mixed farms with a multitude of complex farming systems.

History of farming systems analysis

There are probably few research approaches in the agricultural world today that have received so much attention and popularity amongst scientists, particularly those working in the developing world, as the farming systems approach. This can be concluded from the enormous amount of literature that has appeared since the mid-1970s. A systems approach implies

> studying the system as an entity made up of all its components and their interrelationships, together with relationships between the system and its environment. Such a study may be undertaken by perturbing the real system itself (for example, via farmer managed trials or by pre- versus post adoption studies of new technology) but more generally it is carried out via models (for example, experiments, researcher and/or farmer managed on-farm trials, unit farms, linear programming and other mathematical simulations) which to varying degree simulate the real system (TAC, 1978).

Over the years, various approaches concerning a systems approach in agricultural research have been published. A specific group of approaches characterised with key words like 'farmer orientated' or 'bottom-up' is commonly referred to as 'farming systems research and development' (FSR&D). Shaner et al. (1982) define FSR&D as

> an approach to agricultural research and development that (1) views the whole farm as a system, and (2) focuses on the interdependencies among components under the control of members of the farm household and how these components interact with the physical, biological, and socio-economic factors not under the households' control. The approach involves (1) the diagnostic phase: selecting target areas and farmers, identifying problems and opportunities; (2) the development phase: designing and executing on-farm research; and (3) the implementation phase: evaluating and implementing the results. In the process, opportunities for improving public policies and support systems affecting the target farmers are also considered.

Expectations were high in the 1970s; however, looking back it becomes clear that FSR&D has not always been able to live up to these expectations.

Farming systems analysis (FSA) is the initial and crucial stage of FSR&D and comprises the above step (1) and, partly, step (2). FSA is the understanding of the structures and functions of farming systems, the analysis of constraints on agricultural production at farm level, and ways to translate this understanding into adaptive research programmes (Fresco, 1988). In other words, FSA is a tool that may be used to set a research agenda. The basic steps in FSA are (1) diagnosis: the analysis of farming systems and the identification of constraints; and (2) design: the step from diagnosis to research, both on- and off-station.

Critique on farming systems analysis

Methodologies used in FSA are documented among others in Byerlee and Collinson (1980), Conway (1985) and Collinson (1987). The diagnostic phase usually includes a study of background information, an informal survey (rapid rural appraisal/sondeo) and a formal verification survey. FSA as it has been practised over the years has been subjected to the following criticism and problems.

(1) *FSA can be vulnerable to subjectivity*. Strong emphasis is laid on the participation of the farmer in determining the main constraints, i.e. a bottom-up approach. In practice, however, FSA may be scientist top-down biased. This can be the case when the scientist perceives the problems of the farmer and decides the priority for problem solving, often not considering the interaction between the various activities being practised by the farmer.
(2) *FSA has been too qualitative*. This has made it a difficult tool for policy-makers and scientists to accurately assess (for example, economically) the problems in a region. Determining the order in which problems would have to be addressed was also obscure as a result of this.
(3) *FSA is mainly farmer orientated*. It should be, but it should not *only* be farmer orientated. The farming systems lie in a region and the region will be administered by policy-makers. These policy-makers have certain development scenarios for the region. The instruments they may chose to use (subsidies, taxes, infrastructure, etc.) will influence the 'operational space' of the farming systems. While in most countries it will be the farmer who eventually decides which crop will be grown, at the same time the farmer will strongly be influenced by his environment and the policy-makers.
(4) *FSA has mainly been crop oriented*. To date, research has been mainly confined to crop production processes and rarely has the approach been applied to livestock processes. Other areas generally omitted from consideration to date are more explicit consideration of off-farm enterprises and a more holistic systems approach, which goes beyond the farm gate and attempts to endogenise, for example, the marketing process.
(5) *FSA has suffered from institutional problems*. Recommendations may be rejected because they are inappropriate to the institutional setting for which they were designed. Programmes would become more realistic, appropriate and acceptable if they took account of the capabilities, resources and past activities of the host institutions. Only with the active and constructive support of the local staff and farmers can there be a self-sustaining problem-solving research system.
(6) *FSA is confronted with time conflicts*. FSA is confronted with time conflicts in two ways: (1) in the FSA approach a conflict exists between short-run private gains and long-run social costs. If only the farmer is allowed to indicate the constraints in his system, these will tend to be biased towards the former, which could exacerbate the latter; (2) there is inevitably a time lag in the recognition of a problem, the finding of a relevant solution and its adoption by farmers.
(7) *FSA lacks gender differentiation*. Numerous studies have pointed out that many household activities are gender specific. Consequently this will have an impact on

the type of activities as well as on the extent to which they are practised by members of the household. Certain solutions proposed on the basis of FSA may, therefore, not be feasible as they do not conform to the realities of on-farm circumstances.

(8) *Lack of unification of FSA methods*. In the literature one comes across many different descriptions of how FSA should be conducted. If standardisation of these methods would be possible, this would reduce costs for future FSA studies.

Quantitative farming systems analysis: a new method

With the aid of new developments in research techniques, an attempt is made to improve the methodologies used so far for FSA. The improved methodology, called quantitative farming systems analysis (QFSA) will be described in detail in the following sections. It is envisaged that QFSA will overcome some of the above shortcomings of traditional FSA. When it appears successful, QFSA will become more than a tool for cropping and livestock systems optimisation because development options can be explored and the social and economic feasibility of these options examined at the farm level.

An interdisciplinary research training project (INRES) has developed these improvements and is testing the procedure at present in the limestone area south of Malang on the eastern part of the island of Java, Indonesia (see Figure 23.1). The research team comprises seven staff members of the Malang University representing five disciplines and two Dutch scientists, with support of interdisciplinary task groups of the Brawijaya University in Malang, Wageningen Agricultural University and the University of Leiden.

Quantification of cropping and livestock subsystems is needed and this will be described in the next section where attention will also be paid to the interactions of those components in mixed systems. Later sections will focus on the exploration of development options at farm level, making use of a synthesis of technical and socio-economic information, and finally, will discuss the experiences with QFSA at the Brawijaya University.

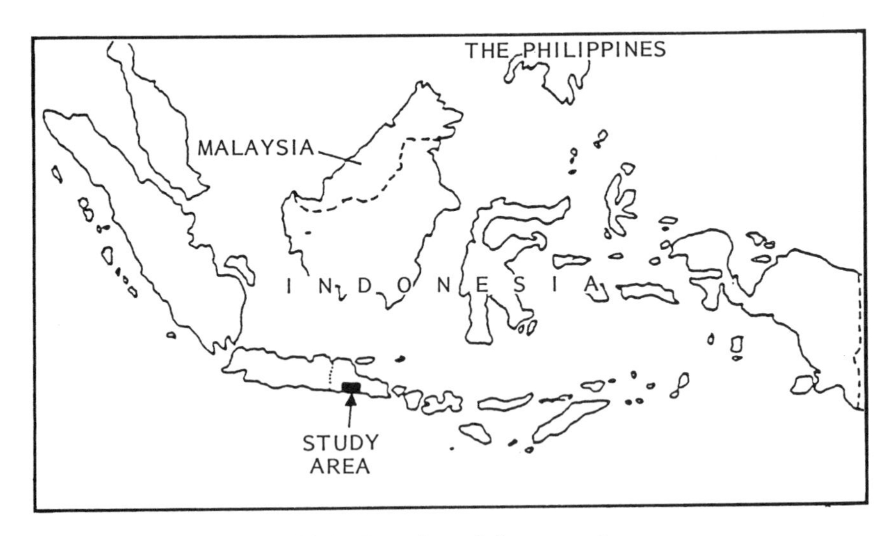

Figure 23.1 Location of the research area

Study area and data gathering

Research started in January 1990 with a sondeo covering 184 households in four different villages in the limestone area south of Malang (see Figure 23.1). Two villages, Putukrejo and Kedungsalam, were selected for further study, mainly on the basis of differences in landform and soil characteristics. A total of 35 farmers participated in a one-year Intensive Farm Household Survey (IFHS) mainly for the purpose of studying in-depth a number of ongoing technical as well as socio-economic processes (for example, decision-making). The sample for the IFHS is a non-random sample which was confined to certain farm size classes (between 0.25 and 1.0 ha), which had to be involved in rearing ruminants, and from which farms with certain crops (flooded rice and sugarcane) were excluded.

As the IFHS survey progressed, serious shortcomings in the sondeo results came to light and insight in the complexities of the mixed farms was largely improved. As a result it was decided that a survey would be conducted in which more households would participate. A Rapid Rural Appraisal (RRA) was performed including 556 households. Out of the 556 households, 150 households were chosen, randomly, for the Expanded Farm Household Survey (EFHS) based on a much larger random sample of farm households and is, therefore, likely to reflect reality to a much greater degree. Apart from these general surveys, disciplinary surveys and data gathering were also performed in the fields of soil science, agronomy, animal science, development economy and rural sociology.

It should be mentioned here that the above-detailed data gathering was very intensive and costly, the reason being that the data gathering had more of a training (new staff) and scientific purpose (new methodology) than a development objective.

QUANTITATIVE ANALYSIS OF CROP AND LIVESTOCK PRODUCTION

Introduction

This section provides the technical coefficients for crop and livestock activities which are needed as input in the activity matrix used in the following section. Use is made of a technique known as Land Use Systems Analysis (LUSA), since crop and livestock activities occur in certain combinations in the research area. LUSA combines data of a Land Unit (LU) with a Land Utilisation Type (LUT) (Van Diepen et al., 1989; Van Lanen, 1991; Driessen and Konijn, 1992). These combinations are called Land Use Systems (LUSs). LUs are internally uniform tracts of land and are defined by their physical characteristics, notably weather data, and soil and terrain information. LUTs are typified by the livestock kept and the crop(s) grown. LUTs can reflect the present situation but also future or potential use. LUSA is directed towards determining the relation between inputs of means of production and corresponding outputs of certain LUTs. The analysis starts with the investigation of physical production potentials, resulting in quantitative estimates of the potential production of the major LUSs. The quantitative estimates are based on the use of crop-growth simulation models (Spitters et al., 1989) and a livestock simulation model. Hereafter the required inputs of various means of production are defined for different production techniques. The selection of the means of production to be considered is based on the goals defined in the IMGLP model. In LUSA a Farming System (FS) is defined as a combination of different LUSs practised by one household on the basis of decisions made in response to physical factors, own priorities and external incentives. The cropping and livestock component of a FS affect each other, both on the output as well as the input side and the interaction will be established by linking the crop-growth and livestock simulation models.

LUSA considers production potentials and ignores socio-economic relations between LUTs within the context of the farm. LUTs compete with each other for external input resources. Farmers will optimise production at the farm level given their own specific objectives, instead of maximising the productivity of each land utilisation type. Therefore, with interactive multiple goal linear programming the socio-economic and the technical analysis is combined in order to explore options for development (see p. 312 and further).

The following technical information, for the IMGLP model, will be required:

(1) potential and attainable production levels for various product groups on well-defined land units in the considered region; product groups are represented by: (a) single cropping, e.g. cassava and maize, and (b) intercropping, e.g. a combination of annuals with perennials;
(2) the technologies related to the attainable and potential yields; variation in agronomic methods, such as fertiliser use, soil and water conservation, pest and disease control, etc.
(3) analysis of the reasons for the yield gaps between potential and attainable, and between attainable and actual yields for the various product groups;
(4) concepts of the ways to sustain production potential: soil erosion, soil degradation, maintain structure and depth of soils;
(5) per livestock type, actual and potential technologies with related inputs and outputs.

Land units

Farmers operate one or more different pieces of land or parcels. A parcel is divided into several subparcels if it is not uniform in land quality or if different LUTs can be distinguished. For each subparcel the land quality is assessed by five land quality variables. Since weather conditions do not vary between LUs, the five land quality variables observed or measured for each subparcel are: (1) soil depth, (2) soil texture (including stoniness), (3) level of terracing, (4) slope, and (5) position. These have led to the classification of farmer's landholdings into four major LUs (Widianto, 1993, personal communication) (Table 23.1).

The individual units are usually very small in size. There seems to be a very strong correlation between the land quality variables slope, soil depth and stoniness. With an increase in slope the percentage of area with deep soils decreases, as does the percentage of area of soils without stones. The distribution of LUs is given in Table 23.1. In the study area, 56% of the farm households operate landholdings smaller than three-quarters of a hectare. The average farm size for the entire survey area is 0.84 ha.

Table 23.1 The classification and distribution of LUs among the land operated by the farmers in the study area (data from EFHS)

	Land class			
	LU1	LU2	LU3	LU4
Distribution (ha)	21.62	52.28	19.45	40.24
Soil depth (cm)	>75	>75	50–75	<25
Soil texture[1]	+ + +	+ +	+	–
Terracing[2]	+ + +	+ +	+	–
Slope (%)	<3	3–50	15–50	>50
Position	valley	slope	upper slope	hill crests

[1]Soil texture: heavy clay with no stones (+ + +) to clay with many stones (–).
[2]Terracing: fully terraced (+ + +) to not terraced at all (–).

Crop activities

To date, a total of 33 Crop Production Activities (CROPs), relevant in the research area, were defined and input–output relations determined. CROPs are combinations of Crop Cultivation Type (CCT), a dominant Annual Crop (AC) and a Crop Production Level (CPL). CCTs are combinations of ACs with tree crops (TCs). For example, intercropping of maize and cassava is characteristic for upland annual cropping systems on Java. Other ACs grown in the area are upland rice and sugarcane while groundnut, soybean, chillies, taro and cowpea can be found incidentally. The main TCs are: banana, coconut, acacia, gliricidia and teak. The CCTs differ from each other in tree density, species composition, vertical structure, function, and ACs. These differences result in variation in the interactions between trees, and between trees and ACs, in terms of light, water and nutrients.

Five CCTs are defined as follows (Sunaryo, 1993, personal communication):

- CCT1: irrigated and non-irrigated land with sparse to no trees
- CCT2: tegal (non-irrigated agricultural land) with trees on the border of the field, at the edges of terraces or randomly distributed
- CCT3: multi-storey systems
- CCT4: woodlands
- CCT5: '*bongkor*', i.e. land not used for annual or perennial cultivation

For these complex CCTs, quantitative estimates have to be determined, either empirically (by on-farm and on-station research) or by making use of (field validated) crop simulation models. Simulation models are only available for a limited number of crops. They hardly exist for AC + TC combinations where the production of an AC must be quantified taking into account interactions/competition with the TC. So, in the farm modelling (see the next section, p. 312 and further) a combination will be used of field data and computed estimates. To date, these are available for five ACs in the major cultivation types CCT1 and CCT2. These are: (1) maize, (2) cassava, (3) upland rice, (4) sugarcane, and (5) maize and cassava intercropped.

Three CPLs are considered. Multiple levels need to be considered to investigate future prospects of increased productivity. CPL1 pertains to an environment in which all biophysical constraints are eliminated; crop growth is determined by temperature and radiation income. The crop is not hindered by pests, diseases, weeds and limiting factors as water and nutrients. CPL1 represents the production potential (PP). In CPL2, the actual availability of water is additionally taken into account and the effects of limited water availability on crop production are calculated. This yields the Water Limited Production (WLP). In CPL3, the limited nutrient availability is additionally considered; production is determined by the actual temperature, radiation, water availability and nutrient availability, leading to the so-called Water and Nutrient Limited Production (WNLP).

The 33 Crop Production Activities (CROPs) distinguished to date, are marked as follows:

$$\text{CROP}_{(a,b,c)}$$

where

a = CCT (1 = land with sparse to no trees; 2 = tegal with trees randomly distributed; 3 = multi-storey systems; 4 = woodlands; 5 = *bongkor*),
b = AC within CCT1 and CCT2 (1 = maize; 2 = cassava; 3 = sugarcane; 4 = upland rice; 5 = maize and cassava intercropped), and
c = CPL (1 = PP; 2 = WLP; 3 = WNLP).

Land Use Systems

Land Use Systems (LUSs) are combinations of Land Units (LUs) with Land Utilisation Type (LUT). A LUT includes multiple crops and animals. However, in order to allow the activity matrix to choose any combination of animal and crop activity, these two activities should not yet be linked with each other. Therefore, our definition of a LUS, within QFSA, will be limited to a combination of a Crop Activity Type (CROP) with a LU only. Only those combinations that occur at a reasonable scale in the study area are used as input in the activity matrix. Such combinations were used because otherwise the number of data for the activity matrix (see p. 312 and further) becomes excessive.

In the study area (Table 23.2) CCT1 mainly exists on LU1 and LU2, and CCT2 and CCT3 are found on all four LUs. CCT4 and CCT5 are mainly cultivated on LU4.

Table 23.2 The distribution of Crop Cultivation Types (CCTs) over the four Land Units (LUs) for a total of 624 subparcels operated by 143 households (data from EFHS)

	LU1 (ha)	LU2 (ha)	LU3 (ha)	LU4 (ha)
CCT1	10.79	11.43	1.12	0.13
CCT2	8.85	34.31	10.54	3.21
CCT3	1.72	4.67	4.60	2.47
CCT4	—	0.35	0.29	4.68
CCT5	0.26	1.52	2.90	29.65

This restricts the number of combinations possible. A total of 96 LUSs are distinguished. These will be marked as follows:

$$LUS_{(a,b,c,x)}$$

where

a = crop cultivation type (1 = land with sparse to no trees; 2 = tegal with trees randomly distributed; 3 = multi-storey systems; 4 = woodlands; 5 = *bongkor*);
b = annual crop only on CCT1 and CCT2 (1 = maize; 2 = cassava; 3 = sugarcane; 4 = upland rice; 5 = maize and cassava intercropped);
c = annual crop production level (1 = PP; 2 = WLP; 3 = WNLP); and
x = land unit (1 = LU1; 2 = LU2; 3 = LU3; 4 = LU4), CCT1 on LU1 and LU2, CCT2 and CCT3 on all the LUs, and CCT4 and CCT5 on LU4.

In the farm model in the next section (p. 312 and further), the landholdings operated by farmers can be varied and all the combinations of land use systems possible can be selected.

Animal activities

At present, a total of 29 Animal Production Activities (ANIM), relevant in the research area, were defined and input–output relations determined. ANIMs are combinations of a Herd Type (HERD), an Animal Type (AT) and an Animal Diet (AD).

HERDs consist of cattle, sheep, goats or a combination. Total benefits of animals include returns from production of meat as well as manure and draft power. Moreover, rearing animals serves to accumulate capital and maintain reserves for greater economic security. Table 23.3 shows the average area of land operated per farm for the different HERDs distinguished. Farmers who keep no animals on the farm operate less land than farmers who keep animals. Keeping cattle is associated with a bigger area of land than keeping only sheep, goats, or a combination of sheep and goats. A herd of cattle combined with either sheep or goats requires about the same land size as a farmer who keeps only cattle.

Table 23.3 Relation of land with type of the herd (data from RRA)

	Type of herd						
Variables	HERD 1 (without animals)	HERD 2 (cattle)	HERD 3 (sheep)	HERD 4 (goat)	HERD 5 (cattle and sheep)	HERD 6 (cattle and goats)	HERD 7 (cattle, sheep and goats)
Land (ha)	0.57	0.97	0.48	0.75	0.97	0.95	1.85
No. of cattle	—	1.89	—	—	1.53	1.92	2.00
No. of sheep	—	—	2.97	—	2.28	—	3.40
No. of goats	—	—	—	2.93	—	2.10	3.20
No. of farms	152	232	31	54	32	48	5

Stall-feeding is predominant. Most of the feed is collected by household members. Feeds include residues of annual crops, grasses, weeds and tree leaves (particularly gliricidia). Concentrates are not fed to the animals. Forage cultivation is not practised. Part of the feeds are produced on land managed by the farm household. However, a significant part—including crop residues—is obtained from outside the farm.

As with crops, a number of input–output relations are determined by using simulation models. However, existing livestock production models usually deal with specific biological processes. Only a few simulation programs embody a whole tropical production system. Examples are the TAMU (Texas A&M University) model for beef production systems (Sanders and Cartwright, 1979), the model from Kahn and Spedding (1983), the simulation of beef cattle systems in the Llanos of Colombia (Levine et al., 1981) and the ILCA model (Konandreas and Anderson, 1982). These models simulate herd dynamics and individual animal careers and do not consider production in relation to the feed resource. Therefore, a new livestock production and feed model at farm level, LIPROFE (Efdé and Van Rheenen, 1992), was developed which includes major farm household influences—particularly labour and land in relation to fodder availability—on livestock production. LIPROFE provides the following options for analysis: (1) monthly feed availability is exogenous and livestock production is endogenous, or (2) livestock production is exogenous and monthly availability of fodder is endogenous. The first type of calculation results in number of livestock units which can be kept on a certain amount of feed. The second type of calculation results in the amount of feed necessary in order to meet the requirements of the livestock units.

In LIPROFE, ANIMs are defined on the basis of AT and AD. The following ATs are distinguished:

- AT1: an adult cow, giving birth to a calf. The feed requirements for the growth of both the cow and the calf are included.
- AT2: an adult cow, producing draft power in the agricultural season and giving birth to a calf. The feed requirements for the growth of both the cow and the calf are included.
- AT3: an adult, non-reproducing and non-lactating cow, kept for meat production and producing draft power in the agricultural season.
- AT4: an adult goat, giving birth to a kid. The feed requirements for both the goat and the kid are included.
- AT5: an adult ewe, giving birth to a lamb. The feed requirements for both the ewe and the lamb are included.

The eight most common feeds fed to the animals in the area are: (1) leguminous and (2) non-leguminous tree leaves, (3) native grasses, (4) elephant grass, (5) maize stover, (6) cassava leaves, (7) sugarcane leaves and (8) rice straw. Feed availability is calculated for two Animal Diets (ADs). In AD1 the monthly composition of the fodder mix is obtained from actual field data. These data are used to calculate the actual amount of each fodder available for the animal. AD2 uses actual data for the monthly composition of the fodder mix but assumes maximum availability for each fodder. LIPROFE matches the feed availability data with the feed requirements of the animals.

The 29 Animal Production Activities (ANIM) distinguished to date are marked as follows:

$$\text{ANIM}_{(d,e,f)}$$

where

d = type of the herd (1 = without animals; 2 = cattle; 3 = sheep; 4 = goats; 5 = cattle and sheep; 6 = cattle and goats; 7 = cattle and sheep and goats);
e = animal type, only within herd 2, 5, 6 and 7 (1 = AT1; 2 = AT2; 3 = AT3); and
f = animal diet (1 = without selection, 2 = with selection).

Technology and management

Agricultural production technologies are well defined ways of converting inputs into agricultural products (Van Keulen, 1992). The technology applied in the study area is strongly determined by the objectives of the producer(s). For this study, two agricultural technology types have been distinguished: 'yield-oriented agriculture' (YOA) strives for the highest production possible under the prevailing conditions whereas 'low external input-oriented agriculture' (LIA) seeks to restrict the inputs obtained from outside the farm. For the YOA, production of crops will take place according to the 'best technical means' (BTM). BTM implies that only those techniques that lie on the production possibilities frontier will be considered (Scheele, 1992). BTM techniques are those techniques that use inputs in the technically most efficient way. This implies that the input of each resource is minimised to the point that the other inputs can be used to their best effect. This point is then determined for all inputs. Crop production adopting an LIA approach seeks to restrict the use of inorganic fertilisers to the minimum by applying manure, crop rotation, etc. LIA is labour-intensive and produces well below calculated crop production potentials.

For the animal production, two feeding management systems have been defined. In the on-farm fodder collection all the feeds fed to the animals are derived from the land operated by the farmer. In the off-farm fodder collection all the feeds fed to the animals are collected from outside the farm with on-farm available labour.

Production situations

Production situations are the combination of levels of output and production technologies on well-defined land units. Analysing a production situation yields one set of (user-defined) inputs and the associated output (potential). Inputs are divided into two classes: (1) primary inputs, e.g. nitrogen and water, with little to no substitution possible, and (2) secondary inputs where substitution is possible, e.g. use of pesticides and type of fodder fed. The level of substitution depends on the goals defined. Crop inputs comprise pesticides, irrigation water, nitrogen and labour. Crop outputs comprise crop yields. Animal inputs are fodder and labour and animal outputs are meat, manure and draft power.

Table 23.4 Input–output coefficients for potential maize and cassava production (LU = 1; CCT = 1; CPL = 1) for two production technologies

	Maize		Cassava	
	YOA	LIA	YOA	LIA
Seed/seedlings (kg/ha)	40	40	10 000	10 000
N fertiliser (kg/ha)	90	—	135	—
P fertiliser (kg/ha)	20	—	20	—
K fertiliser (kg/ha)	50	—	50	—
Manure (t DM/ha)	—	10	—	10
Draft power (h/ha)	—	20	—	20
Yield (t DM/ha)	2.22	1.83	10.65	7.01

Examples of input–output coefficients are presented in Table 23.4 for the potential production of maize (LUS1111) and cassava (LUS1211) grown within CCT1 and on LU1, are given for two agricultural production technologies (YOA and LIA).

Examples of input–output coefficients for a herd with animal type 2 and a 'no selection' animal diet are presented in Table 23.5 for two feeding management systems (all coefficients are expressed per cow, including the requirements of a calf).

EXPLORING OPTIONS FOR DEVELOPMENT AT FARM LEVEL

Introduction

Rural development has long been a major issue with policy-makers. Several, often conflicting, objectives are pursued when considering rural development; the most

Table 23.5 Input–output coefficients per year, for a herd (AT2) with two feeding management systems

	On-farm, no selection	Off farm, no selection
Production:		
Meat (kg)	50	45
Calf (No)	1	1
Manure (kg DM)	1000	1000
Draft power (hours)	100	100
Feed requirements:		
Leguminous tree leaves (kg DM)	602	366
Non-leguminous tree leaves (kg DM)	149	164
Native grasses (kg DM)	312	440
Elephant grass (kg DM)	92	—
Maize stover (kg DM)	—	619
Cassava leaves (kg DM)	154	44
Rice straw (kg DM)	170	239
Sugarcane leaves (kg DM)	253	340
Labour for feeding (hours)	100	390

important being increasing levels of income for people in rural areas while at the same time ensuring that this increase will be sustainable. Issues that have to be considered when pursuing rural development are: (1) which options are technically feasible; (2) are the endeavours of policy-makers also consistent with the objectives of farm household members; and (3) what are the environmental spill-over effects resulting from directions of development?

This section will describe the farm level model that is part of the newly developed decision-supporting tool QFSA and shows how socio-economic and biophysical information can be synthesised. Different directions of development as well as trade-off effects can be studied. The farm level model that has been developed is an Interactive Multiple Goal Linear Programming (IMGLP) model. IMGLP has been utilised for explorative studies at regional level (WRR, 1987; De Wit et al., 1988; Van Keulen, 1992; Van Keulen and Veenenklaas, 1992; WRR, 1992). However, the author is not aware of applications of IMGLP at farm level, in developing countries, within the context of FSA.

Farm level model

The method

To describe the IMGLP method the author has slightly adapted the illustration given by Rabbinge (1991). Two goal variables are used for this purpose: monetary inputs (MI) and gross margins (GM). These two goal variables have been placed along the axis of the graph (Figure 23.2). The line OA shows the relation between MI and GM. When MI = 0, then GM = 0 also; however, an increase in MI will be accompanied by an increase in GM. The goals decreasing MI and increasing GM are, therefore, conflicting. On MI a number of additional constraints are imposed, shown by the top horizontal lines. The optimum values for MI and GM are 0 and 1500, respectively (Figure 23.2(a)). The next step is to determine minimum or maximum acceptable values. If for GM the demand is that GM $\geqslant$ 1000, the number of combinations decreases from OABC to EABD (Figure 23.2(b)) and the best value that can be obtained for MI is 300. If additionally a demand were placed on MI (e.g. MI $\leqslant$ 400), the feasible area would further be reduced to EFG (Figure 23.2(c)) with optimum values for MI and GM being 300 and 1300, respectively. In this very simple example only two goal variables have been used; however, it is possible to include many more goals.

The system

It is important to clearly define the system for which the explorative study is to be conducted because this will determine the type of and the extent to which goals can be pursued. For the explorative work that is being conducted for the limestone area of south-east Java, the chosen system is the farm. Attention will primarily be focused on soil-related activities. As income generation from off-farm activities—utilising farm household labour—is prevalent in the research area, off-farm income-generating activities are also included, although in less detail than the soil-related activities. Activities in the system that take place or could realistically be expected to take place within the explorative time horizon (20 years) are included.

Scenarios and scenario variants

A *scenario* is a direction for development that is anticipated in the future, and a *scenario variant* is a way to distinguish the utilisation of different instruments to accomplish a scenario. The *time horizon* is the period for which the model is considering directions for development. Three major scenarios were explored, namely: (1) technical potentials

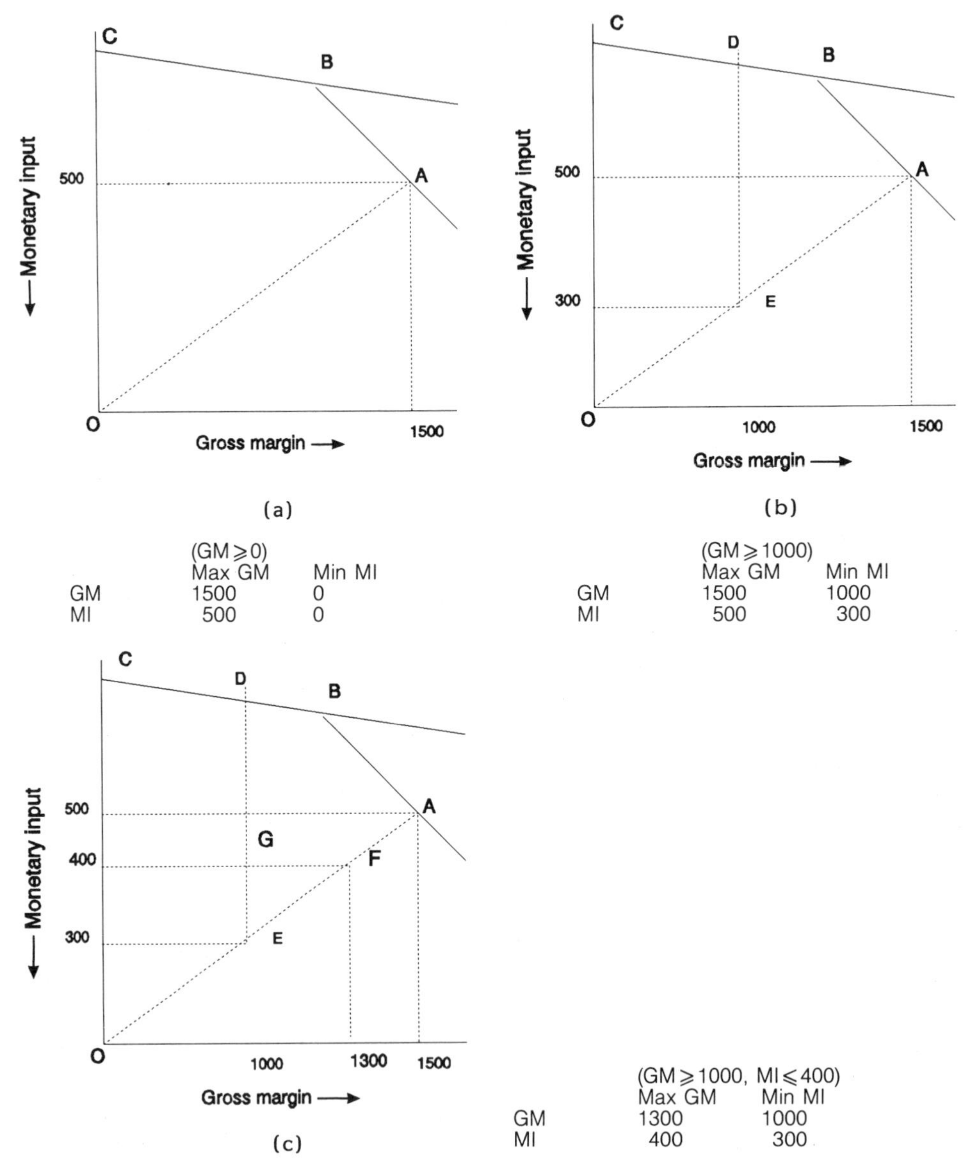

Figure 23.2 Interactive optimisation of two goal variables (monetary input and gross margins, arbitrary units)

Table 23.6 Scenarios and the scenario variants

Scenario	Variant
A Technical	A1 Soil productivity
	A2 Cost effectiveness
B Household	B1 Monetary input
	B2 Employment
	B3 Income
	B4 Diversification
C Environmental	C1 Nitrogen
	C2 Biocides

(*A-Scenario*), (2) household socio-economic wishes (*B-Scenario*) and (3) environmental concerns (*C-Scenario*). Table 23.6 summarises the scenarios and their variants.

Technical scenario

Techniques are explored that will enable farmers to maximise productivity from their land by using the appropriate management practices in the most efficient manner. This will be done by maximising soil productivity in terms of gross returns (*Variant A1*) or by minimising the costs per output (*Variant A2*). In the latter variant, all costs are included except those resulting from the utilisation of hired labour.

Household socio-economics scenario

Goals that can be recognised as being compatible with those of farm household members are pursued. These are: minimising monetary inputs (*Variant B1*), maximising labour utilisation (*Variant B2*), maximising income from any activities (*Variant B3*), and maximising income through a strategy of activity diversification (*Variant B4*). The first scenario is relevant because many options are often not possible for farm household members as they are not in a position to obtain access to the capital necessary to realise them. A characteristic of farm households is that there is a surplus of labour during certain periods of the year. One could argue that this could be considered as a comparative advantage and that for that reason attention should especially be focused on those activities that are labour intensive. This, however, should only be done if certain levels of income can be generated, because maximising labour utilisation will in itself never be an objective. Maximising gross margins from any activities, as will be done in the third variant, can be risky. Households may become too dependent on one type of activity and, therefore, in the fourth variant the same goal is being pursued through a strategy of activity diversification.

Environmental scenario

If soil productivity were to be increased in the future, it seems likely that this will be accompanied by higher levels of inputs, such as nitrogen and biocides. Especially in

Table 23.7 Primary and secondary goal variables

Primary goal variables	Secondary goal variables
1 Gross returns	1.1 Gross returns for annual crops
	1.2 Gross returns for perennial crops
	1.3 Gross returns for livestock activities
	1.4 Gross returns for non-soil related activities
2 Gross margins	2.1 Gross margins for annual crops
	2.2 Gross margins for perennial crops
	2.3 Gross margins for livestock activities
	2.4 Gross margins for non-soil related activities
3 Monetary inputs	3.1 Monetary inputs for annual crops
	3.2 Monetary inputs for perennial crops
	3.3 Monetary inputs for livestock activities
	3.4 Monetary inputs for non-soil-related activities
4 Labour utilisation	4.1 Labour utilisation for soil related activities
	4.2 Labour utilisation for non-soil-related activities
5 Nitrogen utilisation per hectare	
6 Biocides utilisation	
7 Costs per unit product	7.1 Inorganic fertiliser per unit output

areas where both the population density as well as the dependence on agriculture is high, the consequences of certain agricultural practices should be given a prominent place in an explorative analysis. In the environmental scenario thc effects of soil related agriculture on the environment are reduced by minimising the utilisation of nitrogen (*Variant C1*) or biocides (*Variant C2*).

Goal variables

Each scenario and scenario variant has an endeavour or objective to be pursued. In the farm model these are called goals. A *primary goal variable* is a variable (e.g. gross returns) that will be optimised, i.e. they may be maximised or minimised. A *secondary goal variable* is part of a primary goal variable (e.g. gross returns for annual crops) and can be optimised but will usually be used to set certain pre-determined values. Table 23.7 gives an overview of the goal variables.

The iterative procedure

In the IMGLP procedure the eight scenario variants are explored in an iterative manner. After the first iteration cycle, eight optimisations will have been conducted. During each of these optimisations minimum or maximum demands are set on one of the primary goals while demands may be placed on the other goal variables. Table 23.8 shows the demands that were used during the first iteration cycle.

After the first iteration cycle we could conclude that no better values could be obtained for each of the primary goal variables. If there was only one goal to be fulfilled (only one stakeholder) we could stop the exploration of development. However, there are often more (conflicting) goals. In such cases, the following steps during the iteration

Table 23.8 Scenarios, scenario variants and the demands that are placed on the goal variables during the first iteration cycle

	Scenarios[b]							
	A		B				C	
Goal variables[a]	A1	A2	B1	B2	B3	B4	C1	C2
1	MAX	GEs	GEs	GEs	GEs	GEs	GEs	GEs
1.1	U	U	U	U	U	U	U	U
1.2	U	U	U	U	U	U	U	U
1.3	U	U	U	U	U	U	U	U
1.4	U	U	U	U	U	U	U	U
2	GEs	GEs	GEs	GEs	MAX	MAX	GEs	GEs
2.1	U	U	U	U	U	U	U	U
2.2	U	U	U	U	U	U	U	U
2.3	U	U	U	U	U	U	U	U
2.4	U	U	U	U	U	U	U	U
3	LEs	LEs	MIN	LEs	LEs	LEs	LEs	LEs
3.1	U	U	U	U	U	U	U	U
3.2	U	U	U	U	U	U	U	U
3.3	U	U	U	U	U	U	U	U
3.4	U	U	U	U	U	U	U	U
4	GEs	GEs	GEs	MAX	GEs	U	GEs	GEs
4.1	U	U	U	U	U	GEs	U	U
4.2	U	U	U	U	U	GEs	U	U
5	LEs	LEs	LEs	LEs	LEs	LEs	MIN	LEs
6	LEs	LEs	LEs	LEs	LEs	LEs	U	MIN
7	U	MIN	U	U	U	U	U	U
7.1	U	U	U	U	U	GEs	U	U

[a]For a definition of the scenarios, see Table 23.6.
[b]For a definition of the goal variables, see Table 23.7.
Demands that can be placed on goal variables: MAX, maximise; MIN, minimise; U, within the feasible area the goal variable is not constrained; GEs, the goal variable will have to be greater than or equal to a certain set value; LEs, the goal variable will have to be smaller than or equal to a certain set value; Sl, the goal variable is set to its lowest possible value; Sh, the goal variable is set to its highest possible value; S, the goal variable is set to a value between Sl and Sh.

(preferably with participation of the stakeholders) may arrive at an acceptable compromise. During further iteration cycles more specific demands can be placed on the various goal variables. So, those goal variables that were left unconstrained ('U' in Table 23.8) during the first iteration may now be used to more accurately accommodate the specific wishes.

IMGLP at farm level: an illustration

The approach can be illustrated by making use of tentative data (actual data are still being processed). Guestimates were used for the cropping activities maize and cassava monocropped (CCT1) and intercropped (CCT2) for the three CPLs on the four LUs. The assumption is that the perennial crop in CCT2 is coconut. Other activities are limestone burning (four different production means), and off-farm activities (three levels of income-generating activities). Livestock activities are not included in this illustration.

Table 23.9 TCs (YOA) for cassava monocrop with and without coconuts

	LUS1211	LUS1234	LUS2211	LUS2234
Inputs				
Cassava (seedling/ha)	25 000	25 000	25 000	25 000
Coconut (seedling/ha)	—	—	4	4
Inorganic fertilisers: (kg/ha)				
Urea (45% N)	280	330	292	344
TSP (20% P)	50	55	56	63
Biocides (litres/ha)	3	3	3	3
Labour (man-days/ha)[b]	141	168	131	159
Draught power (cow-days/ha)	20	—	20	—
Monetary inputs (Rp/ha)	298 750	112 625	311 400	126 325
Outputs				
Cassava tuber dry matter (kg/ha)	20 000	6554	16 000	5243
Fodder dry matter (kg/ha)	20 000	6554	16 000	5243
Coconut (fruit/ha)	—	—	1800	922
Coconut (wood/ha)	—	—	8	8

[a]The assumption is that every year, 1/15th of the coconut trees are replaced.
[b]In the IMGLP model the labour requirements have been specified according to the month when an operation took place.

Guestimates for the soil-related activities include yields that could be obtained with YOA, where there is a high dependence on external inputs, although no use is made of herbicides. The technical coefficients could be considered to be a reflection of the goals being pursued in scenario variants A1, A2, B3 and B4. Consequently, the impact on goal variables when shifting from one scenario variant to another variant is rather limited, compared to the situation where guestimates reflecting the goals of all scenario variants would have been used.

Input–output coefficients

Technical coefficients

TCs make explicit the extent to which a certain activity draws on the available (limited) resources in order to achieve a certain level of output(s). The *activity matrix* in the IMGLP model contains agricultural as well as non-agricultural activities from which a selection can be made.

For a complete overview of the TCs, the reader is referred to Van Rheenen (1993). Table 23.9 presents the TCs as for cassava with and without coconut as they are for the PP situation on LU1 (LUS1211 and LUS2211) and for the WNLP situation on LU4 (LUS1234 and LUS2234), i.e. the two extremes.

Potential yields without water and nutrient constraints (PP) on LU1 are very much higher than the water- and nutrient-limited yields (WNLP) on LU4, while nutrient and labour demands on LU4 are either equal or higher than on LU1. This is because LU4 is much less suited for crop cultivation than LU1. Consequently, more nutrients are required but also more labour is required for land preparation.

Table 23.10 TCs for limestone burning with different production means

	LB1	LB2	LB3	LB4
Inputs:				
Labour (man-days)	146	122	122	98
Material inputs:				
Limestone (m^3)	7	7	7	7
Firewood	80	80	80	80
Monetary inputs				
Operational costs (Rp)	—	28 000	10 500	38 500
Capital costs	1000	1000	1000	1000
Output:				
Lime	4500	4500	4500	4500

Table 23.10 shows the TCs for limestone burning (LB) that have been included in the model. There are no differences in outputs between LB1,2,3 or 4. The main difference is the manner in which inputs are acquired. With LB1, limestone and wood are collected by members of the household, while in the case of LB4 both these inputs are purchased, thus increasing the monetary inputs necessary for the production process, and reducing labour requirements.

Three categories of off-farm activities were defined: (1) high-income-generating activities (OH), Rp.10 000 per man-day; (2) middle-income-generating activities (OM), Rp.5000 per man-day and (3) low-income-generating activities (OL), earning Rp.2500 per man-day.

Constraints

Constraints that are imposed on the model can be classified into two categories: (1) *technical constraints* which are considered as absolute (e.g. landholding) and (2) *standard constraints* which allow a certain degree of arbitrariness (e.g. the number of days that household members can participate in off-farm income-generating activities). Table 23.11 summarises both categories of constraints.

Results after the first cycle of optimisations

After the first round of optimisations the best and worst values that can be expected for the different goal variables have been made explicit, and consequently the space that the user of the model has to vary the value of the goal variables. Table 23.12 shows the computed results after following the optimisations that are indicated in Table 23.8.

After this first iteration cycle the results can either be accepted or rejected. In the latter case one may wish to tighten certain goal variables. Sooner or later a stage will be reached where it will no longer be possible to tighten a goal variable without sacrificing another goal variable.

The IMGLP technique considers farm household activities from different perspectives. The values of the goal variables need not necessarily be different for the different scenarios. For example, similar goal variable values can be observed for scenarios A1

Table 23.11 Constraints

	Constraints			
	Technical		Standard	
	Min	Max	Min	Max
Land for soil-related activities (per ha)				
LU1	—	0.25	—	—
LU2	—	0.25	—	—
LU3	—	0.25	—	—
LU4	—	0.25	—	—
Household labour (man-days/month)	—	72	—	—
Hired in labour (man-days/month)	—	—	—	36
Limestone burning (no. of burnings/month)				
LB1	—	—	—	2
LB2	—	—	—	2
LB3	—	—	—	2
LB4	—	—	—	2
Off-farm activities (man-days/month)				
OH	—	—	—	4
OM	—	—	—	8
OL	—	—	—	12
Minimum crop requirements (kg/year)				
Cassava	—	—	500	—
Maize	—	—	350	—

Table 23.12 Scenarios and goal variables after the first cycle of optimisations

Scenario[a]	Goal variables[b]								
	1 (Rp × 10^3)	2 (Rp × 10^3)	3 (Rp × 10^3)	4 (Md)	4.1[c] (Md)	4.2[c] (Md)	5 (kg)	6 (litres)	7 Ratio (Rp/Rp)
A1	9394	8894	500	967	100	867	100	1.95	0.73
A2	7000	6713	287	939	71	868	60	1.50	0.02
B1	7000	6926	74	872	71	801	58	1.50	0.14
B2	7871	7371	500	1096	18	1078	9	0.39	0.83
B3	9394	8894	500	967	100	867	100	1.95	0.73
B4	8842	8342	500	1010	110	900	100	2.00	0.83
C1	7000	6500	500	892	13	878	8	0.34	2.50
C2	7000	6500	500	892	10	882	13	0.22	2.42
Min[d]	7000	6000	—	864	110	350	—	—	—
Max[d]	—	—	500	—	—	—	100	2	—

[a]For a description of the scenarios and scenario variants, see Table 23.6
[b]For a definition of the goal variables, see Table 23.7.
[c]These secondary goal variable demands are only imposed in scenario B4.
[d]These are pre-determined minimum and maximum values.
Rp = Rupiahs, Md = man-days.

and B3. This is because gross returns appear to be strongly correlated with gross margins. If TCs were included that were a reflection of the household socio-economic or the environmental scenarios, such a correlation would not necessarily be so explicit. Monetary inputs in scenario variant A2 are higher than in scenario variant B1 because there where costs are minimised per output (scenario variant A2), i.e. the cost effectiveness is being maximised, the costs of hired labour are not included. These are included when monetary inputs are minimised (scenario variant B1). When determining the cost effectiveness per activity it is difficult to pre-determine the quantity of hired labour that will be utilised per activity. The results show that maximising labour gives relatively low levels of gross returns and gross margins. When gross margins are maximised with a strategy of activity diversification, levels of income will decrease compared to the situation with no specific diversification demands. So, although income decreases, so do the risks of depending too much on a certain type of activity. In this way, the price that has to be paid for a less risky existence can be computed. Nitrogen and biocides utilisation can be brought to very low levels, as farm households are in a position to generate a large part of their income from limestone burning and off-farm activities.

Considering, however, the very tentative nature of the data that have been used for this illustration, the practical implications and conclusions that can be drawn on the basis of the results that are presented in Table 23.12 are, at best, very limited.

EXPERIENCES WITH QUANTIFIED FARMING SYSTEMS ANALYSIS AT THE BRAWIJAYA UNIVERSITY

Introduction

The University of Brawijaya (UNIBRAW) in Malang, East Java, Indonesia, consists of 10 faculties: economics, law, administrative science, agriculture, animal husbandry, fishery, medical, basic sciences, technical sciences, and post graduate. UNIBRAW is a young (30 years old) and growing university, with approximately 15 000 students and 1212 academic staff members including 421 MSc and 65 PhD holders. Of the latter group, 41 are members of the Agricultural Faculty (UNIBRAW, 1993). Due to the limited number of PhD holders among the academic staff, UNIBRAW seeks cooperation with other universities. In the Higher Education system of Indonesia, all staff members participate in the execution of the three main tasks of the university, called '*Tri Dharma Perguruan Tinggi*': teaching, research and public service.

Each university in Indonesia is free to select its 'Main Scientific Focus'. In 1977, UNIBRAW selected Rural Development (RD) as its main focus for public service and research. The reasons are: (1) the major part of the population lives in rural areas and generates income from agricultural activities. The present government policy gives priority to those who are below the poverty line. (2) Malang is a small town, which has close ties with the surrounding rural area. (3) UNIBRAW's staff are considered to be sufficiently qualified to support RD endeavours. Later, it was realised that this mission is too broad since it includes all aspects of life in the rural areas and all sectors of RD. A more specific goal had to be set. In 1991, UNIBRAW's main scientific focus was reformulated as 'Development of Science and Technology in Support for Rural

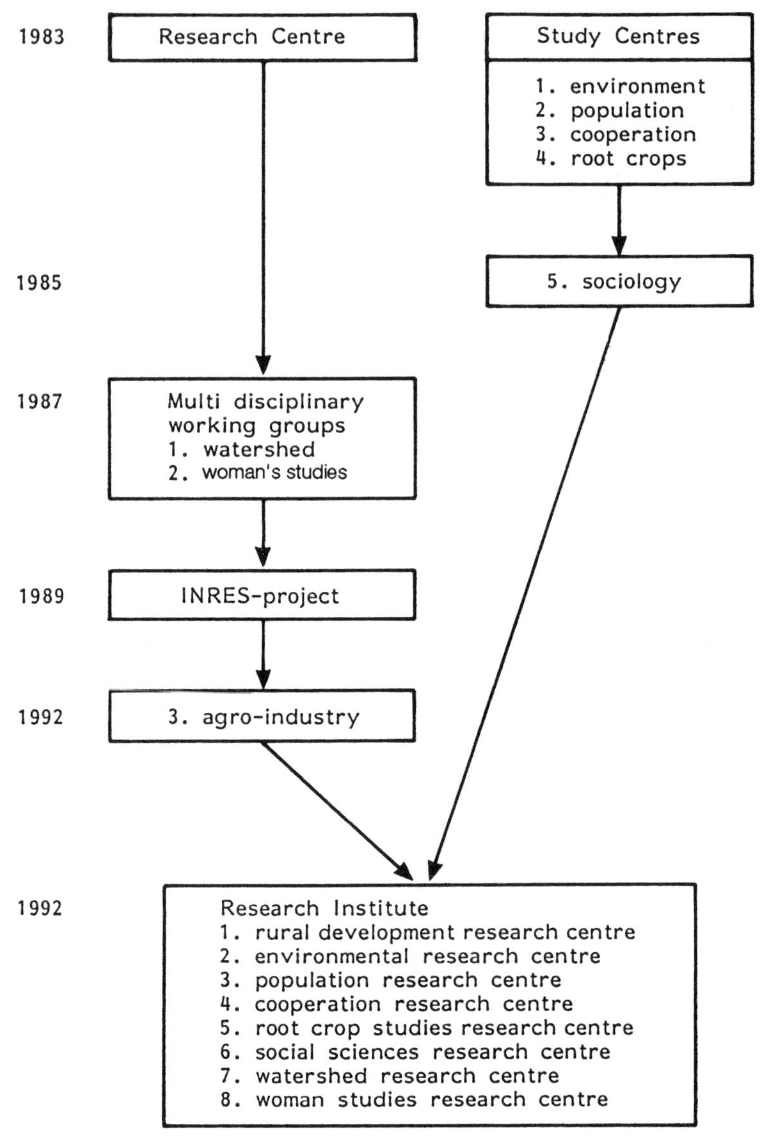

Figure 23.3 Institutional development of UNIBRAW's commitment on research into rural development

Industrialisation'. UNIBRAW's regional mandate is for rainfed agriculture in east Java and eastern Indonesia.

UNIBRAW was aware that a tool was needed to: (1) analyse the existing rural conditions, (2) identify and describe problems, (3) formulate problem-solving strategies and opportunities and (4) help to select alternative scenarios relevant to farmer's regional and national goals.

Historical overview of FSA at UNIBRAW

During the period 1975–89 serious difficulties had been encountered while searching for the appropriate methodology for RD, because it includes so many aspects (Semaoen et al., 1991).

A Research Centre was established in 1983 (see Figure 23.3), and its role was to deal with the coordination, administration and monitoring of research activities. In the same year, four Study Centres were established for: (1) environment, (2) population, (3) cooperation, and (4) root crops (UNIBRAW, 1983). These Study Centres were not part of the Research Centre and it was observed that they worked largely independently of each other. Little was done to establish integrated research that could support RD. Developing policies for RD require information and synthesisation on a number of topics, e.g. agronomy, livestock production, sociology and economics. Consequently, the research approach should be of an interdisciplinary nature.

In 1984, the Research Centre started to coordinate the scattered research by organising a workshop on agro-ecosystems. Staff from several faculties participated in that workshop. The objective of the workshop was to introduce agro-ecosystem analysis. As a follow-up to this workshop, KEPAS, coordinated by the Central Research Institute for Agriculture (CRIA), Bogor, together with some UNIBRAW staff implemented the Agro-Ecosystem Analysis approach in several kabupatens (second level of local government) in east Java. This activity was funded until 1985 by the Ford Foundation. UNIBRAW at that time was lacking the capacity to carry out research on social issues because there was no social science faculty. In 1985, cooperation between UNIBRAW and Leyden University (The Netherlands) led to the establishment of a fifth Study Centre on Social Sciences. Again, this Social Sciences Study Centre was not under the coordination of the Research Centre, but directly under the responsibility of the Rector.

In 1987, the Research Centre started to coordinate two multi-disciplinary research working groups: (1) the watershed and (2) the Woman Studies Working Group. These working groups started working in isolation and there was no adequate integration of research activities.

In 1989, UNIBRAW started to conduct the first stage of an interdisciplinary research project (INRES) under full coordination of the Research Centre. INRES was the final part of a long-term cooperation (> 15 years) between UNIBRAW and the Wageningen Agricultural University and the Leyden University in the Netherlands. The objective of the project was to train staff and develop QFSA with a focus on development options for small farmers in the limestone area South of Malang. The five major disciplines involved in this project are agronomy, soil science, agricultural economics, sociology and animal husbandry.

As a consequence of the narrowing of the main scientific focus towards rural industrialisation, the Research Centre started in 1992 a new multi-disciplinary working group, namely the Agro-Industry Working Group. In the same year the Research Centre of UNIBRAW increased its status and became a Research Institute, which now consists of eight Research Centres including all previous study centres and working groups. One of these research centres is for RD studies (UNIBRAW, 1992), which is a combination of the INRES project and the Agro-Industry Working Group. This Research Centre is expected to continue to play a firm role in RD studies. In 1993, this Research Centre on RD started a second stage of interdisciplinary research, funded by the Indonesian Directorate of Higher Education.

Contribution of QFSA to UNIBRAW

While studying farm households, INRES developed a more objective and quantitative FSA methodology (Stroosnijder and Van Rheenen, 1993) which is known under the acronym QFSA (= Quantified FSA). INRES started to train UNIBRAW staff ('learning by doing') in interdisciplinary FSA and improved in the mean time this research methodology. UNIBRAW, having gained experience with QFSA for a number of years, is now in a position to look back and evaluate the methodology. Such an exercise is part of UNIBRAW's continuing effort to upgrade its research and education. Such an evaluation will enable the university to capitalise on its successes. The various contributions of QFSA—some of which are inevitably closely related—towards UNIBRAW are as follows:

Research questions could be addressed in a quantitative manner

Many research questions had, until the introduction of QFSA, only been addressed in a qualitative way. Some examples of such research areas are: (1) the contribution of annual and perennial crops to farm households; (2) the contribution of livestock to farm households (ruminants appear to contribute little to the regular cash flow, however, they play a very important and valuable role for farm household as money savers); and (3) determination of alternative farm household development scenarios.

Realising the importance of an interdisciplinary approach

As work on the development of QFSA progressed, UNIBRAW staff became more aware of the necessity to look at RD issues in an interdisciplinary manner.

Interaction with scientists outside the University

The QFSA approach has provided ample opportunity for INRES staff to present their results and participate in academic symposiums, e.g. the National Symposium of Farming Systems Research and Development, Malang Indonesia (INRES, 1991). Participating in QFSA has also enabled INRES staff to establish contacts at an international level; they are now all registered as members of the Asian Farming Systems Association (AFSA). Several staff members have participated in international symposiums. All these opportunities enhance the endeavour of UNIBRAW to improve its research activities and to become acquainted with research that is being conducted in other parts of the world.

Contribution of QFSA in the field of academic building

When QFSA was first introduced at UNIBRAW, the initiators of the programme were well aware that the concept would be new for the staff members who would be involved in the programme. Realising this, it was decided that it would be of crucial importance to organise several training sessions, during which project members would be introduced to several techniques that would be utilised within the framework of QFSA. These training sessions were organised for both nucleus (including senior UNIBRAW staff who were involved in the project on a full-time basis) and plasma staff (UNIBRAW staff who participated in the project on a part-time basis) (INRES, 1990) of all the disciplines that were involved. The training sessions could be categorised as preliminary, main and supporting training sessions. A preliminary training session introduced the concept of FSA. Participants were acquainted with the more classical FSA methodologies and the way in which the INRES programme would divert from these. The main training sessions emphasised the methodologies that would be used in the fields of quantified land evaluation, theoretical production ecology, sociology and economics. In these sessions the links between the different disciplines were also highlighted. The supporting training sessions had a more disciplinary character, i.e. quantitative evaluation of soil fertility (QUEFTS), agro-forestry, farm household survey, and application of a database. The knowledge that was accumulated by INRES staff was not only useful in their own research endeavours but was also disseminated to (post)graduate students of the University.

Setting research agendas

When studying the very complex farming systems south of Malang, many research topics surface. Time and again decision-makers at the University have had to decide which topics were to be given priority. One simply did not know in advance which topics should be given more priority. Considering the fact that UNIBRAW has a crucial role to play in the field of RD, it is of the utmost importance that it has suitable tools with which it will be possible to decide which research topics should be given priority. QFSA—as a decision-supporting tool—has made it easier to set research agendas (Van Rheenen et al., 1992). QFSA enables decision-makers at the University to clearly see the gains that can be achieved when considering RD from different perspectives and this can be brought in line with government policies. For example, the Government of Indonesia has subsidised the utilisation of fertilisers. This was done with the aim of increasing the contribution of crops to farm household incomes. With this policy in mind the following questions become relevant: (1) in which ways can fertilisers be used with optimal efficiency, (2) which crops should be stimulated, and (3) which macro-economic measures should be taken so that the income-generating capacity of farm households is stable? Answering these types of questions requires: (1) thorough disciplinary research, and (2) the utilisation of a tool in which the various insights that were gained in (1) can be synthesised.

Changing attitude towards research and teaching

QFSA was developed in the INRES project and with this project many groups were involved at different levels. To all those involved it became apparent that what was necessary in the context of conducting research for RD was that: (1) a quantitative approach was essential if the aim was to evaluate the merits of different possible government policies; (2) the quantitative approach gave the opportunity to gain more insight in the very complex farming systems being studied, and made it possible to formulate research problems in a very concrete manner; and (3) disciplinary research and interdisciplinary research can very well go hand in hand.

UNIBRAW members have become more competent discussion partners in the field of RD

On frequent occasions UNIBRAW staff are required to actively participate in discussions concerning RD with other researchers, extension services, and policy-makers. The work that has been done at the INRES project has enabled them to take part in these discussions in a more concrete manner. Now they can not only talk about RD issues in a qualitative fashion, but they can also substantiate their arguments with quantitative information.

Problems encountered during the implementation of QFSA at UNIBRAW

One does not only learn from the success of a certain approach but also (probably more) from the problems that were encountered while implementing it. Some of these were as follows.

The introduction of new modelling techniques has been time consuming

The approach that was developed at the INRES project was very dependent on several modelling techniques. These techniques were applied at all levels of the farming systems, ranging from crop and livestock simulation models, to decision-supporting models at farm level. Prior to the initialisation of the QFSA at UNIBRAW, the staff had hardly been acquainted with these techniques. The successful application of QFSA depends on the competence of all parties involved, including students, plasma and nucleus staff, project management and external supervisors. It appeared that a great deal of time elapsed before all participants were able to effectively participate in the project.

Data requirements for QFSA are extremely high

For the successful development of QFSA it became apparent that data would be required with a high level of accuracy. These data had to be collected making use of secondary as well as primary sources. When it became apparent that the available secondary data did not reach the degree of accuracy that was required, it became necessary for project members to collect these data. This proved to be extremely time consuming. Now, being in a position to look back, we may wonder whether all the data that initially were thought necessary, really were indispensable. This question is essential, because it will give an idea of the usefulness of the approach in the future. Obviously, if the approach will be so time consuming in terms of data collection, each time it is to be executed, its usefulness will be reduced. Here, it is important to realise that the QFSA methodology had to fulfil several demands. It had to be useful for RD and at the same time had to provide the University with an increased academic insight in the underlying complexities of rural development. If the approach had only contributed towards RD it probably would have been possible to have utilised fewer data and the whole exercise could have been completed in a shorter span of time.

Introducing interdisciplinary thinking is time consuming

Most participants of the INRES programme had been involved in research which carried a strong disciplinary character. The switch from disciplinary thinking to one with a more interdisciplinary orientation was not always easy and needed time. Learning to listen to practitioners of other disciplines and accepting their contributions as valuable input is essential for RD. It is important to realise that this is something that will not happen overnight and during this investment of time, the researcher loses the opportunity to demonstrate his skills in his own specialisation.

Little involvement of stakeholders

It proved simply impossible to involve different stakeholders in RD at an early stage. Ideally stakeholders are involved with this type of research—in an interactive way—at a very early stage. However, involvement for stakeholders to participate in an interactive manner becomes really interesting when they can see what the effect will be of a shift in their objectives. Before this effect can be demonstrated quantitatively, the data necessary for the models should have been collected and the development of the models should have been completed.

The expected contribution of QFSA to UNIBRAW in the future

UNIBRAW has invested a great deal of time and money in QFSA. The results that have been gained so far have given research managers of the University confidence that the investments have been justified. This also became apparent when the project responsible for the implementation of QFSA was subjected to an external evaluation (De Boer et al., 1992). Long-term strategic thinking for a university is of great importance. In this regard the expected contributions that can still be expected from QFSA in the future are as follows.

PhD dissertations

It is envisioned that the QFSA approach will give thorough academic insights in the functioning of farming systems in the limestone area of South Malang. It was considered of great importance that this pioneering work would remain accessible for future generations of researchers at UNIBRAW as well as elsewhere. Also, senior supervisors of the project agreed that the work should be of a high academic standard. The agreement was therefore reached that the work would be published in the form of PhD dissertations. PhD dissertations will be published with the following topics: (1) quantified land evaluation in farming systems analysis, a case study in east Java; (2) the role of perennial crops in farming systems in the limestone area of South Malang; (3) the role of ruminants in the limestone area of South Malang; (4) decision-making processes of 12 farm households in the limestone area of South Malang; (5) the economic consequences of land tenure systems on dryland farming, a case study in the LS area of South Malang; (6) quantifying the physical production potential of farming systems in the limestone area of East Java; and (7) exploring options for development for farming systems in the LS area of South Malang. These PhD dissertations will be supervised by promoters and co-promoters of the Universities of Wageningen, Leyden, Malang and Bandung.

Formulation of a follow-up project using the experience gained while developing QFSA

A result of the INRES project is the computation of several alternative development scenarios. These scenarios, however, only indicate potential options for development and cannot be used as blueprint plans from which development paths can be deduced. If one desires to achieve the computed results it will be necessary to formulate development paths, i.e. which steps have to be taken to leave the actual situation and reach the computed results. In a follow-up project (which has already been approved by the Indonesian government) the following objectives will be pursued: (1) the computed results will be compared with the actual situation; (2) the observed differences in step (1) will be analysed; (3) the tools used will be re-evaluated and weaknesses will be removed; (4) institutionalisation of QFSA in the research and education systems of UNIBRAW; and (5) bridging the gap between agricultural research and the extension services, by showing both the importance of a detailed analysis of farmer constraints and opportunities (INRES I) and the usefulness of an ongoing dialogue with farmers (farmers involvement in planning, operating and evaluating).

Simplification of the QFSA

Recently, discussions in Indonesia have intensified as to how the welfare of the rural population can be increased. Questions have included: (1) how can people in the rural areas be characterised, (2) what are the constraints for RD, (3) what are the potential options for development and (4) how can programmes be planned and implemented while having a high chance of success? By using the up-to-date literature and UNIBRAW experiences, UNIBRAW in cooperation with WAU, will attempt to formulate simple methodologies using QFSA relevant to the actual conditions and disseminate it to local government staff, extension services, researchers and students.

DISCUSSION AND CONCLUSIONS

In this chapter a tool has been presented with which various options for development can be explored and which has the great advantage of providing great flexibility. It is very easy to examine the impact of variations in certain parameters (e.g. labour availability, input–output prices, land resources) on the various goals being explored. Analysis of options for agricultural development in a region, as a basis for planning, should result in the selection of the 'best' development option (Van Keulen, 1992). The same applies at farm level; also at farm level one can conclude that the so called 'best' option will depend on the various stakeholders in rural development, whose objectives may very well be conflicting. The tool described in this chapter can be of assistance in exploring various options and showing the impact that objectives of different stakeholders have on each other.

Is the approach and the tool described in this chapter a panacea for rural development? Certainly not, and one should not want to use it for that purpose. It is not—and does not pretend to be—anything more than a decision-supporting tool. It shows the various stakeholders in rural development which options are, in principle, possible—but it does not present a blueprint plan as to how one can reach these options. Before these options can be realised several constraints will have to be removed and these might very well lie outside the spheres of influence of the actors in the system being studied. A clear example of this is prices; individual farmers are price takers.

ACKNOWLEDGEMENTS

The authors gratefully acknowledge the INRES project, a cooperative project of Universitas Brawijaya Malang, Wageningen Agricultural University and University Leyden in training on interdisciplinary agricultural resarch, for its financial support. The authors would also like to express their gratitude to all staff members of the INRES project, whose cooperation and stimulating scientific guidance made this study possible.

REFERENCES

Byerlee, D. and Collinson, M. P., 1980. *Planning Technologies Appropriate to Farmers: Concepts and Procedure*. CIMMYT, El Batan, Mexico.

Collinson, M. P., 1987. Farming systems research: procedures for technology development. *Experimental Agriculture* **23**: 365–386.

Conway, G. R., 1985. Agroecosystem analysis. *Agricultural Administration* **20**: 31–55.

De Boer, S. J., Makken F. A. and Sarsidi Sastrosumarjo, 1992. *Evaluation Report Interdisciplinary Agricultural Research Training at UNIBRAW University, Malang*. UNIBRAW, Malang, Indonesia.

De Wit, C. T., Van Keulen, H., Seligman, N. G. and Spharim, I., 1988. Application of interactive multiple goal programming techniques for analysis and planning of regional agricultural development. *Agricultural Systems* **26**: 211–230.

Driessen, P. M. and Konijn, N. T., 1992. *Land-Use Systems Analysis*. Wageningen Agricultural University, Department of Soil Science and Geology, Wageningen, The Netherlands and INRES, Malang, Indonesia.

Efdé, S. L., and Van Rheenen, T., 1992. Incorporating livestock in quantified farming systems analysis. Paper presented at The International Seminar on Livestock Services for Small Holders, Yogyakarta, 15–21 November.

Fresco, L. O., 1988. *Farming Systems Analysis, An Introduction*. Tropical Crops Communication No. 13, Dept. of Tropical Crop Sciences, Wageningen Agricultural University.

INRES, 1990. *Interdisciplinary Agricultural Research Training at UNIBRAW University*. INRES Project Document No. 1, January 1990. INRES, Malang, Indonesia.

INRES, 1991. INRES farming systems analysis. In: Iksan Semaoen et al. (eds). *Prosiding Simposium Nasional Penelitiandan Pengembangan Sistem Usahatani Lahan Kering yang Berkelanjutan*, pp. 61–77. UNIBRAW, Malang, Indonesia.

Kahn, H. E. and Spedding, C. R. W., 1983. A dynamic model for the simulation of cattle herd production systems: Part 1—General description and the effects of simulation techniques on model results. *Agricultural Systems* **12**: 101–111.

Konandreas, P. A. and Anderson, F. M., 1982. Cattle herd dynamics: an integer and stochastic model for evaluating production alternatives. ILCA Research Report No. 2, International Livestock Centre for Africa, Ethiopia.

Levine, J. M., Hohenboken, W. and Gene Nelson, A., 1981. Simulation of beef cattle production systems in the Llanos of Colombia—Part 1. Methodology: an alternative technology for the tropics. *Agricultural Systems* **4**: 37–48.

Rabbinge R., 1991. Afwegingen op lange termijn. In: Brussaard, W. and Stortenbeker, C. W. (eds), *Milieu en Ruimte in de Jaren '90*, pp. 55–73. Wageningse ruimtelijke studies 5, Landbouwuniversiteit, Wageningen.

Sanders, J. O. and Cartwright, T. C., 1979. A general cattle production systems model. I: Structure of the model. *Agricultural Systems* **4**: 217–227.

Scheele, D., 1992. Formulation and characteristics of goal. Working Document 64, Netherlands Scientific Council for Government Policy, The Hague, The Netherlands.

Semaoen, I., Liliek, A. and Zemmelink, G., 1991. Farming systems research: development of interdisciplinary research at UNIBRAW University, Malang, Indonesia. Paper presented at The FAO Expert Consultation on Institutionalization of Farming Systems Development, 15–17 October 1991, Rome, Italy.

Shaner, W. W., Philipp, P. F. and Schehl, W. R., 1982. *Farming Systems Research and Development: Guidelines for Developing Countries*. Westview Press, Boulder, Colorado.

Spitters, C. J. T., Van Keulen, H. and Van Kraalingen, D. W. G., 1989. A simple and universal crop growth simulator: SUCROS87. In: Rabbinge, R., Ward, S. A. and Van Laar, H. H. (eds), *Simulation and Systems Management in Crop Protection*, pp. 147–182. Simulation Monographs. PUDOC, Wageningen.

Stroosnijder, L., 1992. La désertification en Afique sahélienne. *Le Courrier ACP-CEE* **133**: 36–39.

Stroosnijder, L., 1993. Systems approach for the quantitative analysis of traditional soil and water conservation techniques in the African Sahel. Research Trainee Project Proposal, Dept. Irrigation and Soil & Water Conservation, Wageningen Agricultural University, Wageningen.

Stroosnijder. L. and Van Rheenen, T., 1993. Making farming systems analysis a more objective and quantitative research tool. In: Penning De Vries, F. W. T., Teng, P. and Metselaars, K. (eds), *Systems Approaches to Agricultural Development*, pp. 341–353. Kluwer Academic, Dordrecht.

TAC (Technical Advisory Committee), 1978. Review Team of the Consultative Group on International Agricultural Research. Farming systems research at the International Agricultural Research Centres. The World Bank, Washington, DC.

UNIBRAW, 1983. Perincian Tugas Bagian, Sub Bagian, Urusan di Lingkungan Universitas UNIBRAW, Malang, Indonesia.

UNIBRAW, 1992. Status Universitas Brawijaya. UNIBRAW, Malang, Indonesia.

UNIBRAW, 1993. Tiga Puluh Tahun Universitas Brawijaya, UNIBRAW, Malang, Indonesia.

Van Diepen, C. A., Wolf, J., Van Keulen, H. and Rappoldt, C., 1989. WOFOST: a simulation model of crop production. *Soil Use and Management* **5**: 16–24.

Van Keulen, H., 1993. Options for agricultural development: a new quantitative approach. In: Penning De Vries, F. W. T. and Teng, P. (eds), *Systems Approaches to Agricultural Development*, pp. 355–367. Kluwer, Dordrecht.

Van Keulen, H. and Veenenklaas, F. R., 1993. Options for agricultural development: a case study for Mali's fifth region. In: Penning De Vries, F. W. T. and Teng, P. (eds). *Systems Approaches to Agricultural Development*, pp. 367–381. Kluwer, Dordrecht.

Van Lanen, H. A. J., 1991. Qualitative and quantitative physical land evaluation: an operational approach. PhD thesis, Wageningen Agricultural University, The Netherlands.

Van Rheenen, T., 1993. Overview of guestimates used for the 'Future of the Lands' symposium (August 1993). Department of Development Economics, Department of Theoretical Production Ecology, Wageningen Agricultural University, Wageningen, The Netherlands.

Van Rheenen, T., Efdé, S. L. and Salyo, S., 1992. Using interactive multiple goals linear programming to set research priorities. Paper presented at The Asian Farming Systems Symposium, 15–17 November, Colombo, Sri Lanka.

WRR (Wetenschappelijke Raad voor het Regeringsbeleid), 1987. Ruimte voor Groei. Kansen en bedreigingen voor de Nederlandse Economie in de komende tien jaar. Rapport no. 29, WRR, Den Haag, The Netherlands.

WRR (Wetenschappelijke Raad voor het Regeringsbeleid), 1992. Grond voor Keuzen. Vier perspectieven voor de landelijke gebieden in de Europese Gemeenschap. Rapport no. 42, WRR, Den Haag, The Netherlands.

CHAPTER 24

Future Land Use Planning and Land Options in Eastern Europe

J. W. M. Hardon,[a] A. M. Yemelianov[b] and J. W. Erdman[a]

[a]Euroconsult, Arnhem, The Netherlands. [b]President's Advisory Council, Moscow and Agricultural Economics Department, Moscow State University, Moscow, Russia

INTRODUCTION

The state of food markets is crucial for economic reform and the democratisation of Russian society. Negative trends of food production contribute to social and political instability, with social outbreaks and with undesirable effects both for Russia and for the whole of the world community.

The roots of today's food crisis in Russia can be traced back in history. After 1917, in accordance with the Boshevik ideology, the state land monopoly was introduced and the natural development of the rural sector was interrupted. Farmers were alienated from property and land, and from free economic activities, thus turning the farmer into a churlish hired worker. Farmers' community and land, and the Russian villages on the whole, were driven to a state which actually required reanimation.

Of great significance are both the restructuring of the agricultural sector and the redefining of priorities. First and foremost, this implies the reduction of losses of produce through accelerated development of storage, processing and transporting sectors, as well as technical and technological conversion at the farms.

LAND REFORM

As part of the process of land reform, private land ownership, life-long proprietorship and leasing were legally acknowledged more than two years ago. However, there were many obstacles. The mechanism of implementing new forms of land relationship has not been elaborated. Often local authorities offer strong resistance, while centralised bodies do not exercise the necessary control. A regulated land market is

The Future of the Land: Mobilising and Integrating Knowledge for Land Use Options
Edited by L. O. Fresco, L. Stroosnijder, J. Bouma and H. van Keulen.

admitted. As for free marketing, there is much more politics than commonsense in the approaches. The task is to give land free of charge to farmers and city dwellers wishing to start farming. This should of course be done considering the current real norms. People do not have the capital to buy land at high market prices.

Forming a rural economy with multiple systems of ownership leads to the development of a farming economy, encouraging various co-operatives and other innovative forms of economic activities. The task is to make the farmer the true owner, a holder of property. The first steps have already been taken. Over 200 000 private farms have been established. However, in the near future these farms will not be able to supply Russia with adequate amounts of food. Nowadays, they are responsible for about 3% of agrarian production. This figure can be expected to reach the level of 5–10%. The problems are not only the lack of machinery, equipment and initial capital, nor the difficulties involved in obtaining the land; the major problem is that most farmers are not psychologically and socially ready to start an independent farm, running the risk and assuming the responsibility inevitably involved. Therefore, while giving all possible assistance to new farmer's initiatives, *kolkhozes* and *sovkhozes* should be reorganised simultaneously on a democratic basis. A state farm should be transformed into a cooperative or association of private farmers. This process has been going on, taking various forms and encountering considerable obstacles. In 1992, the Russian authorities tried to speed up the establishment of private farms and to restructure all state farms during a period of one year. In reality, this implied decollectivisation and the campaign caused much damage. A signal to retreat was sounded. The strategy of state farm transformation became less radical.

FARM RESTRUCTURING

Agricultural policy-makers, farmers and agribusiness managers in Central and Eastern Europe face fundamental difficulties in the transition of collective and state farms (SC farms) to farms and firms that are led on the basis of commercial principles, whether cooperative or private. In the past existing cooperatives and 'combinats' have not organically complemented farming, but were created to cover the whole process of production, processing and wholesale distribution. In the near future agriculture and the food chain have to be reorganised: new firms and integration of markets will make up a more market-oriented structure.

The starting point for new farming structures is a set of state laws. Legal adjustment is putting an end to the systems of collective and state farms. Simultaneously, major changes occur in (i) macro-economic policy, (ii) decentralisation of government authority, and (iii) agricultural policy instruments. The changes in agricultural and food policy deserve special attention. In general, these imply (1) deregulation, (2) termination of subsidies, and (3) structural change in international trade regimes.

At the same time, demand from the domestic market is declining and competition from foreign food producers is increasing. Demand from and access to traditional export markets in Central Europe and CIS have severely declined due to the conversion to hard currency trade and political changes. Even a resumption to barter-deal arrangements is complicated by the present disarray. Thus agriculture, though on its way to market orientation, is still in a crisis.

Large-scale farming has, by now, extended over more than one generation; SC farms determine the economics of agriculture and its infrastructure. Investment activities have been geared to this particular mode of production over many years. It has brought about a capital stock, notably with regard to buildings, machinery and processing facilities, which suits the past scale of farming in those countries. Subdividing the land over a number of smaller private units, although far from easy because of property claims that are submitted or the desire to restore old land boundaries, probably remains the easier part. It is considerably more difficult to decide in which way to reform ownership and management of the capital stock invested in large-scale buildings and equipment.

Finding solutions to this problem is of major importance as it will determine to a large extent the efficient use of available capital stock and other resources in agriculture. Moreover, it has an effect on sectors downstream in the market chain. Taken together, these account for a sizeable part of the economy of the countries concerned. Therefore, efficient resource use in these sectors will have a recognisable effect on overall productivity measures and trends for years to come. There can be no falling back on previous ownership rights as these investments date from the time of large-scale farming. Nor can it be expected that existing capital goods can be replaced in a short space of time by new investments of a more appropriate type for use on subdivided units, due to the persisting overall paucity of investment resources in most countries. Under these circumstances the issue is how to ensure that the most productive use is made of the available capital stock. In the case of a decision to maintain its collective ownership for some years to come, the risk of exploitation at unduly low prices is a threat to its proper use and maintenance and also to the formation of agricultural product prices that will reflect true costs. At the other extreme, when these capital goods are transferred to new owners, their monopolistic position in servicing farmers will tend towards high prices of these services at the expense of farmers' incomes and discourage private farming. There is a need to identify efficient ways to deal with this problem and there is reason to doubt whether a pure approach through markets will generate the higher productivity of the agricultural sector which must be achieved.

Although literature on efficient contracts and guidelines for land reform provides a useful background, the current situation in Eastern Europe is unique due to the pervasive and economically wide nature of the reforms. Surveys have been carried out to describe the agricultural operations. They will basically try to find out how farmers are managing in particular areas, and whether they are at present finding ways of operating holdings of reasonable size without free-riding on available productive capacity. The case studies will study selected farms in more detail and depth.

TECHNOLOGY TRANSFER

While implementing the agricultural reform, Russia should make more effective use of its possibilities. Hereby, assistance from the West can play an important role; not in terms of humanitarian aid, which is ultimately unable to solve the food problem, but rather assistance in transforming the production on the basis of advanced technologies. In this respect, activities started by Dutch experts in the Kolomna district of the Moscow Region are of considerable interest.

Three Dutch companies formulated a general agricultural support programme for four districts in the Kolomna area: the so-called KASP programme. The main characteristics of the KASP programme are:

(1) a programmatic approach, which implies a long-term involvement of the donor and recipient organisation with an outlined development scenario in a medium- and long-term perspective and a detailed plan and budget for annual short-term interventions;
(2) improvement of the total production chain of a limited number of crops, which means that improvements are supposed to ameliorate the total production chain from primary production, storage, processing to distribution and marketing;
(3) in the first year (1993) a focus on potatoes and vegetables (cabbage, carrot and red beet) produced at large-scale production units, aiming at higher productivity per unit and an improved quality of the produce as main objectives.

With the above-mentioned guidelines in mind, the first project within the KASP overall support programme is the establishment of a supply and marketing cooperative, based on the experience of Dutch cooperatives acquired in the last 50–100 years. The objective of the cooperative is to sell improved agricultural produce on markets outside the traditional state-controlled market channels on behalf of its members, existing *sovkhozes* and *kolkhozes*. Furthermore, the cooperative will provide its members, at a charge, with the necessary local and imported inputs from the Netherlands, such as seed potato and vegetable seeds, farm machinery and storage equipment. The price of these imported inputs will be realistic under the prevailing Russian pricing system.

The Russian staff and members of the cooperative, being the managers of the privatised *sovkhozes* and *kolkhozes*, will be trained in the functioning and operation of Dutch supply and marketing cooperatives, keeping in mind that these cooperatives are private enterprises, not controlled by any state organisation. Emphasis will be laid on operating in a free market economy, in which prices are determined by supply and demand. In this respect, production of quality products will be essential in the marketing policy of the cooperative. For 1993, four *sovkhozes* were selected, one from each of the four participating districts. During 1993, further study will be carried out, investigating the feasibility of establishing a centrally organised storage facility and workshop for the cooperative.

The work carried out by Dutch experts has more than local significance. It demonstrates how Western countries can actively contribute to the revival of the Russian rural sector and help overcome the current food crisis.

CHAPTER 25

MSBB—A Concept and Microcomputer Application for Agro-environmental Land Use Analysis and Planning

K. W. Knickel

Institut fuer laendliche Strukturforschung, Johann Wolfgang Goethe-Universitaet, Frankfurt/Main, Germany

CONCEPTUAL FRAMEWORK

MSBB (Modelle Standort- und Betriebstypischer Bodennutzung) is a systems-oriented concept and PC/AT-based software for the analysis of agricultural structures and farming patterns in micro-economic and ecological terms. It facilitates the estimation of structural changes and possible or probable developments in land use resulting from varying economic, technological and political conditions. The concept underlying the MSBB System was developed in the framework of a three-year research project funded by the German Federal Ministry for Research and Technology. The prototype software is presently being tested.

In an agricultural context, systems-oriented means that relevant aspects of and interrelationships between external conditions, farm household resource use, farming system and patterns of land use are simultaneously taken into account—including interactions between livestock and crop subsystems. By providing a systematic framework for data collection and synthesis, MSBB helps to cope with the considerable complexity encountered when trying to understand and describe agricultural systems.

Concept, implementation and use of MSBB are characterised by multidisciplinarity. The two main dimensions are ecological and technological cause–effect relations, and socio-economic determinants. The two corresponding and, under the existing EC market and price policy, contrasting objective functions are environmental protection (natural sciences) and agricultural production (social sciences and managerial economics).

The Future of the Land: Mobilising and Integrating Knowledge for Land Use Options
Edited by L. O. Fresco, L. Stroosnijder, J. Bouma and H. van Keulen.

FIELDS OF APPLICATION

Both dimensions—environmental protection and agricultural production—are assessed simultaneously, either at the level of a representative range of region and farm types or for individual farm-holdings. Linked with that are two different fields of application:

(1) *planning- and policy-oriented* evaluation of developments and measures on the regional, national or European level, e.g. assessing the environmental impact of the reform of the Common Agricultural Policy by simulating the potential responses to individual measures at the level of a representative spectrum of region and farm types (Figure 25.1);
(2) *farm management-oriented* counselling of individual or groups of farmers with similar resources and farm organisation, e.g. by analysing the present situation and simulating adjustments in farm plans and type, scale and intensity of land use, and by evaluating the effects in terms of both farm income and impact on environmental quality.

In the first field of application, a spectrum of regions is selected with sufficiently homogeneous agro-ecological and socio-economic conditions. On the basis of variation in key structural variables, models of typical farm households are defined for each type of region. In the second field of application, data are used for individual farm-holdings.

Independent of the particular aims, the analysis is based on thc cconomic, social and organisational aspects at the level of the farm. Attributes of the natural, socio-cultural, economic and institutional environment are taken into consideration in the formulation of production and objective functions and, apart from that, only as far as they affect farmers' decision-making.

IMPLEMENTATION

MSBB runs on PC/AT microcomputers and is based on standard (commercial) software packages: linear programming (XA), spreadsheet and databank management (SYMPHONY, LOTUS 1-2-3). Fundamental are three databanks: standard planning data (STAN), farm structural data (BETR) and production systems data (VERF). Synthesis of data is achieved by means of linear programming (LP) in combination with various spreadsheet and databank applications. The advantage of using LP is that it allows a relatively realistic analysis and simulation of resource use and of input–output relationships at the farm level. A module for input data (matrix generator) and a module for the interpretation of output data (matrix interpreter) provide uniform frameworks for data handling (Figure 25.2).

There are clearly defined interfaces in MSBB for the simulation of changes in external conditions such as adjustments in producer and/or input prices, set-aside or extensification schemes and nitrogen taxes. MSBB is a problem-oriented and, at the same time, open concept. It has a modular structure. Upstream and downstream bridges can be developed, a linkage with other types of planning is easily possible. Specific problem areas can be formulated in a more detailed way. Data demand depends on the particular question

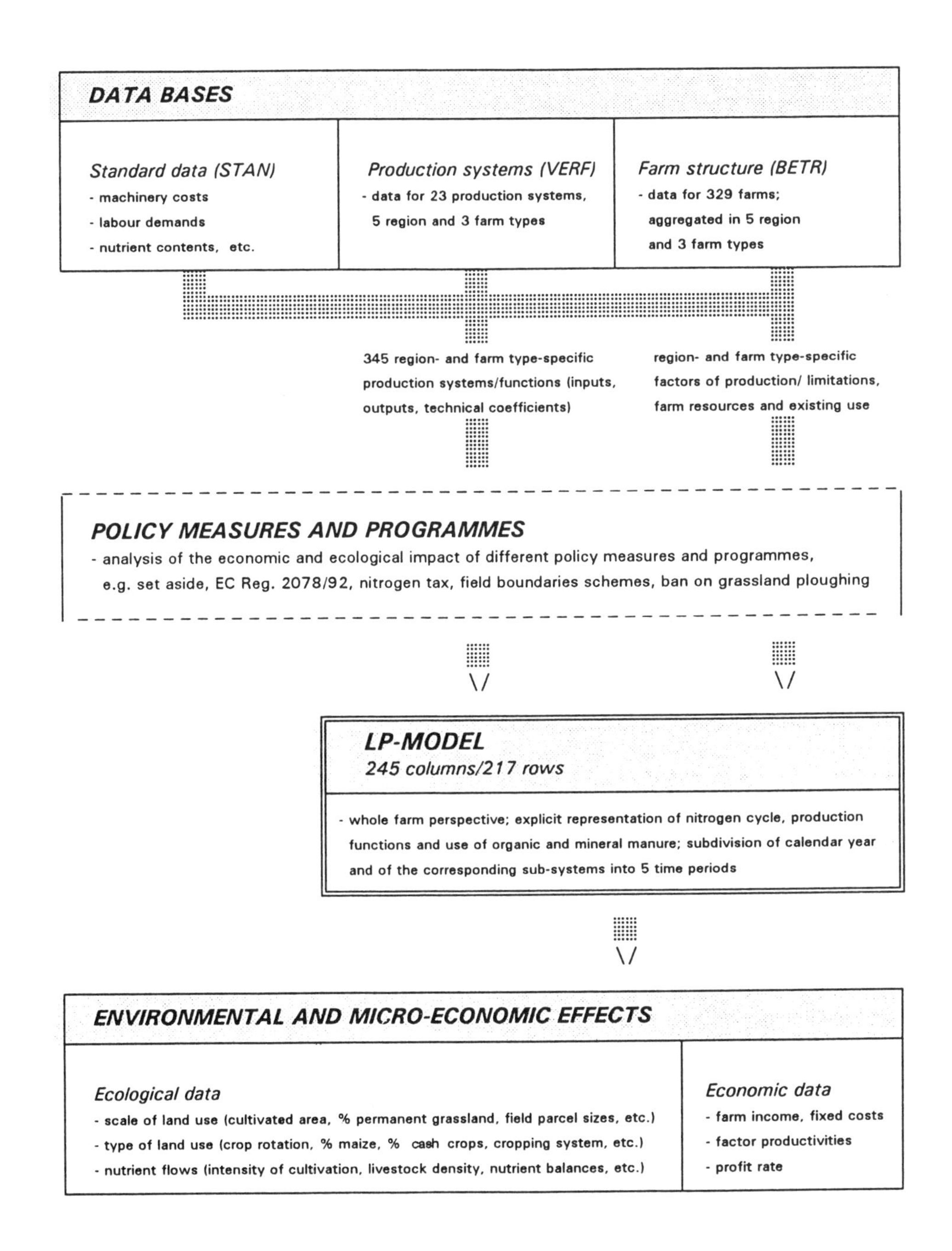

Figure 25.1 Using MSBB for assessing the potential effects of policy measures at the level of a representative spectrum of region and farm types

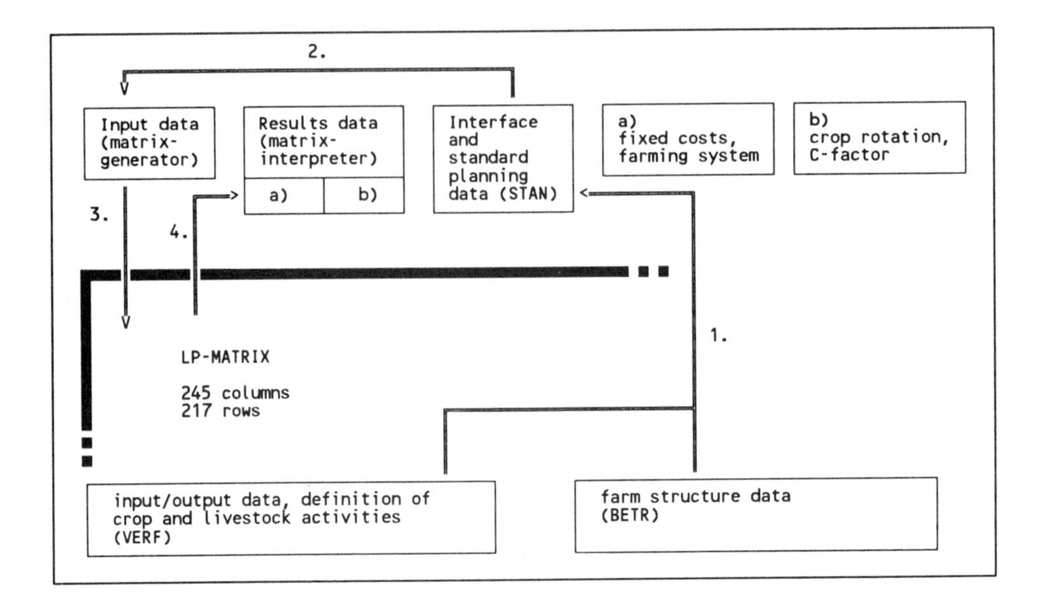

Figure 25.2 Flow of data in the MSBB spreadsheet

studied, and data availability has been a key aspect in the development of the model.

MSBB is closely tied to actual land use conditions in the regions studied, normally having a size of about 1000–2000 ha UAA managed by 50–150 farms. For the simultaneous assessment of production patterns and their environmental impact, selected ecological models are integrated within a more comprehensive whole farm model. In order to cope with the vast amount of data to be handled, a set of production systems-related and agro-environmental indicators is used.

Potential environmental impact is quantified in terms of erosion (C- and P-factor in the Universal Soil Loss Equation), pollution (livestock density, intensity of fertiliser use, $N_{org/min}$-balances, P_2O_5 and K_2O-balances, pesticide use), and scale and type of land use (farm and field parcel sizes, livestock numbers, grassland, arable, irrigated and fallow land, crop rotation, etc.). With the help of these indicators a description can be made of the potential environmental impact of a particular farm system and land use pattern. The region-specific potentials of the natural environment and protection requirements (e.g. the soil erosion risk, characterised by soil type, topography and precipitation or the buffer capacity of the soil and nitrate (NO_3) pollution trends) are compared with these land use related data. Key questions then are: the compatibility of existing land use with protection requirements, the need for adjustments and action (e.g. for soil conservation measures) and, related to that, development constraints and opportunities (Figure 25.3).

Table 25.1 Example of the structure and contents of a user-defined results table in MSBB

RESULTS TABLE	MSBB-3	Location Type: lowland/arable	Farm Type: int/FT*	AWU/farm: 1,00	OPT/I	1989/90

CROP PRODUCTION:	WWHEAT	WRYE	WBRLY	SBRLY	OATS	MIXGR	SBEETS	POTAT	OSRAPE	MAIZE	FBEETS	CCR,F/M	GRASS,W	GRASS,A
N 1 (kgN/ha)	70	50	40	20	20	0	60	60	80	20	40	10	40	70
N 1	0,0	0,0	0,0	0,0	0,0	0,0	0,0	0,0	0,0	0,0	0,0	0,0	0,0	0,0
N 2 increasing	0,0	0,0	0,0	0,0	0,0	0,0	0,0	0,0	0,0	0,0	0,0	11,4	0,0	0,0
N 3 intensity	46,6	0,0	0,0	0,0	0,0	0,0	0,0	0,0	0,0	0,0	0,0	0,0	0,3	0,1
N 4 (+20)	0,0	0,0	4,2	0,0	0,0	0,0	4,9	0,4	0,0	0,2	0,0	0,0	0,0	0,0
N 5	0,0	0,0	0,0	0,0	0,0	0,0	0,0	0,0	0,0	0,0	0,0	0,0	0,0	0,0

LIVESTOCK:	
Dairy cows	0
Heifers	0
Other heifers	0
Bulls	1
Sucker cows	0
Store pigs	21
Other pigs	47
Sheep	0
Live- (LU/haUAA)	0,19
stock (DM/LU)	0,00
density (LU'/haFA)	1,69
MVPs (DM):	
Milk quota (l)	0,00
Cow places (Pl)	0
SugarB quo (dt)	0
Arable ld (ha)	943
Grassld (haGw)	463
Grassld (haGa)	0

CROPPING PATTERN:	
Cereals (% AA)	90,3
Maize (% AA)	0,4
Root (% AA)	9,3
Fodder (% AA)	0,4
CCrop,f (% AA)	1,4
CCrop,m (% AA)	18,9
Grassld (% UAA)	0,7
CRotCon (DMsum)	0
FALLOW/SET-ASIDE:	
Set aside (% AA)	0
Grass ro (% GA)	0
(DM/haGA)	0
PESTIC/DM/haAA)	221
DRAINAGE/CONVERSION:	
Drainage (%Grl)	0
Conv (% Grl)	0

PRODUCTION INPUTS:	
Fert (DM/haUAA)	487
P205 (kg/haUAA)	119
K20 (kg(haUAA)	154
N-Qu (DM/kgN)	0,00
CFeed (1000 MJ)	15
(% purch)	100,0
PFeed (1000 GN)	0
(% purch)	0,0
MARKET COSTS:	
(DM/haUAA)	657
(% FI)	50
Grain s (% GU)	91,9
(t p.a.)	306
Milk s (t p.a.)	0
SLURRY/M Surplus:	
(kgNmin/haUAA)	0
SLStCap (m3)	221

INTENSITY PARAMETERS:	AF	Gw	Ga	LF
N-Applicat -Fertiliser	194	78	82	193
(kgN_{min}/ha)-Slurry/M	13	33	49	13
-other	143	61	71	142
N-Uptake	241	141	159	240
$N\text{-}_{min}$ Balance	108	31	43	108
$N\text{-}_{org}$ Balance (kgNorg)	0	0	0	0
UAA <80 kgNm,o/ha (%)	0,0	0,0	0,0	0,0
Nmin-Per.Bal (kg/haAA)	Sept–Oct:	0	Nov–Feb:	14
EROSION PARAMETERS:	Part-C-Factor	0,04	0,00	
	C-Factor	0,04		
	C-UAA	0,04		

FARM INCOME PARAMETERS:	DM/farm	DM/haUAA	DM/AWU
Standard Gross Margin (SGM)	122 272	2 157	122 272
Fixed Costs/Overheads	47 647	840	47 647
Specific Costs	136 242	2 403	136 242
Farm Income (48,5%)	74 626	1 316	74 626
" with compensation payments	74 626	1 316	74 626

FARM SYSTEM: field crops	ROTATION TYPE: cereals/sugar beet	26.05.93 10:56

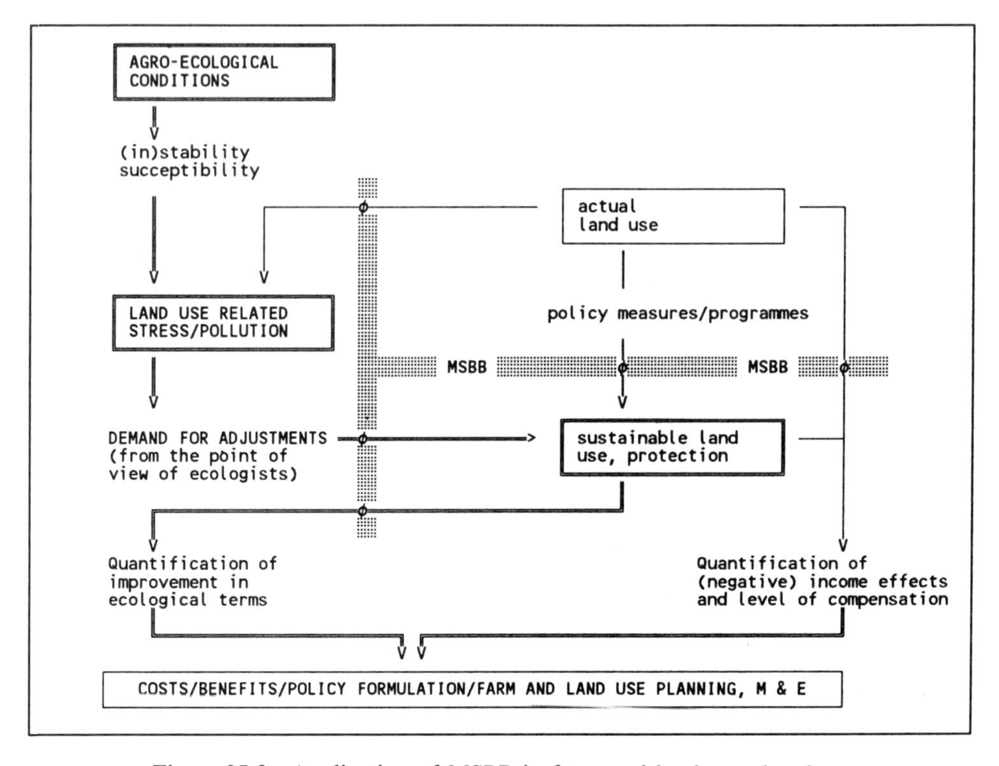

Figure 25.3 Application of MSBB in farm and land use planning

Table 25.2 Comparison of the effect of a producer price reduction by 15% (P-15) and a nitrogen tax of 150% (N-150) on farm income and risk of soil erosion (C-UAA) in four different region and farm types (A–D)

REGION AND FARM TYPE				Farm income			Soil erosion risk		
	Altitude (m)	% Grassland	Farm type	Actual (ECU/AWU)	P-15 (%)	N-150 (%)	Actual (Part-C)	P-15 (%)	N-150 (%)
A	150	1	fc/sb	74 626	−18	−16	0.04	+25	0
B	260	13	fc/os	62 212	−27	−18	0.06	0	0
C	220	40	mixed	58 658	−17	−1	0.10	−10	+10
D	630	100	extgr	39 658	−15	0	0.00	0	0
ϕ				—	−19	−9	—	+4	+3

Source: MSBB.
fc/sb = Field crops/sugar beet; fc/os = field crops/oil seeds; mixed = mixed farm: extgr = extensive grazing.

RESULTS MODULE

The results module of MSBB presents a user-defined range of farm type and region-specific data (Table 25.1). They include indicators of farm organisation and production systems; indicators for income effects; ecologically relevant data; and finally, incentive

Table 25.3 Impact on level of N fertiliser use, N_{min} balance and the proportion of extensively cultivated land (<80 kg total N) in four different region and farm types (A–D)

	N fertiliser use			N_{min} balance			% UAA<80 kg total N		
	Actual (kg/ha UAA)	P-15 (%)	N-150 (%)	Actual (kg/ha UAA)	P-15 (%)	N-150 (%)	Actual (% UAA)	P-15 (%p)	N-150 (%p)
A	193	−8	−21	108	−15	−34	0	0	+1
B	158	−1	−17	105	−4	−23	9	+3	+3
C	32	−12	−81	64	−16	−31	4	+14	+42
D	0	—	0	31	—	0	65	0	0
ϕ	—	−5	−30	—	−9	−22	—	+4	+12

Source: MSBB.

and compensation rates for agro-environmental policy measures (e.g. costs arising from specific measures and restrictions; marginal costs connected with crop rotation regulations; shadow prices of different types of land).

The data on potential environmental impact and the corresponding micro-economic effects inform policy analysts and planners. Immediate discussion of results and feedback are possible, allowing a close collaboration between analyst, planner, farmer and advisory services. Sets of farm-level data provide a basis for a more qualitative extrapolation and assessment of the impact at a regional level. Examples are given for two alternative policy measures and four region and farm types in Tables 25.2 and 25.3. The two measures compared are a producer price reduction by 15% (P-15) and a nitrogen tax of 150% (N-150).

CHAPTER 26

Non-agricultural Land Use among Farm Households in Scotland

M. Mitchell

SAC-Auchincruive, Ayr, Ayshire, UK

BACKGROUND

This chapter presents some of the results of the 'Pluriactivity in the Agricultural Sector in Scotland' project funded by the Economic and Social Research Council (Mitchell and Doyle, 1993) as part of the Joint Agriculture and the Environment Programme. It was an interdisciplinary project combining the modelling expertise of geography, microeconomics, socio-economics and ecology to explain the phenomenon of pluriactivity, defined as the engagement of one or more members of a farm household in either off-farm work or on-farm non-agricultural enterprises. Previous studies have looked at the incidence of pluriactivity, which is not a new departure for farm households, but there was a need to look deeper: why do households become pluriactive, what effect does it have on farm management, how does it impact on the environment and what implication does it have for the local and regional economy?

Pressures on the agricultural sector are increasing due to falling incomes, the need to reduce production levels, policy reforms and the environmental lobby. It has been predicted that this will lead to large structural changes. Strak (1989) says that in the UK the 'rural economy will become more and more dependent on non-farm activities in the future' and 'over the next decade . . . we can expect a further 34 000 full-time employees to leave the industry and another 18 000 farmers to go'. Arkleton's study (1988) found that this is reflected throughout the EC. Changes in farm management will inevitably lead to changes in land use, whether it means the countryside reverting to a 'natural' state, land being used for non-agricultural enterprises or city industrialisation spreading outwards. One aspect of this change is the farm household's decision to become pluriactive, and the following looks at which internal and external factors may influence that decision.

The Future of the Land: Mobilising and Integrating Knowledge for Land Use Options
Edited by L. O. Fresco, L. Stroosnijder, J. Bouma and H. van Keulen.

INTRODUCTION

A detailed survey of 506 farm households was carried out in three regions of Scotland during 1991. The regions chosen were illustrative of the situation on the mainland. Grampian Region has the influence of a city, Aberdeen; Fife Region is a major tourist area; and Dumfries & Galloway is a remote agricultural area which exhibits a low incidence of pluriactivity. They also reflect a cross-section of farm types from Less Favoured Areas to the arable lands of Grampian and the dairy farms of Dumfries & Galloway. All farms in the three regions were stratified according to farm type and British Size Unit (BSU) which is a measure of size based on income. Farms less than 4 BSUs were excluded from the study as they are, by definition, part-time farms. A random sample of 7% of each cell was then drawn and a face-to-face interview of the sample was carried out.

As well as gathering physical data on the farm and farm household, the questionnaire covered the household's background, education, expectations of succession, and a whole series of questions to try to assess their attitude to change, risk and how they form their opinions. The data were stored in an Oracle relational database and it is this information which was used to assess the internal factors that may influence a farm household's decision to become pluriactive.

It was also decided to investigate which external factors would influence the opportunity to become pluriactive. Several non-agricultural enterprises were investigated to assess which geographic, physical, locational, social and economic factors may influence their success. This can be used to identify areas which are as yet undeveloped in terms of these enterprises, but exhibit apparently appropriate conditions.

INTERNAL FACTORS

As stated above, the survey contained many questions to do with the household's characteristics and attitudes. Several of the questions asked the farmer to indicate on a scale of one to six what he thought of innovations, what influenced decisions, his willingness to try new ideas, to what lengths capital would be risked on a possibly profitable enterprise, his attitude to diversity, etc. Answers to these questions were combined to give a risk score, which was in turn used as one of the variables when discriminant analysis was carried out, comparing the pluriactive and non-pluriactive households. The variables used for the analysis are listed in Table 26.1.

SPSS Discriminant Analysis (SPSS, 1990) was then used to test whether the means of the variables for the two groups of households (pluriactive and non-pluriactive) were significantly different. If the observed significance level is small (i.e. less than 0.05) the hypothesis that all group means are equal is rejected. The results of the analysis are listed in Table 26.2.

It can be seen from Table 26.2 that the variables GET and HA do not have a significance level less than 0.05. Therefore it was concluded that the pluriactive sample means were not significantly different from the non-pluriactive group, so the hectare measure of farm size and how the farm was acquired did not influence the decision. The other factors, however, did seem to be significant. Pluriactive households have a

Table 26.1 List of variables used in discriminant analysis between groups of pluriactive and non-pluriactive farm households

Variable	Description
RISK	Risk score
AGE	Age of the farmer
QUAL	Post-school qualifications
GET	How the farm was acquired: bought, inherited, rented, managed, etc.
BSU	British Size Unit, a measure of size based on income
HA	Hectares
FTYPE	Farm type
FFAMILY	Whether the farmer and spouse have farming backgrounds
NO	Number of people in the household
NO16	Number of people over the age of 16 in the household

Table 26.2 Results of discriminant analysis on the groups of pluriactive and non-pluriactive households

Variable	Significance level
RISK	0.0053
AGE	0.0004
QUAL	0.0013
GET	0.3708
BSU	0.0148
HA	0.1490
FTYPE	0.0109
FFAMILY	0.0013
NO	0.0000
NO16	0.0000

higher mean risk score; they have more members, including those over 16 years; they are better qualified; the farmer is younger; they live on farms with smaller BSUs; and they tend to be arable or smaller horticultural enterprises.

EXTERNAL FACTORS

If farm households are to be encouraged to diversify into non-agricultural enterprises it would be useful to be able to provide some kind of guidance as to what that enterprise may be. Therefore the next stage of investigation was to look at a variety of activities and to devise some method for assessing their viability. The activities were chosen either because they are very common activities for farm households to take up, or because, although they are very rare, they provide much higher levels of income for farm households than the frequent, smaller enterprises. The activities chosen are listed in Table 26.3 along with the reasons they were started and the annual average income they provide. It shows that in the majority of cases the enterprise was started to maintain or increase the household's income, and that the highest income levels are obtained from the least common activities, namely shooting and livery stables.

Table 26.3 Reasons for starting on-farm enterprises and average annual income from the activity on farms surveyed in Grampian and Dumfries & Galloway

	Why was the enterprise started?	Average annual income
Bed and breakfast	82% to remain in farming	£3279
Farm shop	60% to remain in farming	£1417
Caravan sites	67% to remain in farming	£2010
Self-catering cottages	88% to remain in farming	£4600
Clay pigeon shooting	60% to remain in farming	£13000
Livery stables	various	£14062

The first step was to identify where existing enterprises were, regardless of whether or not they were carried out on a farm. The presence of these enterprises was measured by a variety of geographical, physical, economic and locational factors outside of the farm, though existing farm shops and livery stables are obviously more likely to be situated on or close to farms.

Once the coordinates of the enterprises had been fixed, the characteristics of the parishes within which they had been established were identified. This was done using SPSS Factor Analysis on a list of variables that exhibited a high correlation with each other, and the presence of these factors seemed to coincide with the presence of the enterprise. It is important to note that factor analysis does not explain a causal relationship, merely the varying degrees of existence of the factors in parishes which contain the enterprise. However, the factors identified in one region may be used to predict where that enterprise would be expected in another, and compared with the actual locations. If the predictions are reasonably accurate, the model may be translated to other areas.

The starting set of variables used were:

- coastline
- main road
- employment
- distance from a town
- number of towns
- centres of population
- tourist attractions
- tourist attraction visitors
- tourist information office enquiries
- farm type
- ITE Land Classification
- population

Tests within SPSS Factor Analysis showed which set of variables should be included for each of the enterprises being investigated, and reduced these to three or four factors. The SPANS Geographical Information System was then used to produce maps which spatially represent areas that show potential for development. An example of this is given in Figure 26.1, which shows the areas of Dumfries & Galloway that exhibit potential

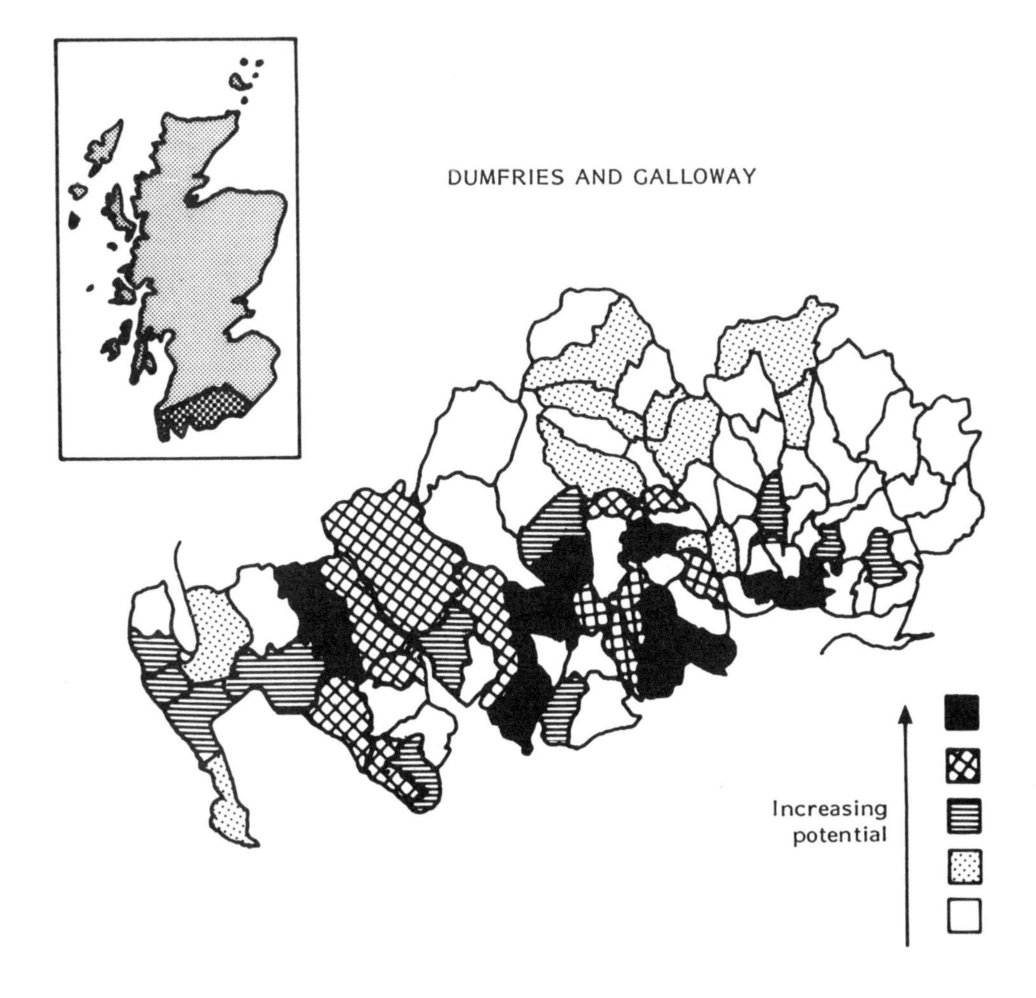

Figure 26.1 Areas of potential for developing caravan sites in Dumfries & Galloway

for developing caravan sites. In this instance the variables grouped into four factors: tourist activity in the area, employment opportunities, remoteness and ITE Land Classification. Looking at the map, the parishes surrounding tourist travel routes and major roads come out as being the main potential areas, with the remote, inaccessible hill areas showing the least.

CONCLUSIONS

The results of this study indicate that there are identifiable factors which describe pluriactive farm households. Although the results of the discriminant analysis may be as expected, confirmation that there are indeed differences between pluriactive and non-pluriactive households means that measures aimed at encouraging farm households to use their land and other factors for uses other than agriculture could be presented in a way that makes them attractive to households with similar characteristics. This, combined with the information about areas that have potential for development, would provide the opportunity for policy-makers to have some influence on land management and land use by providing incentives for particular development in targeted areas.

REFERENCES

Arkleton Research, 1988. *Rural Change in Europe.* Arkleton Trust (Research) Ltd., Oxford, UK.

Brun, A. H. and Fuler, A. M., 1991. *Farm Family Pluriactivity in Western Europe.* Arkleton Trust (Research) Ltd., Oxford, UK.

Dalton, G. E. and Wilson, C. J., 1989. *Farm Diversification in Scotland.* Economics Report No. 25, Scottish Agricultural College, Aberdeen.

Dent, B., Doyle, C. J., Heal, O. W., Crabtree, J. R., Dalton, G. E. and McGregor, M., 1993. *Pluriactivity in the Agricultural Sector in Scotland.* Economic and Social Research Council, London.

McInerney, J. and Turner, M., 1991. *Patterns, Performance and Prospects in Farm Diversification.* Agricultural Economics Unit, University of Exeter.

Mitchell, M. and Doyle, C., 1993. *Pluriactivity on Scottish Farms—Regional Socio-Economic Impacts and Prediction of Potential.* Economic and Social Research Council, London.

SPSS Inc., 1990. *SPSS/PC+ Advanced Statistics 4.0.* SPSS Inc., Chicago, Illinois.

Strak, J., 1989. *Rural Pluriactivity in the UK.* National Economic Development Office, London.

CHAPTER 27

Land Use Optimisation along the Sukuma Catena in Maswa District, Tanzania

J. A. Ngailo,[a] J. M. Shaka,[a] Ph. A. Kips[b] and F. van der Wal[a]

[a]National Soil Service, Tanga, Tanzania. [b]The Winand Staring Centre for Integrated Land, Soil and Water, Wageningen, The Netherlands

INTRODUCTION

Optimal use of an area of land is arrived at when land use goals have been realised with a minimum of negative effects. In many developing countries, land use optimisation is concerned with boosting agricultural production and is usually a process instigated by overall government policies and development objectives such as food self-sufficiency. The direct role of government in land use optimisation is especially evident in large-scale projects of land development, e.g. in irrigation. By contrast, land use optimisation can also be a more or less spontaneous process from within the farming community and led by the objectives and needs of that community. Here the role and influence of government is at best supportive or facilitative, by favourable price policies, credit and marketing facilities and infrastructural provisions.

In this chapter we present a case of 'spontaneous' land use optimisation from within the farming community. The objective is to show how this process has led to a diversified land use system, optimally adapted to the available land resources. The case presented serves to illustrate the importance of looking at land use and land use changes from a systems point of view. Understanding the complicated nature of land use systems with all the intricate relations between soil potentials and the spatial arrangement of soils on the one hand, and land use and economic changes on the other hand, is a prerequisite to any outside intervention aiming at a more efficient and sustainable land use. A particular point in focus is to what extent soil degradation is allowable in land use optimisation.

The area is Maswa District (3564 km^2) in Shinyanga Region, located some 120 km south-east of Lake Victoria in Tanzania. Socio-culturally the district is part of a larger

The Future of the Land: Mobilising and Integrating Knowledge for Land Use Options
Edited by L. O. Fresco, L. Stroosnijder, J. Bouma and H. van Keulen.

geographic zone called Sukumaland, i.e. the home area of the Wasukuma people. The present study is based on data collected during a reconnaissance soil survey of the district (NSS, in prep.) and also draws upon the results of more specific soil research work in the area (Ngailo, 1992; Shaka, 1992). The physical data are integrated with the published results of socio-economic and farming systems research (Brandström, 1985; FSRP, 1989b; Ebong et al., 1991; Meertens and Ndege, 1993), leading to a rather complete picture of the land use system in place.

GENERAL DESCRIPTION OF THE STUDY AREA

Maswa District, like most of Sukumaland, has an open landscape of gently undulating plains with long slopes to wide flat valley bottoms. Characteristic are the granite tors ('kopjes') standing out here and there on the otherwise smooth crests. The altitude in most of the district is in the range 1200–1300 m.

The district has a dry savanna climate and is situated in agro-ecological zone P8 (medium altitude plains with soils of moderate fertility; short crop growing periods of 3–3.5 months with unreliable onset dates; De Pauw, 1984). The average annual rainfall varies from about 700 mm in the south-east to about 900 mm in the north-west, i.e. in the direction of Lake Victoria. The rainfall pattern is pronounced mono-modal. According to long-term average data, the rainy season starts in November–December and ends in early May. About 85% of the annual rainfall is recorded in this period. Within the rainy period, January and February are somewhat drier months. Mean rainfall data, however, give a distorted picture. Rainfall in Maswa District, and in Sukumaland at large, has a cellular character. It tends to fall in heavy, localised rainstorms, separated by dry spells. As a result, rainfall in any one part may differ considerably from year to year and both onset date and length of the rainy season are unreliable. Because of the short dependable growing period, its unreliable onset date and the unfavourable infiltration and moisture storage capacities of the major soils (see below), the district is generally considered rather marginal for agriculture. The Government therefore recommends the cultivation of drought-tolerant crops such as sorghum, millet, sweet potatoes, cassava and pigeon peas, besides cotton as cash crop (TFNAP, 1985). But in reality a large variety of crops are grown in Maswa District, including rice and maize. In addition to arable agriculture, livestock (mainly cattle) is an important element of the land use system.

The key feature of agriculture in Maswa District is coping with the limited amount and unpredictability of rainfall. One way to do so is planting as early as possible. Land preparation and planting normally begin in October for maize, sorghum and rice, and in November for cotton. Clearly the food crops are given priority and the other crops are planted later because of insufficient labour for land preparation. Spreading of plantings of the same crop in time (staggered planting) is also practised, especially in the drier areas (FSRP, 1989a).

The people of Maswa District, the Wasukuma, are agro-pastoralists growing crops and rearing cattle. Traditionally cattle is the main form of wealth. Brideprice is commonly paid in cattle, and herd size largely determines one's social status. Cattle also provide a buffer against food shortage (FSRP, 1989a).

In recent decades the population of the district has increased substantially (Kurji, 1985), mainly by migration from Mwanza Region and other areas further north around Lake Victoria. Population density in the western part of Maswa District (Sengerema Division) has now reached comparatively high levels (about 70 persons/km^2 according to the 1988 population census; FSRP, 1989a). Lack of firewood is now a serious problem (cow dung is widely used as an energy source).

Up to the 1950s the staple grains in the district were sorghum and bulrush millet (Rounce, 1949). Since the 1950s these have largely been replaced by maize, although sorghum is still grown as a drought- and waterlogging-resistant crop in the south-east. Another important change was the rapid expansion of cotton cultivation, particularly after the Second World War. As a result of this increase in cash cropping, Sukuma agriculture advanced into a state of semi-commercialisation. At the same time shifting cultivation was replaced by fallow systems and permanent systems due to the increasing population density and the introduction of the plough. Room for grazing was reduced. As earnings from cotton were frequently invested in cattle, the growing cattle population had to depend on a diminishing range (Von Rotenhan, 1966). The threat of overgrazing was in fact a major factor behind the migrations into Maswa, later on into Mwagilla and Meatu in the east, and lately from there to far-away places such as Mbeya and Rukwa in south Tanzania (FSRP, 1989a).

Since the 1970s cotton production has steadily declined. Because of persisting low producer prices against increasing input prices, farmers became less and less interested in growing the crop and shifted attention to cultivating wetland (paddy) rice. This development, largely fostered by the farming community itself (Meertens and Ndege, 1993), has led to the present situation in which rice is as important a cash crop as cotton. Today good rice land is in short supply. Although there is still room for developing rice fields, the land is claimed by long-term residents, so that newcomers have to buy (FSRP, 1989a).

THE SUKUMA CATENA

The land use pattern in Maswa District is very much linked to a particular toposequence of soils from ridge crest to valley bottom. This recurrent toposequence is called the Sukuma catena and was first described by Milne in the legend of his Provisional Soil Map of East Africa at scale 1:2 000 000 (Milne, 1936). The potentials of the different soils in the catena are well understood by the Wasukuma and reflected in a unique and rich indigenous soil nomenclature. The Sukuma soil names imply a certain suitability for crop cultivation, mainly based on soil depth, workability, and susceptibility to waterlogging (Milne, 1947). This local terminology has much practical relevance (Acres, 1984) and is completely in line with the concept of soil series as developed in the USA.

The typical Sukuma catena is developed on granite, although Milne (1947) expresses some doubt whether this is really the case in all parts of Sukumaland. He suggests that possibly in part of the area the soils on the lower slopes and in valley bottoms are developed in, or derived from, old lake (Victoria) deposits. In Maswa, NSS (in prep.) and Shaka (1992) found further evidence for this suggestion.

SOILS

Name	Luguru	Luseni	Itogoro	Mbuga
Depth	—	shallow to moderately deep	deep to very deep	very deep
Drainage	—	well drained	imperfectly drained	poorly drained
Texture	—	sandy	loamy to clayey	clayey
Reaction	—	strongly acid	very strongly alkaline	moderately alkaline
$CaCO_3$-content	—	non-calcareous	strongly calcareous	moderately calcareous
Sodicity	—	non-sodic	very strongly sodic	mod.to strongly sodic
Mineralogy	—	quartz, kaolinite, haematite, feldspar	smectite, calcite	smectite, calcite
Classification	(Rock outcrop)	Eutric Regosols, petroferric phase	Calcaric Phaeozems, sodic phase	Calcic Vertisols, sodic phase

LAND SUITABILITY

	Luguru	Luseni	Itogoro	Mbuga
Cotton	—	moderate	moderate	low to moderate
Maize	—	low	unsuitable	low
Rice	—	unsuitable	moderate	moderate
Grazing	—	low	moderate	high

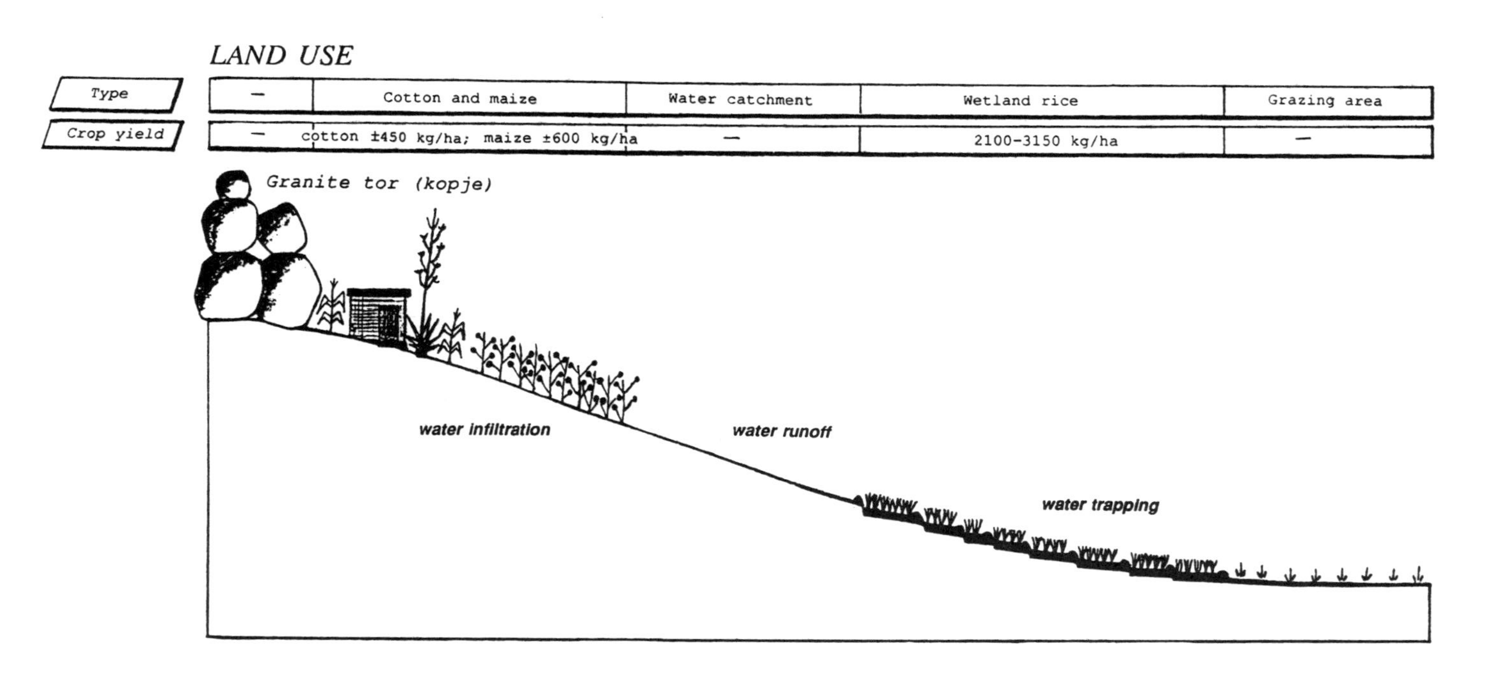

Figure 27.1 The Sukuma catena and the typical land use system in Maswa District

The soils making up the Sukuma catena are, from ridge crest to valley bottom: Luguru, Luseni, Itogoro, and Mbuga (Figure 27.1). The Luguru are in fact the granite outcrops (the 'kopjes') and will not be considered further. The other soils are described below.

Luseni

The Luseni soils are yellowish brown, well-drained sandy soils located on the ridge crests and upper slopes (Figure 27.1). They are formed in granite and classify mostly as Eutric Regosols, petroferric phase (FAO, 1988). Most of these soils are moderate deep over a continuous ironcrust. Typically, on top of this crust a thin horizon of iron gravel is present. Water infiltration into these soils is good but they have a very low capacity to store water. They are strongly acid and infertile, with very low CEC and organic matter levels (Table 27.1). The soils, however, are easy to work. Generally the Luseni soils are used for growing cotton and maize. In the drier areas in the southern part of the district, sorghum is grown instead of maize. But even in the northern part with more rainfall maize yields are marginal (Fig. 27.1), except in old kraals where yields up to 2.4 tonnes/ha are obtained (FSRP, 1989a). Farmers, however, take the risk of growing the crop because it is highly valued (food preference).

Itogoro

The Itogoro soils are dark grey, imperfectly drained, loamy to clayey soils located on the long very gentle middle and lower slopes. Most of these soils are formed in lacustrine deposits related to former high levels of Lake Victoria. The soils are calcareous throughout and classify mostly as Calcaric Phaeozems, sodic phase. Some classify as Mollic Solonetz (FAO, 1988). Typically the surface soil is rather sandy (surface wash from the Luseni upslope). Because the more clayey subsoil is massive and rather compact these soils were coined 'hardpan soils' (Milne, 1947). Water infiltration into these soils is difficult, the massive subsoil impedes internal drainage. Soil reaction is mostly very strongly alkaline (pH above 9.0). The soils are very strongly sodic with ESP values above 25. The Itogoro soils on the mid-slope are mostly used for grazing; on the lower slope they are used for cultivating wetland rice (in bunded fields). In places they are used for cotton, which is grown on ridges or broadbeds to avoid waterlogging and to enhance rooting. Maize gives a poor yield on Itogoro, mainly because of its low tolerance to sodicity.

Mbuga

The Mbuga soils are black, poorly drained, cracking clays and heavy clays located in the wide flat valley bottoms. They are formed in colluvial and fluvio-colluvial deposits. They are calcareous and classify as Calcic Vertisols, sodic phase (FAO, 1988). These soils are difficult to work. They have a very narrow moisture range for easy tillage; when dry they are very hard, when wet they are very sticky. In the rainy season these soils are susceptible to waterlogging and ponding. Soil reaction increases with depth from mildly alkaline in the surface horizons to moderately or strongly alkaline in the

Table 27.1 Selected data of soils making up the Sukuma catena

Soil name	Horizon	Depth (cm)	Texture[a]	pH-H_2O	CEC (cmolc(+)/kg)	EC (dS/m)	ESP	Organic C (%)
Luseni	Ap	0–15	S	5.9	2.4	0.02	4	0.5
	C	15–56	LS	5.1	2.5	0.01	2	0.4
	Cc	56–75	grLS	5.2	3.1	0.01	2	0.2
Itogoro	Apk	0–10	SCL	8.4	24.2	0.14	6	1.0
	AB	10–30	CL–SC	9.3	31.9	0.57	37	1.4
	Btkn1	30–45	CL	9.3	31.5	0.64	35	1.3
	Btkn2	45–65	CL	9.4	31.1	0.61	34	1.3
	Btkn3	65–90	CL	9.2	32.2	0.81	33	—
	2Ckn	90–100	grCL	9.2	31.1	0.91	34	—
Mbuga	Ahn	0–15	CL	7.8	28.7	0.11	8	0.9
	AC	15–60	C	8.4	30.2	0.19	12	0.8
	Ckn1	60–90	C	8.7	28.9	0.24	17	—
	Ckn2	90–110	C	8.8	32.4	0.35	17	—
	Ckn3	110–140	C	9.0	31.1	0.40	20	—

[a] S = Sand; LS = loamy sand; SCL = sandy clay loam; SC = sandy clay; CL = clay loam; C = clay; gr = gravelly.

deeper subsoils. Likewise surface horizons are slightly sodic, and deeper subsoils are strongly sodic. The Mbuga soils are mostly used for grazing. There is comparatively little cropping on the Mbuga because of the adverse physical conditions.

LAND USE OPTIMISATION

The mid-slope Itogoro soils are kept bare intentionally by (over-)grazing. With the first rains these soils slake and develop a surface crust, thereby reducing infiltration so that maximum water runoff to the rice fields is achieved. This water harvesting for rice is necessary because of the relatively low and unreliable rainfall. So, degradation of the soil of one slope segment is allowed so as to maximise use of the adjacent lower segment. This kind of land use optimisation, i.e. increasing the potential of one soil area at the expense of another (nearby) one, is also known from Bukoba District (west of Lake Victoria). Here the kibanja–rweya system entails the use of farmyard manure and grasses from the rweya (grassland with marginal soils) to improve the fertility of the kibanja (FSRP, 1989b; Floor et al., 1990). Actually the well-documented plaggen system in western Europe (The Netherlands) falls in the same category of land use optimisation.

The bare slope segments (or catchment intervals) on the mid-slope are made big enough to ensure that a large amount of rainfall runs off and is collected in the rice fields downslope through openings in the bunds. In places shallow channels are made to lead the water to these openings. The catchment intervals are nowhere made too long, so that rill and gully erosion are avoided and no damage is done to the rice fields. On very long Itogoro slopes farmers have constructed two or more segments of bunded rice fields separated by water catchment intervals, thus controlling erosion. The technique as such is described as 'micro-catchment runoff irrigation' by Tauer and Humborg (1992), working in the West African Sahel. It is also known from the Negev Desert (Evenari et al., 1982).

The land use system currently in place in Maswa District is a reflection of the constant drive of the Wasukuma for food security. Cash income is less important than food security. The Sukuma farmer will try to get land in all soil-slope segments of the catena. The most ideal situation is to have a continuous strip of land from ridge crest to valley bottom (Mbuga are mostly communal lands for grazing with access for all farmers having cattle). In favourable rainfall years cotton yields will be relatively good, as well as maize and rice yields. The farmer will have a good income from the sale of cotton and the surpluses of maize and rice. He will use most of this money to buy one or more cows. Cattle are seen as an insurance against food shortage in bad rainfall years when cotton is providing insufficient income and maize and rice yields are not enough to survive. In these circumstances he will sell one or more cows to buy food. In intermediate years cotton may still produce, whereas maize and rice will be just enough to feed the family (no surpluses).

The increased importance of wetland rice in the area has greatly improved farmers' food security. Before, farmers had grown (i.e. were forced to grow) cotton (on Luseni mainly) whereas maize and sorghum were the main food crops. Maize yields, however, have never been high because of the erratic and overall low rainfall and the adverse soil conditions (low fertility, low water-holding capacity, high sodicity). Rice provided the needed diversification of cultivation as cattle and cotton (as non-food crops) had often proved to be too meagre a basis for long-term food security. Unlike cotton, rice could be used as food and as cash crop. Also, rice offered the possibility to exchange food directly (rice for maize). Another advantage of rice cultivation is that recurring investments to grow the crop are much lower than those for cotton. The latter crop can hardly be grown without pesticides and hence renders farmers more dependent on (availability and prices of) material inputs.

Contrary to cotton and maize, wetland rice cultivation offers the possibility to retain surplus water at low cost (i.e. in the bunded fields). Before, where cotton was grown on Itogoro soils, it was mostly planted along the slope to get rid quickly of excess water in case of heavy downpours, thus avoiding waterlogging. Today, in parts where cotton is still grown on Itogoro soils, cultivation is often still top-down, but now also with the aim to provide water to the rice fields located further downslope. In this way all surface water, and eroded soil (containing plant nutrients), are retained in the system. No water and soil are lost to the Mbuga. Consequently the Mbuga nowadays hardly get ponded in areas with rice fields on the bordering lower slopes and in these places more Mbuga soils are being cultivated because the risk of ponding has reduced substantially. Another effect is that drinking water reservoirs situated in the Mbuga do not get silted up as quickly as in the past when this was identified as a problem (DHV, 1975).

CONCLUSIONS

In Maswa District three kinds of land use optimisation can be recognised: (1) land use optimisation by crop diversification in order to obtain maximum food security; (2) land use optimisation by improving the potential of one soil area at the expense of another; and (3) land use optimisation by creating a near-closed system without (or with only very minor) water, soil and nutrient losses. The first element is contrary to what land use optimisation entailed in the days of cotton expansion, i.e. a move towards monocropping. The second element is in line with what is/was done in some other agricultural areas with severely constraining environmental conditions. The third element is important from the viewpoint of sustainable agriculture.

Recommendations on soil and water management (and conservation) in the area and related land use changes must be based on a thorough understanding of the land use system in place and the properties and potentials of the Sukuma catena. A good insight into the socio-economic drives of the local population (i.e. food security and the related desire to have land on all segments of the catena; social status through cattle ownership) is important, too. In Maswa District, farming systems research and soil research together provide the necessary basic information for comprehensive, community-based land use planning.

Indiscriminate implementation of erosion preventive measures will result in reduced rice yields and thus jeopardise farmers' food security. So, safeguarding the future of the land does not automatically mean that all lands need to be protected and conserved maximally. The Itogoro water catchment intervals should not be planted with trees, nor should contour farming be practised there. Instead it is recommended trees suitable for firewood be planted at field boundaries on the Luseni soils (on the ridge crests), and along the Luseni–Itogoro border where subsoils remain moist longer.

REFERENCES

Acres, B. D., 1984. Local farmers' experience of soils combined with reconnaissance soil survey for land use planning—an example from Tanzania. *Soil Survey and Land Evaluation* **4**: 77–86.

Brandström, P., 1985. The Agro-pastoral Dilemma: Underutilisation or Overexploitation of Land among the Sukuma of Tanzania. Working paper No. 8, African studies programme, University of Uppsala, Sweden.

De Pauw, E., 1984. *Soils, Physiography and Agro-ecological Zones of Tanzania. Crop Monitoring and Early Warning Systems Project.* Ministry of Agriculture and FAO, Dar es Salaam, Tanzania.

DHV, 1975. Shinyanga Regional Integrated Development Plan, Final Report. DHV Consulting Engineers. Prime Ministers Office, Tanzania, Ministry of Foreign Affairs, The Netherlands.

Ebong, C., Gijsman, A., Husson, O., Kivunge, K., Yun Li, X., Rusamsi, E. and Wongchanapai, P., 1991. Food security, livestock and sustainability of agricultural systems in Sukumaland, Tanzania. Paper presented at the final seminar, 10th International Course for Development Oriented Research in Agriculture, IAC, Wageningen.

Evenari, M., Shanan, L. and Tadmor, N., 1982. *The Negev—The Challenge of a Desert*, 2nd edn. Harvard University Press, Cambridge, MA.

FAO, 1988. *FAO–Unesco Soil Map of the World*, revised legend. World Soil Resources Reports No. 60, FAO, Rome.

Floor, J., Kimaro, D. N., Vlot, J. E. and Van Kekem, A. J., 1990. Soil fertility aspects and sustainability of the traditional farming system in Bukoba district, Tanzania. Paper presented at the 10th Annual General Meeting of the Soil Science Society of East Africa, Arusha, Tanzania.

FSRP, 1989a. Diagnostic Survey of Maswa and Meatu Districts (Phase 1: Informal Survey). Tanzania/Netherlands Farming Systems Research Project Working Paper No. 4, District Agricultural Office, Maswa. Agricultural Research Institute Ukiruguru, Mwanza, Tanzania, and Royal Tropical Institute, Amsterdam, The Netherlands.

FSRP, 1989b. Diagnostic Survey Bukoba District. Tanzania/Netherlands Farming Systems Research Project Working Paper no. 3, Agricultural Research Institute Ukiruguru, Mwanza, Tanzania.

Kurji, F., 1985. *Population and Conservation in the Serengeti-Maswa Area, Part I—The Demography*. Research Paper No. 9, Institute of Resource Assessment, Dar es Salaam, Tanzania.

Meertens, H. C. C. and Ndege, L. J., 1993. *The Rice Cropping System in Sengerema Division, Maswa District*. Tanzania/Netherlands Farming Systems Research Project Field Note No. 37, District Agricultural Office, Maswa, Agricultural Research Institute Ukiruguru, Mwanza, Tanzania, and Royal Tropical Institute, Amsterdam, The Netherlands.

Milne, G., 1936. *A Provisional Soil Map of East Africa (Kenya, Uganda, Tanganyika and Zanzibar) with Explanatory Memoir*. Amani Memoirs, East African Agricultural Research Station, Amani, Tanzania.

Milne, G., 1947. A soil reconnaissance journey through parts of Tanganyika territory, Dec. 1935–Febr. 1936. *Journal of Ecology* **35**: 192–265.

Ngailo, J. A., 1992. An investigation of physical properties and water management techniques in a semi-arid environment. MSc Thesis, Department of Soil Science, University of Reading, UK.

NSS, in prep. *Soils of Maswa District (Shinyanga Region, Tanzania) and Their Agricultural Potential*. National Soil Service, Agricultural Research Institute Mlingano, Tanga, Tanzania.

Rounce, N. V., 1949. *The Agriculture of the Cultivation Steppe of the Lake, Western and Central Provinces*. Department of Agriculture, Tanganyika Territory. Longmans, Green & Co., Cape Town, South Africa.

Shaka, J. M., 1992. Nature and patterns of soils in Maswa District, Tanzania. MSc Thesis, Department of Soil Science, University of Reading, UK.

Tauer, W. and Humborg, G., 1992. *Runoff Irrigation in the Sahel Zone; Remote Sensing and Geographic Information Systems for Determining Potential Sites*. Technical Centre for Agricultural and Rural Cooperation (CTA) ACP-EEC, Margraf Publishers, Weikersheim, Germany.

TFNAP, 1985. *The Tanzania National Agricultural Policy (Final Report)*. Task Force on National Agricultural Policy, Ministry of Agriculture and Livestock Development, Dar es Salaam, Tanzania.

Von Rotenhan, D. F., 1966. *Bodennutzung und Viehhaltung im Sukumaland, Tanzania; die Organisation des Landbewirtschaftung in afrikanischen Bauernbetrieben*. Afrika Studienstelle, IFO-Institut für Wirtschaftsforschung (München), Springer Press, Berlin, Germany.

CHAPTER 28

Selected Case Studies at Farm Level

A METHODOLOGICAL CONCEPT FOR THE DEVELOPMENT OF SOIL AND WATER CONSERVATION PACKAGES IN SMALL-SCALE FARM AREAS

H. J. Krüger[a] and U. Wiesmann[b]

[a]Soil Conservation Research Project Ethiopia, Addis Ababa, Ethiopia. [b]Group for Development and Environment, Institute of Geography, Berne, Switzerland

Different degradation processes are destroying the natural resources of African mountains. Large efforts have been made in different parts of Africa to protect soil and water, by application of traditional soil and water conservation (SWC) measures by farmers and by implementation of newly introduced SWC measures by governmental organisations and NGOs (Non-Governmental Organisations). Nevertheless, progress in soil and water conservation is slow and achievements have been few. African mountains are losing considerable quantities and qualities of soil and water resources; an alarming situation given the growing population pressure.

The approach to soil and water conservation used in the past is the subject of a critical review among concerned experts. Experts accept more and more the following elements as keys for an improved approach: participation of the community and the single farmers, low external inputs, and use of integrated soil and water conservation packages.

The term 'integrated' has three aspects:

(1) the consideration of SWC in all parts of land use (livestock management, arable farming, forestry, etc.);
(2) the development and implementation of a strategy which is in the context of the farm environment. The most important question is the use of synergetic effects by combination of different single soil conservation measures;
(3) the newly added soil and water conservation elements should fit into the existing farming systems and should initiate development processes.

The Future of the Land: Mobilising and Integrating Knowledge for Land Use Options
Edited by L. O. Fresco, L. Stroosnijder, J. Bouma and H. van Keulen.

A methodological concept for the design and participatory implementation of integrated soil and water conservation packages considers socio-cultural and technical aspects. The 10 steps of the procedure are as follows:

(1) Survey on key natural resources: preliminary data-collection on key natural resources, e.g. information on soils, topography, hydrology, vegetation cover and land use, in view of an assessment of degradation status and dynamics. The survey includes methodological elements of Participatory Rural Appraisal.
(2) Evaluation of social and economic potentials and limitations for SWC. Through an informal survey which includes methodological elements of Participatory Rural Appraisal, social and economic potentials and limitations for soil and water conservation are evaluated. This evaluation addresses potentials and limitations on four levels: (a) plot and farming system; (b) overall household strategies and decision-making; (c) community organisation and development; (d) relation to the overall economic and political development.
(3) Formulation of demands towards soil and water conservation methods. On the basis of the evaluation of the social and economic potentials and limitations for SWC (step 2) demands are formulated towards the development and testing of measures. These demands mainly include aspects which are thought to increase the social and economic acceptance of measures, e.g. the importance of short-term productive effects of the measures.
(4) Determination of priorities for soil and water conservation. Priorities in location, timing and approaches for soil and water conservation measures are determined on the basis of demands (step 3) and on the assessments of erosion hazards and land use potentials.
(5) Selection of soil and water conservation measures. Soil and water conservation measures for testing and implementation are selected on the basis of the priorities (step 4) and through considering aspects of acceptance within the economic and social system (steps 2 and 3).
(6) Estimation of conservation effects of the measures. Only the selected soil and water conservation measures which are fitting to the priorities and which are thought to attain high acceptance are tested and modelled in view of their soil loss and runoff-reducing effects.
(7) Detailed social and economic studies on specific aspects. Through close scrutiny of the technical and biophysical aspects and collaboration with the local community, problems and tasks are identified which require in-depth evaluation, e.g. labour availability, cropping calendar, community organisation in view of maintenance, etc.
(8) Supporting soil and water conservation measures. Additional measures which support and enhance the effects of the core measures (steps 4 to 6) are selected, designed and tested as components of an integrated package. These measures might not have a high acceptance rate, but through treating them as additional and supporting measures, aspects of non-acceptance can be kept to a minimal level.
(9) Design of points of breakage within the conservation system. Construction of perfect conservation systems is impossible due to limited resources. Therefore, to avoid erosion damage on productive land, breakages of conservation structures are

designed on spots with low productive importance or low importance for the community.

(10) Implementation concept and participatory implementation. The implementation of the suggested specific soil and water conservation package (steps 4 to 9) is planned in a participatory way and the implementation is enhanced through assistance to the organisation of the target groups (single farm households and/or local community). As all measures will be sub-optimal (mainly due to step 2 and 3), the measures are continuously further developed and adapted in the sense of conservation-based rural development. This implies that during implementation one will draw back on steps 3 to 9 (iterative process).

The above-mentioned procedure uses experience from a 10 year soil and water conservation research programme in Ethiopia and Kenya. Collected data and information were transformed to key figures and functions, allowing the estimation of important elements for the assessment of the existing local erosion potential and the determination of the correct objectives for soil and water conservation measures. The presented procedure is an attempt to close the gap between existing knowledge about designing and implementation of single soil conservation measures and missing experiences in the designing and implementing of integrated soil conservation packages suitable for the African context. The presented procedure will be subject to revisions and improvements after tests in trials and pilot projects.

THE EFFECTS OF TILLAGE AND IRRIGATION PRACTICES ON SOILS IN GHANA: A CASE STUDY

N. Kyei-Baffour, E. Y. H. Bobobee and S. A. Koram
University of Science and Technology, Kumasi, Ghana

Market gardening to supply adequate amounts of vegetables to the Accra-Tema region has been the main objective of the Weija Irrigation Project, managed by the Weija Irrigation Company (WEICO). This objective has not been achieved because of factors such as poor water management, waterlogging and the absence of a reliable drainage system. Preliminary results are available for salinity and compaction measurements of the soils monitored on three different plots: uncultivated plots (U); developed and cultivated plots but abandoned for the past three years (DU); and developed and continuously cultivated plots (DC). The textures of the three plots are loamy sand, sandy clay loam, and sandy loam for U, DU and DC, respectively. Bulk densities for all the test plots were in the range 1450–1870 kg/m^3. Higher bulk densities and soil resistances occurred on DC at depths of 20–30 cm, suggesting compaction at plough sole depth.

Electrical conductivity of saturation extracts (EC), pH, and exchangeable sodium percentage (ESP) suggest salt build-up in the soils, which may be partly responsible for the dwindling crop yields. Further studies into the pH, EC, and ESP at different soil depths, and in the irrigation water are continuing. Proper irrigation water management, incorporation of organic matter into the soil, drainage by a series of beddings, subsoiling, and the reduction of vehicular traffic have been suggested as short-term solutions to ameliorate production problems and to improve workability.

THE EFFECT OF THE COMBINATION OF CROPPING PATTERNS WITH GRASS ON FARM LEVEL IN WEST JAVA, INDONESIA

Nurpilihan
Padjadjaran University, Jatinangor, West Java, Indonesia

As a consequence of the Indonesian farmer's limited land resources, he generally does not plant strips of grass with his main crops when he cultivates sloping land. However, planting grass strips on steep areas does have numerous advantages: it reduces erosion and runoff and it could also be used as a source of animal feed.

In West Java, 30–40% of the land has sloping and rolling surfaces with a 0–8% slope and the rest of the land has a larger than 8% slope. The top soil on cultivated land is being increasingly ruined by soil erosion. The combination of cropping patterns with grass is an alternative cropping pattern on the farmer's land in West Java, and could affect the percentage of canopy coverage and lessen the effect of the impact of raindrops on erosion and runoff.

The purpose of the experiment here described was to study the effect of combinations of cropping patterns with and without grass strips on soil erosion, runoff and land productivity. The experimental design employed was a Randomised Complete Block Design with two factors: cropping patterns with/without grass strip (TK) and degree of slope (S). The levels of treatment of TK were as follows:

- TK-1: a single cropping pattern (corn) with a grass strip
- TK-2: a multiple cropping pattern (corn, soybean and upland rice) with a grass strip
- TK-3: a multiple cropping pattern (corn, peanut and upland rice) with a grass strip
- TK-4: a single cropping pattern (corn) without a grass strip
- TK-5: a multiple cropping pattern (corn, soybean and upland rice) with a grass strip
- TK-6: a multiple cropping pattern (corn, peanut and upland rice) without a grass strip

The degree of slope factor consisted of three levels, respectively: S-1, 8%; S-2, 16%; and S-3, 24%. The results showed that the amounts of erosion were 11.8, 22.5 and

28.0 ton/ha on slopes of 8, 16 and 24% respectively, cultivated with corn only. It could be reduced to 6.4, 7 and 15.1 ton/ha on land with slopes of 8, 16, 24% respectively, by using a multiple cropping pattern consisting of corn, peanut and upland rice with a grass strip.

DIGITAL TERRAIN MODELLING FOR THE SITING OF A CENTRE PIVOT IRRIGATION SYSTEM IN SUGAR CANE

I. Jhoty, M. Chung, D. Ah Koon, J. Deville and C. Ricaud
Mauritius Sugar Industry Research Institute, Réduit, Mauritius

A study using digital terrain modelling was undertaken to select the most appropriate site for a center pivot irrigation system, for which minimum land levelling was required. This irrigation system has a total arm length of 335 m to irrigate 35 ha and the ground clearance under the truss is 3.8 m.

The PCARC/INFO PCTIN module for digital terrain analysis was used to evaluate terrain limitations, and situate the centre pivot system. Field boundaries and 2 m contours from 1:2500 maps were digitised separately, and slope maps derived. Slope categories were regrouped to give a better representation of the topography, and sites having depressions or strong ground height differences were pinpointed.

The locus of the pivot centre was defined by fitting the outmost perimeter of the overall radius of 335 m in the delimited project area. The best site was then identified whereby minimum land levelling was required. Analysis of four profiles was performed along radii across the critical areas to reveal the ground height differences that would be encountered by the pivot arms, and by the wheels along defined spanlengths.

The terrain analysis performed, highlights the various constraints of the topography and enables the siting of the centre pivot within the given constraints. The main results of the analysis included the identification of the best site for the pivot, which satisfies the boundary limits whilst avoiding the worst slope of the region. The profile highlighted the areas where land levelling analysis was required, so as to reduce the pressure head fluctuations along the trusses, allowing designers to optimise water application efficiency. Also, the feasibility of the cut and fill had to be weighed in the investment cost.

The results demonstrate the usefulness of DTM (Digital Terrain Modelling) in determining the appropriate site, enabling the modification of truss spans according to terrain constraints for minimum land levelling and investments and optimum performance. Similar questions on selecting best sites may be resolved by this method.

PART VI

THE FUTURE OF THE LAND: LESSONS AND CHALLENGES

CHAPTER 29

Can Land Use Planning Contribute to Sustainability?

R. J. Olembo

United Nations Environment Programme, Nairobi, Kenya

LAND: A RESOURCE UNDER PRESSURE

Land as an entity is generally finite, while the natural resources it supports can vary over time and according to management conditions and uses. As each nation moves to the next level of economic intensity, human requirements and economic activities place ever-increasing pressures on land resources, creating competition and conflicts and resulting in non-optimal use of both land and land resources. Intelligent and sensitive planning of the use of land resources can be an appropriate early tool. By examining all uses of land in an integrated manner, it makes it possible to build consensus and minimise conflicts through informed negotiations among competing uses, to make the most efficient trade-offs and to link social and economic development with environmental protection and enhancement, thus helping to achieve the objectives of sustainable development. The essence of the integrated approach finds expression in the coordination of the sectoral planning and management of activities concerned with the various aspects of the use of land resources.

If planning is approached flexibly, it can provide a broad framework for taking decisions in a forward looking and enlightened manner. Flexibility in land use planning can be guaranteed if the agreed plan: (1) is able to tolerate unexpected disturbances or changes in circumstances, and (2) is able to produce new or amended approaches quickly when necessary. Although flexibility is often thought to be a good thing, it has its own risks. For example, an over-flexible plan endangers consistent and powerful development. If all *ad hoc* ideas can be realised easily, it means that earlier plans (often based upon thoughtful goal setting and extensive basic studies) may be too easily rejected. One advantage of strict and purposeful land use planning has been to show citizens the future

The Future of the Land: Mobilising and Integrating Knowledge for Land Use Options
Edited by L. O. Fresco, L. Stroosnijder, J. Bouma and H. van Keulen.

land use of their natural environment and, thus, make them feel more secure and involved in directing their destiny.

In order to provide a basic framework for understanding the ramifications of sustainable land use planning, concepts should be defined and clarified; the driving forces behind changes identified and the impacts of proposed land use policies made patently clear to all stakeholders.

UNCERTAINTIES IN RESOURCE STATUS: EXTERNALLY INDUCED BIOPHYSICAL STRESSES

Gradual loss of topsoil is one of the major threats to human well-being, but it is not readily apparent, at least not in its early states. To sustain development, its demands must be within the support capabilities of the land, and the environment in general, otherwise degradation will take place.

The world's demand for food, fibre and fuel production is ever increasing and is likely to increase strongly until at least the year 2050 when the world's population is expected to be at its peak. To meet this demand, vast tracts of land are now being farmed more intensively, and new and often marginal land is being brought into production. The result is widespread soil loss both in developed and developing countries through soil erosion, salinisation, sodification, toxicity, fertility depletion and waterlogging.

Whether soil loss is due to erosion, chemical pollution, poor drainage or urban development, it reflects inadequate land use policies and ineffective implementation of soil management and conservation programmes. Continued mismanagement of this most basic resource constitutes a threat for many developing countries (and indeed for the entire world), the magnitude of which is often underestimated. There is a strong need for improved data on soils and terrain resources and for the development of an information system that can deliver accurate, useful and timely information on soils and terrain resources to decision-makers and policy-makers. This will make it possible to focus attention to critical areas so as to channel scarce funds to those areas where they can be used more effectively.

CAUSES AND PROCESSES OF LAND DEGRADATION

While the causes of land degradation are diverse and often complex, there are some clear trends. Three distinct periods of soil degradation are recognised:

(1) early civilisations occurring in the ancient past and up to 250 years ago;
(2) an era of European expansion in the Americas, Australia, Asia and Africa;
(3) the post Second World War period, very much related to human and animal population explosions, particularly taking place in the developing countries.

By 1930 the threat of land degradation was clear everywhere and many governments initiated extensive soil conservation programmes. No region is free from natural disasters. However, some environments are less stable than others or are more likely to suffer disruptions. Areas predisposed to disaster include those with: steep slopes, exposed coasts, proneness to soil salinisation, intense or erratic rainfall, or hurricanes. Increasing demographic pressure results in overuse of high potential land and/or the misuse of marginal, often easily degraded land. The impact is double-edged: an increase in demand made upon the environment and the destruction of the resource base. What really matters is the rate of change in population in critical regions, resulting in exceeding the limits of carrying capacity.

It is not the mere exploitation of marginal land that automatically leads to land degradation. Factors such as fluctuations in rainfall, availability of inputs (including labour), outbreaks of pests or diseases, and law and order make land more or less marginal and bring the risk of degradation. Exploitation perpetuates poverty and poverty creates environmental degradation, leading to increased poverty. The toll on natural resources takes many forms, including soil erosion, loss of soil fertility, desertification, deforestation. Short-term land tenure will induce exploitation of land resources, and landlessness will push the poor towards more marginal land for their livelihood.

The biggest challenge to sustainable development is political vicious circles. The more meagre the resources, the more important becomes good governance and cooperation. Specific measures need to be taken to eliminate and prevent the occurrence of corruptive practices. Many developing countries now spend much of their revenue servicing foreign debts and consequently are less able to afford necessary land management inputs. The result is that the modernised sector of agriculture, which has taken so much of investment in recent years, breaks down, soil conservation is underfunded, and the land is ruthlessly exploited to try and generate desperately needed foreign exchange.

There has been a tendency for Western attitudes and technology to replace established, often successful, Third World production practices—sometimes with catastrophic results. The increased prices of various oil-based imported inputs drastically altered the costs of inputs. Externally dictated demands for cash crops also have led to many *ad hoc* decisions on land use. Human diseases like malaria, bilharzia, and recently AIDS not only kill, but rob countries of billions of working hours per year. Such outbreaks of diseases cause land to be neglected. Where seasonal poverty, malnutrition and seasonal disease outbreaks are widespread, a feedback spiral may begin.

THE CONCEPT OF SUSTAINABILITY

The World Commission on Environment and Development (WCED) defined 'sustainable development' as 'development that meets the needs of the present without compromising the ability of future generations to meet their own needs'. The term has been criticised as ambiguous and open to a wide range of interpretations, many of which are contradictory. The confusion has been caused because 'sustainable development', 'sustainable growth' and 'sustainable use' have been used interchangeably, as if their meanings were the same. They are not. 'Sustainable growth' is a contradiction in terms: nothing physical can grow indefinitely. 'Sustainable use' is applicable only to renewable resources: it means using them at rates within their capacity for renewal. 'Sustainable development' used in the Strategy of Caring for the Earth means: improving the quality of human life while living within the carrying capacity of supporting ecosystems. While the theme sustainability permeated the United Nations Conference on Environment and

Development (UNCED) and forms the backdrop of Agenda-21, the UNCED did not attempt a new or further definitional refinement. Rather, it bestowed political respectability to the notion of sustainable development.

The concept of sustainability is currently applied to land use systems of divergent geographical scales, from individual fields or farms to regions, countries or even the world as a whole. However, at each scale different processes operate, each at its own rate. In the same way, the distinction between renewable and non-renewable natural resources implies a specific time scale. It is argued that ecological sustainability can be adequately defined only with reference to specific spatial and time scales. Fresco and Kroonenberg (1992), discussing the subject of time and spatial scales in relation to ecological sustainability, review the specific points in the processes at which sustainability is most affected in time and space. They identify these as: disturbance of cycles, damage to system components (e.g. topsoil) and complete destruction of the system. Acceptability of damage or disturbance depends on the time and spatial scales considered; for example, deterioration of land qualities may be acceptable in small patches throughout the landscape or when they occur at a slow pace.

For each individual agro-ecosystem, whatever its scale, we might define internal processes, acting within its spatial and temporal scale, and external processes, acting from outside the system. Productivity, efficiency and stability of an agro-ecosystem are functions of its internal dynamics, but resilience is defined with reference to the external processes affecting the system. Therefore, the sustainability of the agro-ecosystem cannot be determined without referring to processes that act at greater time scales than the internal processes.

Sustainability must be translated into a set of boundary conditions defining acceptable levels of outflows of matter and energy from the agro-ecosystem. In other words, sustainability nearly always means limits placed on production potential, whether spatial (decisions on limiting the area and land types taken into production) or temporal (limiting annual outputs). Acceptability implies that an arbitrary limit is set by a group of decision-makers, whether farmers or governments. Here, time and spatial scales interfere as well: what is acceptable to farmers in view of their objectives that do not extend beyond the farm and the next generation, may be not be acceptable to an entire nation. The concept of sustainability can be used in two ways: to evaluate existing systems and to design new land use systems.

If we have to translate sustainability in practical terms, we have also to keep the following tenets upfront: (i) social, cultural, economic and other societal needs must be harmonised and satisfied in an equitable, efficient and environmentally sound manner; (ii) planning should help in doing this by deliberately focusing on the way in which decisions are taken; (iii) broad guiding principles including a broad concept of environment; respecting the closed nature of the world system with finite resources, where it is imperative to live off the 'income' of natural/environmental resources and not the 'capital' assets; respecting the three main ecological principles for sustainability as enunciated in the WCS, namely:

- maintenance of essential ecological processes and life-support systems,
- preservation of biological diversity, and
- sustainable use (of species).

Integrating human development and environmental conservation requires:

- institutions capable of an integrated, forward-looking cross-sectoral approach to decision-making and strategic thinking;
- effective policies and comprehensive frameworks that guarantee the productivity and diversity of the living systems into the future;
- the promotion of technology which not only increases the benefits from a given stock of resources but maintains or even enhances it;
- in-built systems for monitoring and assessing (M&A) of progress, and mechanisms for midstream adjustments and corrections. To be effective, such M&A systems need to be linked closely to the perceived information needs of the people and institutions that determine the course of development. Managers and policy-makers need early warning indicators of potential problems that could lead to unsustainability, and they need them in time for adjustments or corrective action.

DEALING WITH SUSTAINABLE DEVELOPMENT IN PRACTICE

The concept of sustainable development provides little guidance in terms of creating effective development policies and programmes. A useful operational goal may be to avoid non-sustainable development, focusing on what can be done to avoid potentially non-sustainable developments. This is something which can be dealt with in a concrete and practical way by interdisciplinary teams working at the project and programme levels and by decision-makers setting the directions for development at the policy level.

Monitoring and assessment activities are critical in the project development and policy processes associated with avoiding non-sustainable development. Managers and policy-makers need early warning indicators of potential problems that could lead to non-sustainability; and they need them in time to take corrective action. Key indicators can be monitored and assessed to provide information needed by project managers and by policy-makers in setting the directions for future development.

Projects are basic building blocks of development, but they are only a small part of the total human activity affecting the sustainability of development. Fundamental policy changes also are needed to encourage people to think about sustainability issues as they go about their everyday activities quite outside a project context. Thus, we need to consider sustainability issues both in terms of project interventions and in terms of policy interventions. The distinction between project- and policy-level considerations also relates to the need to deal with sustainability at different scales of human and ecological activity. Sustainability is of concern at all levels, and there are strong interactions between the different levels of concern. An aggregation of local-level actions can lead to broader and eventually global effects. Global level concerns must lead to policy changes.

M&A systems need to be comprehensive enough to identify warning signs at different levels of aggregation. If M&A is highly disaggregated, it can miss identifying critical linkages and the changes which are occurring in them. Thus, in dealing with sustainability issues, it is important to deal both at the project and the macro or policy levels, and to do so in a coordinated fashion.

Dealing with externalities

Externalities are effects of an action (project or policy) that are outside the decision context or concern of those taking the action. These effects may be known but disregarded by the project planners or other authorities, or they may be unknown, unexpected and unintended effects. For example, in watershed management projects, externalities often relate to the upstream–downstream relationships: land and water uses upstream affect those who live downstream, e.g. changes in quantity and timing of water flows. Thus, project planners have to be aware that what may be a move toward more sustainable development in the uplands may turn out to contribute to non-sustainable development downstream. As another example, the building of a road in a project area to provide access to a multi-purpose dam may result in an unplanned influx of new settlers on unplanned watersheds, which in turn may lead to increased downstream problems. It is evident that negative externalities can lead to problems of non-sustainability for: (1) people outside the project area (spatial externalities); (2) people living in some future period of time beyond the project life (temporal externalities); and (3) sectors of the economy or individuals outside the context, or defined jurisdiction boundaries, of the project.

Carrying capacity

Many people have defined sustainability in terms of carrying capacity, a concept long used to describe the maximum population size that the environment can support on a continuing basis. The concept of carrying capacity was developed in the field of population biology, and can be transferred to human systems only by analogy. Human carrying capacity has been defined simply as the 'number of people that a given amount of land can support' (WRI/IIED, 1986). Nevertheless, Watt's (1977) reference to an optimum, sustainable quality of life 'within the carrying capacity set by milieu of local, regional, and even international resources' suggests the complexity of this concept when applied to human systems.

Odum (1983) attempted to clarify the meaning of the concept by distinguishing between maximum and optimal carrying capacity. He defined maximum carrying capacity as the maximum allowable population size that, while theoretically sustainable, exists at the threshold and is vulnerable to even small changes in the environment. Optimal carrying capacity, on the other hand, is a smaller, more desirable population size that is less vulnerable to environmental perturbations. Ophuls (1977) writes that a sustainable level of human demand on the environment is 'perhaps as little as half' the maximum carrying capacity.

It is important to recognise that the carrying capacity of any given region is subject to change. It may be increased through investment of capital and technology, or through imports of energy and materials from outside the region. Studies of national carrying capacities are often flawed because they do not consider the exchanges between countries. Urban-industrial zones, in particular, depend on much larger land areas for their maintenance.

The carrying capacity of an area may decrease as a result of declines in the biological productivity of the land. The *Global 2000 Report* describes the feedback loop that leads to continuous declines in human carrying capacity through complex interactions of social and economic factors. On a global scale, human carrying capacity is finite. Although much recent discussion has focused on whether the global human population has exceeded its limit, it is generally agreed that the available data are not comprehensive or accurate enough to precisely calculate the number of people the planet could support. Attempts to blame environmental degradation on overpopulation grossly oversimplify a much more complex problem.

Land productivity and population supporting capacity: getting closer to societal needs

FAO (together with IIASA) has been working on the subject of population supporting capacity estimates since the early 1970s (FAO/UNFPA/IIASA, 1982). The methodology is simply based on an assessment of the land productivity which implies identification of crop, livestock and fuelwood land utilisation types (LUT), and their ecological (climatic, edaphic and landform) requirements. Then, from the agro-ecological cells in the land resources inventory, district by district, land used or required for irrigation, cash crops and for non-agricultural purposes is deducted. The remainder is an inventory of land potentially available for rainfed cultivation, and for productivity assessments.

The next stage in the assessment process allows for development planning applications, which involves the calculation of the quantities of edible calories and protein that could be produced by the different crops, livestock, and products from other land uses, using information on the nutritional composition of the products. The crops, including grassland, that can produce the largest or desired quantity and quality of calories and protein in each agro-ecological cell are then selected, and the results from each cell in each climatic zone in each district are added to determine the maximum potential production. The potential population-supporting capacity is computed as potential population density in persons per hectare, which is then compared with the present and projected population densities.

CONCLUSION

Tropical deforestation and degradation, resulting from increasing population pressures, have fostered the current popularity of the concept of sustainability. However, in practice we need adequate tools for planning sustainable development, such as assessments of carrying capacities or human population supporting capacities. However, there is growing scepticism of development planning, particularly after the collapse of the centrally planned economic system and the widespread acceptance of the market mechanism as a tool for economic development. Much of the debate is a result of misunderstanding of the realities of each system. Central planning has failed, more than anything else, because of a lack of flexibility and because it has relied too heavily on command and control methods, rather than harnessing individual contributions. True, market mechanisms harness this individual energy and self interest to produce goods and services, but often cannot address the common good or long-term perspectives in relation to the environment. Existing examples of successful planning approaches include national socio-economic development planning (e.g. Singapore) or industrial and technological planning (e.g. Japan) or planning of product development and marketing strategies in particular industries (e.g. IBM). But for many poor developing countries, the messages are mixed. The World Bank Structural Adjustment Programmes, with their many demands for policy reforms, including stressing privatisation, raise questions for many planners in the central bureaucracies in these countries. However, it is the very state of the soil and the current pressures on it which prompt us to take all measures to ensure that this essential human survival resource is managed and utilised in ecologically sensitive and sustainable ways.

Several agencies, including UNEP, are known for their attempts to promote the idea that environment should not be separated from development, that both environmental/land resources and economic aspects should be integrated in development planning. The agricultural perspective is too narrow to be able to address such an encompassing subject as the carrying capacity for humans on the earth.

REFERENCES

FAO/UNFPA/IIASA, 1982. *Potential Population Supporting Capacities of Lands in the Developing World*. Technical Report of Project Int/75/P13, FAO, Rome.

Fresco, L. O. and Kroonenberg, S. B., 1992. Time and spatial scales in ecological sustainability. *Land Use Policy* **9**: 155–168.

Odum, E. P., 1983. *Basic Ecology*. Saunders Publishing, New York.

Ophuls, W., 1977. *Ecology and the Politics of Scarcity*. W. M. Freeman, San Fransisco, CA.

Watt, K., 1977. *The Unsteady State*. University Press of Hawaii, Honolulu, HI.

WRI/IIED (World Resources Institute and the International Institute for Environment and Development), 1986. *World Resources 1986*. Basic Books, New York, USA.

CHAPTER 30

The Challenges of the Future of the Land

C. H. Bonte Friedheim[a] and A. H. Kassam[b]

[a]ISNAR, The Hague, The Netherlands. [b]The Secretariat of the Technical Advisory Committee to the CGIAR, Research and Technology Division, FAO, Rome, Italy

LAND, OUR MAIN RESOURCE

Most people live on the land, from the land and with the land. Land has always been:

- a source of minerals, fertilisers and non-renewable energy
- an influence on climate
- a source of fresh water
- a source of biodiversity
- a source of food, fibre, agricultural and forestry products, and renewable energy
- a biological factory for recycling waste
- a living and working space for man, meeting also recreational and infrastructural needs and requirements

While land is important as a resource, it is not necessarily the same land that is required for all purposes. Basically there is biologically fertile or potentially fertile land, as well as land which cannot be used at all, or only seasonally or partially, in any biological process. Light, temperature, energy, available water, and natural or induced soil fertility are the determining factors. It seems necessary to accept or to establish an interrelationship between the different uses of land. Even more important, we must determine the present and future minimum requirements of land quantity and quality for a growing number of people and for a sustainable life as we know it. Such investigations are required on global, regional, and national levels.

There should be some principal guidelines and considerations for addressing the different challenges of the land and setting policies for land use. For example:

- negative effects of man's interference with land should be, as much as possible, reversible;

The Future of the Land: Mobilising and Integrating Knowledge for Land Use Options
Edited by L. O. Fresco, L. Stroosnijder, J. Bouma and H. van Keulen.

- when land can be used for more than one purpose, protection of the land for the most important long-term use must be attempted;
- in future, increasingly global values are likely to take precedence over regional, and regional over national values and uses;
- future generations' needs for land must be considered.

These guidelines can help to determine practical policies for land and for its present and future use. A precondition for sound guidelines and policies is the collection, analysis and use of required data. In many countries and for many national, regional and global purposes, necessary data and time series are still lacking, and estimates and guestimates are used instead of facts. Modern information technologies must be applied in dealing with the challenges of the future of the land. It seems necessary to briefly describe the different uses of land.

Land as a source of minerals, fertilisers and non-renewable energy

Land contains much of mankind's inherited wealth. Mining these resources does not need to decrease the biological potential of land in either quantity or quality. On the contrary, with the right practices, land's biological potential can be improved through mining. Therefore, mining policies and regulations should not only consider exploitation of non-renewable resources but should emphasise the sustainability of the land and its overall usefulness during and after mining.

Land as an influence on climate

Any influence of land on climate is the result of a biological process. Man, as well as animals and plants, lives on the land and produces carbon dioxide, oxygen, methane and other gases. Different land uses produce different gases of different quantities, which not only influence microclimatic conditions but can also affect the global climate and induce changes. Large-scale and very specific land use plans must take such possibilities into account. Recently recorded climatic changes must be studied in order to determine causes and effects, their magnitude and likely future trends. The warming tendencies of the global climate might result from, among other things, land use changes. In turn, these climatic changes will affect new land use practices in large areas.

Land as a source for fresh water

The need for fresh water poses possibly the greatest challenge of all to land and to man. It must be assumed that fresh water will be the limiting factor for life on land for some time to come. In many areas and regions, the quantity of available fresh water and its distribution are directly related to land use in the region and upstream, as well as to the land's potential for sustaining human life and for agricultural production. The capture and storage of fresh water is not yet considered an economic activity. Water is widely considered a free public good. Consumers are generally charged with distribution costs, but not with any production or replacement costs. With ever increasing demand and a limited or fast decreasing water supply potential in many areas, this attitude and its related policies have to change and change soon. Otherwise the collection and supply potential of water will become very limited and even irreversible. In a growing number of countries, the need for sufficient supplies of fresh water and desalinated sea-water requires a new approach. Policies, traditional practices, and real costs and benefits of water supplies for different users must be reviewed, in line with changing demand and supply.

Land as a source of biodiversity

Cultivated land and other land used by man for his different activities, has lost much of its original biodiversity. The quantity and quality of the globe's biodiversity is unknown and its value for the future cannot be estimated. Sufficient land must be set aside in different agro-ecological zones to be used as banks of biodiversity. Even with such banks, the earth will lose much of its biodiversity. Mankind will have to pay for such land and for biodiversity reserves; these cannot be charged to the countries where they are established and where they must be protected. Similarly, biodiversity is to be regarded as mankind's and not as an individual nation's heritage. In view of the past decreases and losses of undisturbed land reserves of high biodiversity value, global action is urgently needed.

The protection of natural forests is an integral part of any programme to safeguard biodiversity. Land under forests in certain environments and landscapes also serves the various tasks of our land better than any other form of land use, with the exception of the production of most agricultural commodities.

Land as a source of food, fibre, agricultural and forestry products, and renewable energy

The loss of productive land through man's farming and other activities should not be underrated. Desertification, the salinisation of irrigated areas, and the deforestation of large areas has critically affected much land, made it largely unsuitable for long-term agricultural production. These are all well-known facts and contribute to the general estimates that about 10% of our land has been degraded and has become unproductive. In addition, man-made or man-instigated soil erosion has increased in most regions of the world, decreasing natural soil fertility, and limiting sustainable production and the possibilities of productivity gains.

We must consider multi-purpose uses of land. It is possible to think of foregoing some of land's production potential and future productivity gains in order to accommodate other purposes as well. Agricultural land and forests can be regarded as contributing factors in the globe's climate and climatic changes, as a source of its biodiversity, and as providing necessary fresh water. Costs and benefits to society of different land uses can be calculated separately and different land uses can be paid for differently, if such a notion of multiple purposes and uses is accepted.

The search for fuelwood and other sources of biomass energy destroys the agricultural potential of the land, especially in many dry areas. Food security and food self-sufficiency are widely accepted political goals. Energy security and energy self-sufficiency in rural areas are closely linked to food security, especially for the poor. However, few energy plans and programmes exist, especially for the rural areas. Traditionally, renewable energy from biomass is considered a free good, easily consumed, but seldom renewed to meet the growing demand.

Land as a biological factory for recycling waste

Without land to recycle large quantities of man's waste in a biological process, most industrialised countries could not survive for long periods of time and would encounter serious air and water pollution problems. Yet land as a biological recycling factory is regarded as a public and free good. More investigations are necessary to determine the costs and benefits to society for using land to recycle waste.

Land as living and working space for man, meeting also infrastructure and recreational requirements

In earlier times—and in large areas even today—farmers lived on their land, selecting their homesteads to meet their needs. But as much as possible they avoided diverting their most productive and fertile land to other purposes. In recent times roads were used which were inexpensive to build and to maintain. Cities grew as centres of activities in agricultural areas or around centres of communication. A growing urban and non-agricultural population with rising incomes increasingly requires space for recreational purposes.

The biological long-term value of the land was seldom taken into account when selecting land for non-production purposes in a biological sense. These historic trends are still in evidence today, despite general knowledge of the more important purposes and uses of high-potential agricultural land and in spite of the availability of low-potential agricultural land or land with no productive use in a biological process. Only present land costs are calculated, not the replacement costs of agricultural land. Individual and short-term benefits determine choices. Future needs, common or public necessities and requirements are seldom considered. The low costs of most agricultural land for other uses can no longer be accepted. Today's agricultural land is subsidised heavily by future generations and their needs for land as the basis for biological processes. Strict land use regulations are required and must be observed and enforced, if food and agricultural production and land's other tasks are to be possible in decades and centuries to come.

THE CHALLENGES

World agriculture so far has met the challenges of a growing population and it will also fulfill its future role in further productivity increases and in the necessary protection of its natural resource base and of the environment. In recognising the achievements of scientists we also want to pay tribute to farmers' knowledge and scientific work, to the breeding of plants and animals, to the improvement of husbandry, and to the large-scale interdisciplinary research between agriculturists and water engineers.

Four sets of challenges are posed by: (1) the future size of the effective demand; (2) the resources needed for the production of the necessary biological output; (3) the need to strengthen the knowledge and technology base for the sustainable development of land; (4) the enabling cultural, institutional and political environment needed for land-based sustainable development. We have looked at the challenges from a global perspective, but we now concentrate on the future of the land in the developing world, recognising that there are considerable differences between and within continents, regions and countries. These differences can be found on the resource side, on the supply side and on the demand side.

The future size of the effective demand

Much has been said and written about the future size of the likely effective demand for agricultural commodities. Historically, the growth in global agriculture, especially food production, has kept pace with population growth. This is also due to the fact that per head of the population, industry is using much less agricultural raw material than earlier in this century. During industrialisation, many agricultural products have been replaced by other raw materials for a growing industry and for a fast rising demand.

An end to rapid growth in population in the developing countries is in sight and a stable global population of some 12 billion people can be expected by the end of the next century, of which some 10 billion will be living in the developing world. Population growth reflects above all the rapid decrease in death rates that most countries have experienced over the last 50 years. Fewer children die and most people have a longer life span than their grandparents, getting closer to the biological limit of human life.

All efforts to decrease population growth rates faster than previously expected will relieve the growing pressure on the land and allow more and more rapid economic growth. In most developing countries children are economically and socially advantageous; they guarantee the survival of their elders. Without them and their responsibilities, other means of support, like social security schemes, must be found. Based on our present knowledge and a common, global political will to overcome hunger, malnutrition and poverty, food production can meet the effective demand of generations to come.

Besides population size, existing poverty and the effective demand for agricultural products will be the major determinants of total demand. The effective demand stems largely from the non-agricultural population and creates and absorbs surpluses in excess of the agricultural, rural subsistence requirements. If the non-agricultural population is not only small but also poor and often provided with external food aid, then national production will not rise. Rising effective demand in the non-agricultural sector will raise output, provide income to agriculture, and overcome rural poverty and the unsustainable farming practices that go with it. A large effective demand is also a sign of an efficient delivery system and infrastructure, which is a precondition for agricultural and rural development.

The resources needed for the production of the necessary biological output

The second challenge deals with land and non-land resources necessary for the required global production level. There is no doubt that additional resources of land, water, nutrients, energy and capital must be found and applied, in addition to improved genotypes of crops and animals, in order to meet the supply challenge. The inputs must be combined in more efficient production systems and must result in production and productivity increases. In the last 100 years the production gains were considerable. Total factor-productivity increases within agriculture were beyond expectations and considerably improved previously very low production levels in most areas. Further increases in productivity in all areas can and must be expected.

With respect to cereal requirements for the future world population, Africa is most vulnerable, but it has the potential to be self-supporting, in contrast to the Near East. Only in the Near East region is the total present or potentially available cultivable land area insufficient to produce sufficient food for its population. In addition to some 2.5 billion hectares of potentially available suitable cultivable land in the developing world, there are over 700 million hectares of rainfed marginal cropland, some of which could be upgraded. However, potentially suitable land can only be used if it is accessible to people. Trans-border migration has come to an end during this century. Therefore, the first choice for each country must be to increase factor productivity, especially land productivity. At the same time the demand for land from non-agricultural interests is growing and national land use policies must maximise the long-term benefits and potentials of land.

Studies have shown that with the right levels of inputs, the land resources of the developing countries can support a population in excess of what is currently assumed as the future maximum level. Again there will be serious regional differences. In some countries, population growth will exceed, or has already exceeded, the potential supporting capacity of the available national land resources, and transfers will have to be made, as they are now, between countries and between regions, but on a larger scale.

The supporting capacity of land will depend on nutrients, energy, and water, as well as on productive plants and animals and appropriate farming systems. There is a need for increased levels of inputs, if sufficient food is to be produced for a larger world population. In most areas the minimum factor will clearly be water. The first approach must be to increase water-use efficiency and reduce waste. Shortage of fresh water is already a serious problem in most of the drier areas of the world and water shortages will often compound the effects of inadequate land resources on food production. Much of the fresh water is an international public good and requires transnational planning and international agreements.

The need to strengthen the knowledge and technology base for the sustainable development of land

A third challenge refers to knowledge and technology. Sources of research capacity are the developed countries, the developing countries, the public sector and the private sector, and finally the international agricultural research work. In the past most of the research work has concentrated on production and productivity gains and not on natural resources management. We must insist on improvements in both sectors. In future, these tasks will be much more inseparable than ever before and will have great impact on land resources and their use. But in the end, developing and harnessing the potential of both intensive and extensive production systems or high-potential or marginal environments will depend first and foremost on the economic and policy framework, at the national as well as at the international level.

The enabling cultural, institutional and political environment needed for land-based sustainable development

Sustainable development will depend on an enabling national cultural, institutional and policy environment. This cannot be transplanted from other countries. Such an environment must be created within each nation. Education, more than any other social instrument, hastens the pace of cultural change and development. In addition, institutional capacity must be developed, maintained and sustained within each country. Each country must be able to generate, absorb and use new technologies (especially in the area of information), master an interdisciplinary approach, deal with the necessary changes, develop national and international policies and programmes, and above all identify choices, options and priorities. This is a tall order and unfortunately is neglected in many countries and by many donors. Among the various national institutions, the universities will have to play a key role. The more developed countries have benefitted from strong universities. In the field of land use, the policy challenge is especially critical and little experience exists in most countries. National policies will affect production but international policies, especially trade policies, might have an even bigger influence on development, especially rural and agricultural development.

Scientists, technicians and experts groups need to link closely with each other in order to mobilise and to integrate knowledge. In the light of the present and future challenges, the objectives of such mobilisation and integration should not be to publish or perish but to advance and to apply. It is the application that counts.

CONCLUSION

Mankind faces many different challenges related to land and its future. Agriculture is only one, admittedly a very major user of land resources. With growing populations, non-agricultural use of land is becoming ever more important. A waste of land resources, above all high-potential resources, can no longer be tolerated. As long as food production requires land, agriculture must have first claim to a large share of a decreasing vital and productive resources.

We have been optimistic in saying that to feed a growing population the necessary physical resources and the land exist and that the knowledge and technologies are or can be made available. With still considerable land reserves, but with some of marginal quality only, and with generally more difficult production conditions, food will become ever more expensive. This is in sharp contrast to the recent experience of food becoming relatively less expensive. More expensive food will push the poverty problem to the forefront. It will not be the farmers who are forced to find solutions but the policy-makers, and the attitude of society tc food and food production will have to change. Are they prepared? Are they being prepared?

We must also make a better case for more research to deal with future problems of food and food production, land use and the environment. Results from new research will not be available for dissemination until the first years of the next century when, according to some projections and under some assumptions, we must expect a food crisis, a rural energy crisis, an environmental crisis, and a rural crisis leading to further increases in migration into mega-cities with their own serious problems. In spite of such warnings, research expenditures for agriculture and for rural problems seem to be decreasing.

We are at a crossroads. Agriculture demands further scientific advances because producers and consumers need production increases and productivity gains. Other sectors compete with agriculture for resources—for human, financial and physical resources, but above all, for land resources.

Universities are challenged to provide scientific knowledge and leadership. When we think about and discuss these challenges let us remember some optimistic advice from Longfellow, who was the most popular US poet in the last century:

> Look not mournfully to the past—it comes not back again;
> wisely improve the present—it is thine;
> go forth to meet the shadowy future without fear—
> and with a manly heart.

CHAPTER 31

Platforms for Decision-making about Ecosystems

N. Röling

Wageningen Agricultural University, Wageningen, The Netherlands

Dedicated to Dr Richard Bawden and his colleagues at the University of Western Sidney ('Hawkesbury'), Australia, who have been among the first to 'liberate' agricultural science and its teaching for post-Newtonian pursuits.

INTRODUCTION

An intriguing question is: can social science be dispensed with at a conference about mobilising and integrating knowledge for land use options? Can the summary of what a major centre for agricultural science has achieved in the past 15 years do without social science? My answer is that without social scientists, the conference only tells part of what agricultural science is all about.

I will explore the limitation of focusing exclusively on building scenarios on the basis of interactive multiple goal planning models (IMGPMs), without paying attention to human decision-making in developing and using those scenarios. In the process, I will hopefully deconstruct the notion that scientists develop options for politicians (Long and Van der Ploeg, 1989). I will try to do all this by developing a perspective on a *coupled system*, comprising:

(1) a 'hard' ecosystem and
(2) a 'soft' platform for decision-making about that ecosystem.

I intend to show that the coupled system provides a better model and tool for integrating agricultural knowledge for sustainable natural resource management than IMGPMs alone, even if IMGPM has made a tremendous contribution by providing a framework for integrating the hard sciences.

The Future of the Land: Mobilising and Integrating Knowledge for Land Use Options
Edited by L. O. Fresco, L. Stroosnijder, J. Bouma and H. van Keulen.

First, I will give an example of an ecosystem and a concomitant platform and demonstrate the relevance of the coupled system. Secondly, I will try to explain the different epistemological foundations of hard ecosystems and soft platforms, which explains why soft social science cannot, and should not, be built into IMGPMs. Thirdly, I will explore the heuristic nature of the coupled system in terms of its capacity to generate relevant research questions for an agricultural science mandated to serve the sustainable management of natural resources. Finally, I will give some examples of the work of soft scientists that is relevant in this respect.

AN EXAMPLE TO DEFINE THE AREA OF DISCOURSE

Yellowstone National Park (Keiter and Boyce, 1991) calls to mind a wilderness of geysers, mountain streams and forests, where the grizzly bear, buffalo and other animals roam free; one of the few places left on earth where the modern visitor can partake in a world unspoiled by people and governed by the 'invisible hand' of nature.

Upon closer examination, this image rapidly crumbles. Outside the Park, tin mines are in operation which provide the State of Wyoming with a large share of its revenues. The mining threatens the aquifers which feed the geysers. Beyond another section of the boundary, the State Forestry Commission has extensive timber plantations. It is interested in the cubic feet per acre and fears the forest fires which the Park's administration considers a 'natural' phenomenon and allows to rage unchecked. Cattle ranchers also share a boundary with the Park. They complain that the buffaloes which roam widely outside the Park carry brucellosis which infects their cattle. Inside the Park, there is heated discussion between people who want to optimise 'nature', and those who want to optimise the opportunity for people to learn, the Park's original mandate. With 3.5 million visitors a year, humankind is clearly its most important species. The facilities to accommodate people seriously affect natural processes.

After the severe fire in 1988, it was recognised that the Park could not be considered a self-contained single-purpose natural-resource management unit. The ecosystem involved obviously covers a much larger area than the Park; the Greater Yellowstone Area comprises actors with diverse and conflicting, but interdependent, claims on the natural resources. This recognition led to the formation of the Greater Yellowstone Coordination Committee, a platform for decision-making which would take into account the various interests and provide opportunities for conflict resolution, negotiation, accommodation and the consensus-building required for concerted action. Far from being shaped and managed by natural forces, the future of Yellowstone is determined by the messy business of stakeholders trying to reach a compromise about fires, the reintroduction of wolves (which were exterminated back in 1930), the migration routes of animals, and the depth of mining operations.

In short, managing the Park, let alone the Greater Yellowstone Area, is not only a question of biophysical information and technical intervention. Its management requires accommodation between human actors (Long, 1984; Long and Long, 1992) who use the same natural environment but with different purposes. They are interdependent in that each affects the desired outcomes of the others. Therefore, environmental management involves a collective 'agency' (the capacity to make a difference) at a platform of decision-making which comprises all stakeholders. The 'hard' ecosystem, which is seen to require unified management, thus cannot be managed except by the development of a 'soft' platform for purposive action among diverse stakeholders.

THE EPISTEMOLOGICAL FOUNDATIONS OF HARD ECOSYSTEMS AND SOFT PLATFORMS

The 'hard' ecosystem, that is, the assumptions about the *nature of human knowledge implicit in hard science* may be illustrated by the definition of 'systems' given by Rabbinge and Van Ittersum (see Chapter 3): 'Systems are limited parts of reality with well-defined boundaries.' Implicit in this definition are the following assumptions:

- the system exists objectively, and
- reality exists independently of the knowing subject.

We can expand this by what I consider to be the other key assumptions of hard science:

- Reality operates on the basis of causal natural laws.
- Through empirical, analytical and experimental methods, it is possible to build factual knowledge which represents objects as they exist in the real world.
- One can build simulation models which integrate the hard sciences and allow exploration of future states of the system, given different human ends. The information provided is objective and true. After all, scientific knowledge is a factual, timeless, universal, lawlike and context-free commodity.

In this view, technology is applied science to be used for instrumental manipulation. Western society is built on the belief in instrumental reason and in the capacity of science to solve problems. This conference bears witness to this approach. It is free from post-Newtonian thinking (Uphoff, 1992).

Assumptions about human knowledge implicit in soft systems thinking (e.g. Checkland, 1981) start from the concept of humans as intentional beings. Instead of unquestionably accepting the ends or the 'why' questions, and focusing only on the means or the technical 'how' questions, soft systems thinkers consider the ends to be the very stuff of discourse. They cannot be assumed because they are the very bone of contention. Secondly, humans are sense-making beings. This means that soft platforms cannot be studied with the premises of positivism. Giddens (1984) explains this by what he calls the 'double hermeneutic'. Natural phenomena do not change as a result of the human effort to know them. Whether we think the sun turns around the earth, or vice versa, does not change the behaviour of these celestial bodies. But knowledge produced about human actors re-enters human society, is interpreted and affects human action. People actively construct their own realities through learning in social processes. Constructivism is an approach to human knowing which is more appropriate for soft platforms than positivism or realism. Hard science can, of course, be seen as one specific way of constructing reality.

As an example of constructivist thinking, I offer an alternative definition of a system: a system is a construct with arbitrarily defined boundaries for discourse about complex phenomena to emphasise wholeness, inter-relationships and emergent properties.

Where Rabbinge and Van Ittersum look at *models* of systems as constructs, I consider the *systems themselves* as constructs. Hard science assumes an objective reality and looks for causes. Soft science assumes that there are as many realities as people, and looks for reasons. Instead of explanation, soft science looks at interpretation. With one objective and true world, disagreement means mutual negation. With multiple constructed realities, disagreement means negotiation and accommodation (Maturana, undated).

The objective of soft science is not to predict and control, but to stimulate self-reflection, discourse and learning. Soft science operates through the double hermeneutic and its achievement should be judged on that basis. Its achievement cannot be judged on the basis of instrumental reasoning, for example, in terms of cubic feet of institution building. Soft and hard scientists operate on totally different epistemologies. Soft science cannot be part of IMGPMs because those models are based on positivist, if not realist, assumptions. Soft science can only play a role in a platform/ecosystem interface dynamic.

THE HEURISTIC NATURE OF THE COUPLED SYSTEM

Research areas which emerge from the coupling between soft platforms and hard ecosystems are as follows:

The goodness-of-fit between perceived ecosystems and platforms for decision-making

This issue is of great interest for the survival of humankind: it is not just a question of indicating that a certain ecosystem at whatever level of aggregation is under threat, but very much also a question of a platform that corresponds with the perceived ecosystem, so that decisions can be made about a more sustainable management of that ecosystem. In many cases, existing platforms for decision-making, such as the household, prefecture, farmer association, car factory, province, or nation, have not been set up for natural resource management and/or do not correspond to the ecosystem to be managed. A great challenge facing us today is to move from human organisation for purposes of production, profit making, etc., to organisation for effective natural-resource management. After having moved from being producers only to being entrepreneurs as well, farmers must now be redefined as natural-resource managers.

Related to this issue are such questions as: can effective platforms be created? How does the identification of an ecosystem under threat affect the development of platforms? Can we learn to develop platforms fast enough to salvage the planet?

The processes occurring at the interface between hard ecosystems and soft platforms

The interface focuses the attention on the role of information and learning. To manage it effectively, the stakeholders in the ecosystem must develop an information system about the ecosystem which allows them to learn and take effective concerted action to manage it in a sustainable manner. In such an information system, the work of hard science can play a crucial role. But this role needs to be carefully studied. So far, rather naive ideas about the use of scenarios by politicians have preempted this research area. But other questions arise as well. For example, stakeholders coming together in a platform to manage an ecosystem must learn from scratch about the ecosystem, agree on its boundaries, share concepts with which to discuss it and decide about its sustainable management, develop new indicators for success, and develop new methods of 'making things visible'.

Increasing the level of social aggregation of the platform/ecosystem combination

This is an issue which is of extreme importance because the level at which interdependence between stakeholders is recognised moves increasingly in a global direction. This means that actors and platforms must increasingly be considered as 'holons' or subassemblies (Koestler, 1967) of larger wholes and must learn to form effective platforms, that is, platforms at which collective agency (Giddens, 1984) with respect to the appropriate ecosystem level can be exerted. Moving up the level of aggregation means that diverse and conflicting objectives must be assembled into rich pictures and mutually accommodated to shared perspectives in painful negotiation. Social dilemmas play a crucial role in these processes (Messick and Brewer, 1983). Brinkman (Chapter 1) has pointed out that, instead of land use planning, it would be better to speak of 'land use negotiation'.

The facilitation of platform formation

How can one move stakeholders in an ecosystem from interacting on the basis of strategic reasoning to communicative reasoning (Habermas, 1984, 1987; Funtowicz and Ravetz, 1990)? How do stakeholders come to the joint appreciation that the sustainability of the ecosystem in which they operate is a priority problem? What coercive and non-coercive interventions are taken with what effects? How does diversity among stakeholders affect these effects? How can one facilitate joint learning of stakeholders on platforms as they move from joint problem appreciation to collective action (Checkland, 1981)? What role does hard science play in such processes and how do such processes affect hard science? Hard science may be used as a tool for social learning, an approach which differs considerably from the assumptions of land use planning by experts.

Sustainability as the emergent property of a coupled system

The key reason for constructing systems is that properties emerge at the system level which cannot be predicted from studying the components making up the system (Checkland, 1981). The total contribution is more than the sum of the contributions of the parts. Hard scientists consider sustainability to be the emergent property of an ecosystem in equilibrium, a point at which the carrying capacity is not exceeded (Jiggins, 1994).

Soft scientists or post-Newtonian biophysical scientists consider sustainability to be the emergent property of a soft learning system (e.g. Bawden and Packam, 1991). Until such time that hard scientists recognise the essentially constructivist nature of human knowledge and move to post-Newtonian perspectives, sustainability may be considered the emergent property of a coupled system between soft platform and hard ecosystem. Such a perspective gives room to the role of human interventions in the ecosystem, to the importance of the multiple perspectives and objectives with which the ecosystem is being used by different stakeholders, to interface phenomena (especially information) between soft and hard systems, and to how platforms learn their way to more sustainable futures and develop agencies to create those futures.

Agricultural science that ignores these issues is bound to become irrelevant. The perspective of the coupling between soft platforms and hard ecosystems seems to imply a role for agricultural research that is different from building expertise which can be utilised by politicians and others according to the linear model of the role of science in society (e.g. Kline and Rosenberg, 1986). Agricultural research can, perhaps, better be defined as an interactive resource for learning our way to more sustainable futures.

ARE SOFT SCIENTISTS DOING SOMETHING ABOUT IT?

Hard scientists do not believe that soft scientists can and do make a useful contribution, partly because of their epistemological assumptions and the focus on instrumental reasoning.

My point of departure is that of the post-Newtonian quantum physicist David Bohm, with whom an interview appeared in the New Scientist shortly after his death (Bohm, 1993): 'Science consists not in the accumulation of knowledge, but in the creation of fresh modes of perception.' Accordingly, our aim is to create perspectives on human activity which can re-enter society as a basis for learning. As explained, this is Giddens' double hermeneutic.

Examples of what soft scientists in Wageningen may contribute are:

- The Wageningen philosophers have been working for years to stretch up the naive realism and positivism with which agricultural science in Wageningen is traditionally pursued and have tried to move it from an instrumental to a communicative perspective in the sense of Habermas (e.g. Koningsveld, 1982). In this respect they have made an important contribution where it counts: after all, agricultural science is applied science.
- Long and Long (1992) have introduced the actor-oriented perspective into Wageningen development sociology, and in doing so, have firmly established the perspective of the intentional, sense-making human actor, be it at the individual level or at a higher level of aggregation. Actors pursue their own projects and clash with others in arenas of struggle. A host of development sociologists have carried out very interesting research on this basis. It focuses on what actors actually do and think, without imposing reified realities from the outside. This refreshing approach has led to painful deconstruction of some cherished perspectives on how the world can be changed (e.g. Long and Van der Ploeg, 1989).

- Van der Ploeg (1993) and his colleagues started with people's basic intentional nature and explored the diversity of human ends and its consequences. It appears that farmers shape their own very different futures, irrespective of the compelling nature of the economic and technical context. This is consistent with the actor-oriented perspective which rejects the notion that people are determined by social systems and other macro forces. The research has important implications for our understanding of the role of people's objectives in sustainable natural resource management.
- An important influence on the Wageningen extension school has been Checkland's Soft Systems Methodology (SSM) (1981) which focuses on taking a set of actors through a process of shared problem appreciation, learning about the problem and taking collective action to improve it. With the collaboration of Checkland, we are developing this methodology, which has been developed into an effective tool since 1979 for corporate situations, for use in platforms for decision-making about ecosystems. Engel (in prep.) has developed SSM into an interactive methodology for actors to learn their way to platforms for effective decision-making about domains of agricultural innovation, a method tested and adapted several times and used across the globe.
- Leeuwis (1993) has deconstructed the conventional perspectives with which the use of information technology is considered and has developed an empirically based soft perspective on 'communication technology' which can provide important insights into the utilisation and development of GISs, IMGPMs and other tools of agricultural science.
- Our Programme for Comparative Research on Knowledge Systems for Sustainable Agriculture has looked at rural people's knowledge in the development of sustainable land use options (Brouwers, 1993); at the way farmers learn their way to more sustainable arable practices under pressure from environmental legislation (Van der Ley and Proost, 1992; Somers and Röling, 1993; Röling, 1993) and at the revolutionary Non-Formal Education approach taken to farmer learning of Integrated Pest Management in irrigated rice in Indonesia, one of the few large-scale and successful deviations from linear transfer of technology approaches to induced technological change (Van de Fliert, 1993). A study of conditions for integrated water level management in the Netherlands (Westendorp and Röling, 1993) showed that, at the level of specialised institutions, the planning emphasis was on official documentation, coercive mandates, segmented expertise, vested interests and boundary maintenance. At the level at which the natural resource is actually managed, stakeholders realised their mutual interdependence through informal interaction which allowed them to take integrative positions. It was these informal mechanisms that proved essential for creating effective water use as a coupled system.
- Finally, Van Woerkum and Aarts (1993) have looked at communication as a policy instrument and are establishing the basically interactive nature of using such 'instruments'. Their contribution goes a long way in providing insight into the processes which characterise the utilisation of hard science knowledge in policy.

The fact that the soft scientists in Wageningen have made such contributions does not diminish the fact that they have, on the whole, operated in narrow disciplinarian confines, and not seen themselves as part of an agricultural science which is a resource for the planet's survival. In that respect, the hard scientists have been much more forward looking.

ACKNOWLEDGEMENTS

Critical comments by Dr Janice Jiggins and Frank Vanclay are gratefully acknowledged. The paper uses some ideas which were originally developed by J. Woodhill and N. Röling.

REFERENCES

Bawden, R. J. and Packam, R., 1991. Systems praxis in the education of the agricultural systems practitioner. Paper presented at the 1991 Annual Meeting of the International Society for the Systems Sciences, Östersund, Sweden. University of Western Sidney-Hawkesbury, Richmond (NSW), Australia.

Bohm, D., 1993. 'Last words of a quantum heretic', interview with John Morgan. *New Scientist* **137**: 42.

Brouwers, J., 1993. *Rural People's Knowledge and its Response to Declining Soil Fertility The Adja Case (Benin)*. Published Dissertation, Wageningen Agricultural University, Wageningen.

Checkland, P., 1981. *Systems Thinking, Systems Practice*. John Wiley, Chichester, UK.

Engel, P. G. H., (in prep.). *Knowledge Management in Agriculture: A Fundamental Requirement for Sustainable Development*. Published Dissertation, Wageningen Agricultural University, Wageningen.

Funtowicz, S. O. and Ravetz, J. R., 1990. *Global Environmental Issues and the Emergence of Second Order Science*. CD-NA-12803-EN-C, Report EUR 12803 EN. Commission for the European Communities, Directorate General Telecommunications, Information Industries and Innovation, Luxembourg.

Giddens, A., 1984. *The Constitution of Society: Outline of the Theory of Structuration*. Polity Press, Cambridge, UK.

Habermas, J., 1984. *The Theory of Communicative Action. Vol. 1: Reason and the Rationalisation of Society*. Beacon Press, Boston, USA.

Habermas, J., 1987. *The Theory of Communicative Action. Vol. 2: Lifeworld and System. A Critique of Functionalist Reason*. Beacon Press, Boston, USA.

Jiggins, J. L. S., 1994. *Changing the Boundaries: Woman-centered Perspectives on Population and the environment*. The Island Press, Washington, DC.

Keiter, R. B. and Boyce, M. S., 1991. *The Greater Yellowstone Ecosystem: Redefining America's Wilderness Heritage*. Yale University Press, Boston, MA.

Kline, S. and Rosenberg, N., 1986. An overview of innovation. In: Landau, R. and Rosenberg, N. (eds), *The Positive Sum Strategy. Harnessing Technology for Economic Growth*, pp. 275–306. National Academy Press, Washington, DC.

Koestler, A., 1967. *The Ghost in the Machine*. Penguin Arkana, London.

Koningsveld. H., 1982. *Het Verschijnsel Wetenschap. Een Inleiding Tot de Wetenschapsfilosofie*. Boom, Meppel, The Netherlands.

Leeuwis, C., 1993. *Of Computers, Myths and Modelling: The Social Construction of Diversity, Knowledge, Information and Communication Technology in Dutch Horticulture*. Published Doctoral Dissertation, Wageningen Agricultural University, Wageningen.

Long, N., 1984. Creating space for change: a perspective on the sociology of development. *Sociologia Ruralis* **24**: 168–184.

Long, N. and Long, A. (eds), 1992. *Battlefields of Knowledge: The Interlocking of Theory and Practice in Research and Development*. Routledge, London.

Long, N. and Van Der Ploeg, J. D., 1989. Demythologising planned intervention. *Sociologia Ruralis* **29**: 226–249.

Maturana, H. R. (no date). *Reality: The Search for Objectivity, or the Quest for a Compelling Argument*. University of Chile, Faculty of Sciences, Santiago, Chile.

Messick, D. M. and Brewer, M. B., 1983. Solving social dilemmas: a review. *Review of Personality and Psychology* **4**: 11–44.

Röling, N., 1993. Agricultural knowledge and environmental regulation: the Crop Protection Plan and the Koekoekspolder. *Sociologia Ruralis* **33**: 212–231.

Somers, B. M. and Röling, N., 1993. Ontwikkeling van kennis voor duurzame landbouw: een verkennende studie aan de hand van enkele experimentele projekten. NRLO, Den Haag, The Netherlands.

Uphoff, N., 1992. *Learning from Gal Oya. Possibilities for Participatory Development and Post-Newtonian Social Science*. Cornell University Press, Ithaca, NY.

Van De Fliert, E., 1993. *Integrated Pest Management. Farmer Field Schools Generate Sustainable Practices: A Case Study in Central Java Evaluating IPM Training*. Published Dissertation, Wageningen Papers, Wageningen Agricultural University, Wageningen, The Netherlands.

Van Der Ley, H. A. and Proost, M. D. C., 1992. Gewasbescherming met een toekomst: de visie van agrarische ondernemers: een doelgroepverkennend onderzoek ten behoeve van voorlichting. Department of Communication and Innovation Studies. Wageningen Agricultural University, Wageningen.

Van Der Ploeg, J. D., 1993. Rural sociology and the new agrarian question: a perspective from the Netherlands. *Sociologia Ruralis* **33**: 240–260.

Van Woerkum, C., and Aarts, M. N. C., 1993. De integratie van communicatie en overheidsbeleid. Department of Communication and Innovation Studies. Wageningen Agricultural University, Wageningen.

Westendorp, J. and Röling, N., 1993. Natuurbeleid en geintegreerd peilbeheer. Report to the Ministry of Agriculture, Nature Management and Fisheries, Wageningen Agricultural University, Wageningen.

CONCLUSION

Planning for the People and the Land of the Future

L. O. Fresco

Wageningen Agricultural University, Wageningen, The Netherlands

Planning for sustainable land use is, in some ways, still in its infancy. This is so because of our limited understanding of the sometimes conflicting goals of land use and sustainability. Also, the long-term effects of land use and hence the inherent constraints on planning for the future, are still largely unknown.

The contributions to this book show clearly that much work still needs to be accomplished in two closely related, yet distinct areas. The first relates to the process of land use planning itself, its goals and the actors involved. The second refers to the science of land use planning in the wider sense, in particular its conceptual shortcomings and the methodology of assessing biophysical and socio-economic options.

THE PROBLEMS OF LAND USE PLANNING AS A PROCESS

It becomes apparent that a considerable rift divides two schools of thought. Simply put, there are those concerned with the future of the land, versus those concerned with the future of the land users. In other words, does land use planning start from an analysis of land use potentials or from an analysis of the needs of the various land users? Obviously, the two are intricately linked, unless we would accept that the land of the future is empty. Yet, these differences in starting point and emphasis, which often mirror differences in disciplines (soil and crop (i.e. technical) sciences versus social science), influence the entire process of land use planning from its very beginning.

In this respect the major question is: where do the goals of land use planning come from? The papers in this volume illustrate clearly that most goals are technology driven, i.e. generated by expectations about the application of new land use technology to agricultural production, but increasingly also by the need to meet environmental

The Future of the Land: Mobilising and Integrating Knowledge for Land Use Options
Edited by L. O. Fresco, L. Stroosnijder, J. Bouma and H. van Keulen.

standards. Goals derived from rural household objectives and constraints are given far more limited attention (this is, in part, a reflection of the selection of contributions). Although such a technology drive is often true of innovations (e.g. car manufacturing), land use planning as a process requires special attention to involve the users for whom land is a primary source of livelihood.

The 'silence of the users' has been the most remarkable feature emerging from the contributions: true interaction with users throughout the process of land use planning is still to come. The need to educate the users in negotiating their futures has been highlighted, suggesting that no effective planning is possible unless users are organised in some fashion, e.g. through platforms. Scenarios provided by researchers can feature as an input in discussions. However, whether all users have the flexibility to 'play along' according to the scenarios, remains to be shown. Many appear to prefer (quick) answers over (debatable) options.

A major constraint in land use planning highlights the differential time horizon between decision-makers and planners or scientists. Should the plan be perfect before it can be implemented, or can development pathways and policy measures already be formulated during the planning process? There is considerable merit in seeing land use planning as an ongoing, continuous process with feedback and feedforward contributions from various stakeholders, so that the 'silence of the users' be interrupted.

The slow pace of land use planning, often taking many years just for the first scenario formulation, is due to the general lack of high quality baseline data and the cost of collecting these. More thought should be put into making use of existing data sets, however limited they are. The need to reduce and justify the generally high investments in data collection and modelling forms an important impetus to the development of land use planning models requiring low data inputs. This could mean the development of expert systems as tools for exploration, to be followed by more sophisticated models requiring more detailed data.

Definition of the interrelations between biophysical potential and socio-economic constraints is shown to be an ambitious venture, leading to pleas from some quarters to separate biophysical exploration from prediction and policy formulation. Several cases have demonstrated the value of guidance provided by socio-economic considerations in the modelling of biophysical processes, e.g. in the selection of cropping and livestock techniques that are modelled.

A final word of caution: in land use planning, as in other promising sciences, the risk of a revolution of rising expectations is real. The problems are great, solutions are called for urgently and interactive land use planning seems the appropriate answer. Yet land use planners must continue to be modest about their contributions, which can only provide some oil in the great wheel of history, for fear of creating a backlash of disappointment.

LAND USE PLANNING AS A SCIENCE

Linking upstream, specialised academics with downstream, impatient implementers remains the great challenge. At some moments, it might seem that the wheel of land use planning has been happily reinvented, notwithstanding decades of work (often published under different key words, but relevant nevertheless). From the chapters emerges a feeling that, in recent years, the debate has not evolved sufficiently in conceptual terms. The discussions point to a potential deadlock between two schools that are emerging among land use planning scientists. There are those who focus on relatively rapid explorative methods, involving simulation modelling of a limited number of, often loosely defined, options ('the explorers'). Yet there are others who emphasise the need for detailed, farm level case studies with a strong participation of social scientists ('the holy grail seekers'). Clearly, we must move beyond the explorative, yet not get swamped with too much detail.

The major difference emerging between land use planning in the 'western' world and in developing countries is that in the latter planning needs to aim primarily (if not exclusively) for increased agricultural output, whereas in the former other considerations (e.g. limitation of discharge of pollutants, reduction of the agricultural area, preservation of landscapes) lead to the inclusion of other goals beyond (sustainable) agricultural production. Increasingly, however, balancing agricultural production and environmental constraints becomes the aim throughout the world.

The challenge (and, occasionally, the tragedy) of an applied science like land use planning is the continuous conflict between the desire for deeper scientific understanding and the need for the design of rapidly applicable methodology. As a consequence, there is always a tension between the collection of 'casuistic' details that are needed for implementation in a given area and the need to identify the underlying patterns and trends.

Interdisciplinarity appears as the two-sided sword of land use planning science. Although there is general agreement on the need to transcend disciplinary boundaries, a large gap remains between the disciplines. Natural and social sciences are governed by different laws, apply different units of analysis and explanatory variables. Their jargons differ, and so do the media through which they communicate. Above all, the degree of quantification differs and it seems unlikely that social factors can really be integrated well into current technical models, with the exception of some economic and demographic variables. On the other hand, it will be worthwhile to attempt to integrate more systematically socio-economic concerns and variables in expert systems and decision-support models, perhaps also through the use of artificial intelligence.

The inherent limitations of modelling require that explicit decisions be made with respect to the total number of variables that can be internal to the model. It seems, from the contributions to this book, that modelling subsystems yields more satisfactory results than handling complex models for entire systems (including their subsystems). However, there should be a balance between the subsystems to avoid a biased representation of the system as a whole. One of the key questions in modelling refers to the constraint of linearity. In nearly all cases where feedbacks are operating, even if they are linear themselves, their interaction is unlikely to be so, requiring model adaptations to deal with substantial non-linearities. This becomes even more urgent because of the need to build flexible models that can deal with ongoing and future land use dynamics.

Furthermore, it has become clear that confidence limits must be formulated explicitly, reflecting data quality and the absence of overall validation of nearly all models. So far, confidence limits are not included when modelling results are reported. Extrapolation, of course, is practically unavoidable, implying that more work needs to be done to develop protocols for extrapolation. This entails questions about the type of data sets required in order to apply a model successfully. Overall, there seems great merit in adapting existing models for application to new situations, rather than developing yet another model. Both concerns suggest a strong move towards an agreed set of minimal data and procedures for model building and sensitivity testing.

There is substantial danger in generating conclusions about other levels of aggregation than the ones the model refers to. The loss of heterogeneity and detail in scaling up from the farm level is difficult to accept for many (social) scientists. Hopefully, this will lead to a more systematic use of intermediate levels of analysis situated between farm and (sub)region.

Finally, the great diversity as well as the high level of the contributions in this book show that the attempt to take stock of the advances in land use planning has been worthwhile, even if the development of a shared framework is still hampered by the lack of a common language.

ACKNOWLEDGEMENT

The author is grateful to all the members of the editorial and organising committees for their contributions to the discussions leading to the concluding chapter.

Index